Applied Computer-Aided Drug Design: Models and Methods

Edited by

Igor José dos Santos Nascimento

Programa de Pós-Graduação em Ciências Farmacêuticas (PPGCF)
Departamento de Farmácia
Universidade Estadual da Paraíba
Campina Grande-PB
Brazil

Applied Computer-Aided Drug Design: Models and Methods

Editor: Igor José dos Santos Nascimento

ISBN (Online): 978-981-5179-93-4

ISBN (Print): 978-981-5179-94-1

ISBN (Paperback): 978-981-5179-95-8

need for a court order if at any point you breach any terms of this License Agreement. In no event will any delay or failure by Bentham Science Publishers in enforcing your compliance with this License Agreement constitute a waiver of any of its rights.

3. You acknowledge that you have read this License Agreement, and agree to be bound by its terms and conditions. To the extent that any other terms and conditions presented on any website of Bentham Science Publishers conflict with, or are inconsistent with, the terms and conditions set out in this License Agreement, you acknowledge that the terms and conditions set out in this License Agreement shall prevail.

Bentham Science Publishers Pte. Ltd.
80 Robinson Road #02-00
Singapore 068898
Singapore
Email: subscriptions@benthamscience.net

BENTHAM SCIENCE

CONTENTS

*Samuel Baraque de Freitas Rodrigues, Rodrigo Santos Aquino de Araújo, Thayane
Regine Dantas de Mendonça, Francisco Jaime Bezerra Mendonça-Júnior, Peng Zhan
and Edeildo Ferreira da Silva-Júnior*

PREFACE

The drug discovery and development process is time-consuming and demands a high financial cost. In this way, it is estimated to take approximately 10 to 17 years, costing around 4 billion dollars. This stimulates the advancement of new methodologies that can accelerate the discovery process and increase the probability of a promising molecule. In addition, constant developments in informatics and computations have led to the routine use of high-performance computing in medicinal chemistry. Thus, Computer-Aided Drug Design (CADD) methods emerge, capable of providing critical information for the design of new molecules, essential in any new drug discovery program [1, 2].

In this context, the book *"Applied Computer-Aided Drug Design: Models and Methods"* appears, presenting the computational methods used by researchers and pharmaceutical companies. Each chapter explains a technique with high precision so that readers can apply it in their research.

This first edition is organized into nine chapters, namely:

Chapter 1 *"Ligand- and Structure-Based Drug Design (SBDD and LBDD)"*: Promising Approaches to Discover New Drugs. Here, the reader will have an approach from a historical perspective on strategies used in designing new drugs until the development of LBDD and SBDD strategies, exemplifying important discoveries of commercial drugs.

Chapter 2 *"Quantitative Structure-activity relationship (QSAR) in studying the biologically active molecules"*. This chapter will bring the principles and methods of this technique based on LBDD. It will present a historical perspective from the first QSAR models to the most current ones like 6D-QSAR. Furthermore, it provides a great read on protocol validation procedures, which are crucial to successful QSAR studies.

Chapter 3 *"Pharmacophore Mapping: An Important Tool In Modern Drug Design and Discovery"*. This chapter approaches a method that can be applied to SBDD and LBDD protocols. The reader will have a historical perspective of the evolution of the method, a presentation of the leading software used, and, in the end, a great background on carrying out a well-validated virtual screening protocol based on pharmacophore. Further, the text addresses successful studies and how their protocols were carried out.

Chapter 4 *"Up-To-Date Developments In Homology Modeling"*. Similar to the previous chapters, the readers will have a theoretical basis on the technique, quite explored when there is information about the target without an experimental structure. Homology modeling is a powerful tool for constructing and applying molecular targets in drug design studies. With this, readers can perform this protocol safely and efficiently.

Chapter 5 *"Anticancer Activity of Medicinal Plants Extract and Molecular Docking Studies"*. In fact, this is the most used tool by drug developers worldwide. Through this technique, new drugs can be safely planned, or even virtual screenings can be carried out to find new drugs. Thus, the authors will bring the technique's theoretical framework, the method's evolution, computational software, and studies in which the application of molecular docking was vital to finding promising molecules.

Chapter 6 *"FBDD & de novo Drug Design"*. In this chapter, the main tools used in Fragment-Based Drug Design (FBDD) and *de novo* Drug Design (DNDD) will be presented, mainly through in silico approaches. It is essential to highlight that these methods control molecules from scratch, generating critical *hits* that later become optimizable *leads*. In addition, all the theoretical frameworks and important discoveries are applied through these strategies.

Chapter 7 *"Molecular simulation in drug design; an overview of molecular dynamics methods"*. Despite being a promising technique, molecular docking has several problems, such as disregarding the flexibility of the active site during simulation. Thus, this chapter will address the molecular dynamics technique, which tries to solve some problems from molecular docking. In fact, with the popularization of computers in drug design, this is the fastest-growing technique, and its application is essential in drug discovery programs. Thus, with great clarity, the authors present the theoretical framework and how to apply it in a design campaign for new drugs.

Chapter 8 *"Quantum Chemistry in Drug Design: density function theory (DFT) and other quantum mechanics (QM)-related approaches"*. The application of quantum chemistry (QM) protocols in predicting biological activity or enzymatic mechanism are highlighted in the current drug discovery process. Increasingly, researchers are adopting these tools in their drug development projects. Thus, in this chapter Rodrigues et al. They explored the entire theoretical foundation of QM, focusing on applying Density Functional Theory, providing new insights to medicinal chemists to use in their projects.

Chapter 9 *"Free energy estimations for drug discovery: Background and perspectives"*. This chapter is one of the most current and essential in this book. Here are shown energy predictions and applications of perturbation theory in drug design. This approach has gained increasing prominence in medicinal chemistry, mainly for solving some limitations related to classic MD simulations. In this way, an excellent theoretical framework and its application in drug design are shown with updated examples.

I hope that with this book, readers will have new insights and be able to safely apply the protocols shown here, providing new trends that help discover new drugs to improve the quality of life of the world's population.

Igor José dos Santos Nascimento
Programa de Pós-graduação em Ciências Farmacêuticas (PPGCF)
Departamento de Farmácia
Universidade Estadual da Paraíba
Campina Grande-PB
Brazil

REFERENCES

[1] Nascimento IJS, de Aquino TM, da Silva-Júnior EF. The New *Era* of Drug Discovery: The Power of Computer-aided Drug Design (CADD). Lett Drug Des Discov 2022; 19(11): 951-5.
[http://dx.doi.org/10.2174/1570180819666220405225817]

[2] Nascimento IJ, de Aquino TM, da Silva-Júnior EF. Drug Repurposing: A Strategy for Discovering Inhibitors against Emerging Viral Infections. Curr Med Chem 2021; 28(15): 2887-942.
[http://dx.doi.org/10.2174/1875533XMTA5rMDYp5] [PMID: 32787752]

List of Contributors

Abha Vyas	Department of Pharmaceutical Chemistry, L.J. Institute of Pharmacy, L.J. University, Ahmedabad 382 210, India
Anwesha Das	Department of Medicinal Chemistry, National Institute of Pharmaceutical Education and Research, Ahmedabad, Palaj, Gandhinagar 382355, Gujarat, India
Arijit Nandi	Department of Pharmacology, Dr. B.C. Roy College of Pharmacy and Allied Health Sciences, Durgapur-713206, West Bengal, India
Blanca Colín-Lozano	Facultad de Ciencias Químicas, Benemérita Universidad Autónoma de Puebla, Puebla, México
Burak TÜZÜN	Plant and Animal Production Department, Technical Sciences Vocational School of Sivas, Sivas Cumhuriyet University, Sivas, Turkey
Dharmraj V. Pathak	Department of Pharmaceutical Chemistry, L.J. Institute of Pharmacy, L.J. University, Ahmedabad 382 210, India
Edeildo Ferreira da Silva-Júnior	Institute of Chemistry and Biotechnology, Federal University of Alagoas, Lourival Melo Mota Avenue, AC. Simões campus, 5587072-970, Alagoas, Maceió, Brazil Laboratory of Medicinal Chemistry, Federal University of Alagoas, Lourival Melo Mota Avenue, AC. Simões campus, 5587072-970, Alagoas, Maceió, Brazil
Emin SARIPINAR	epartment of Chemistry, Faculty of Science, Erciyes University, Melikgazi, 38039, Kayseri, Turkey
Esin Aki-Yalcin	Department of Pharmaceutical Chemistry, Faculty of Pharmacy, Cyprus Health and Social Sciences University, Guzelyurt, Northern Cyprus
Fernando D. Prieto-Martínez	Instituto de Química, Universidad Nacional Autónoma de México, Ciudad de México, México
Francisco Jaime Bezerra Mendonça-Júnior	Laboratory of Synthesis and Drug Delivery, Department of Biological Sciences, State University of Paraiba, João Pessoa 58429-500, PB, Brazil
Hardik G. Bhatt	Department of Pharmaceutical Chemistry, Institute of Pharmacy, Nirma University, Ahmedabad 382 481, India
Igor José dos Santos Nascimento	Pharmacy Department, Estácio of Alagoas College, Maceió-AL, Brazil Pharmacy Department, Cesmac University Center, Maceió-AL, Brazil Programa de Pós-Graduação em Ciências Farmacêuticas (PPGCF), Departamento de Farmácia, Universidade Estadual da Paraíba, Campina Grande-PB, Brazil
Mallika Alvala	MARS Training Academy, Hyderabad, India
Muhammed Tilahun Muhammed	Department of Pharmaceutical Chemistry, Faculty of Pharmacy, Suleyman Demirel University, Isparta, Turkey
Paresh K. Patel	Department of Pharmaceutical Chemistry, L.J. Institute of Pharmacy, L.J. University, Ahmedabad 382 210, India

Peng Zhan

Department of Medicinal Chemistry, Key Laboratory of Chemical Biology (Ministry of Education), School of Pharmaceutical Sciences, Cheeloo College of Medicine, Shandong University, 44 West Culture Road, 250012 Jinan, Shandong, PR China

Rodrigo Santos Aquino de Araújo

Laboratory of Synthesis and Drug Delivery, Department of Biological Sciences, State University of Paraiba, João Pessoa 58429-500, PB, Brazil

Ricardo Olimpio de Moura

Programa de Pós-Graduação em Ciências Farmacêuticas (PPGCF), Departamento de Farmácia, Universidade Estadual da Paraíba, Campina Grande-PB, Brazil

Samuel Baraque de Freitas Rodrigues

Institute of Chemistry and Biotechnology, Federal University of Alagoas, Lourival Melo Mota Avenue, AC. Simões campus, 5587072-970, Alagoas, Maceió, Brazil

Serap ÇETINKAYA

Department of Molecular Biology and Genetics, Science Faculty, Sivas Cumhuriyet University, Sivas, Turkey

Sneha R. Sagar

Department of Pharmaceutical Chemistry, L.J. Institute of Pharmacy, L.J. University, Ahmedabad 382 210, India

Thayane Regine Dantas de Mendonça

Laboratory of Medicinal Chemistry, Federal University of Alagoas, Lourival Melo Mota Avenue, AC. Simões campus, 5587072-970, Alagoas, Maceió, Brazil

Vijeta Kumari

Laboratory of Natural Product Chemistry, Department of Pharmacy, Birla Institute of Technology and Science, Pilani (BITS Pilani), Pilani Campus, Pilani-333031, Rajasthan, India

Yelzyn Galván-Ciprés

Escuela Nacional de Ciencias Biológicas, Instituto Politécnico Nacional, Ciudad de México, México

CHAPTER 1

Ligand and Structure-Based Drug Design (LBDD and SBDD): Promising Approaches to Discover New Drugs

Igor José dos Santos Nascimento[1,2,3,*] and **Ricardo Olimpio de Moura**[3]

[1] *Pharmacy Department, Estácio of Alagoas College, Maceió-AL, Brazil*

[2] *Pharmacy Department, Cesmac University Center, Maceió-AL, Brazil*

[3] *Programa de Pós-Graduação em Ciências Farmacêuticas (PPGCF), Departamento de Farmácia, Universidade Estadual da Paraíba, Campina Grande-PB, Brazil*

Abstract: The drug discovery and development process are challenging and have undergone many changes over the last few years. Academic researchers and pharmaceutical companies invest thousands of dollars a year to search for drugs capable of improving and increasing people's life quality. This is an expensive, time-consuming, and multifaceted process requiring the integration of several fields of knowledge. For many years, the search for new drugs was focused on Target-Based Drug Design methods, identifying natural compounds or through empirical synthesis. However, with the improvement of molecular modeling techniques and the growth of computer science, Computer-Aided Drug Design (CADD) emerges as a promising alternative. Since the 1970s, its main approaches, Structure-Based Drug Design (SBDD) and Ligand-Based Drug Design (LBDD), have been responsible for discovering and designing several revolutionary drugs and promising *lead* and *hit* compounds. Based on this information, it is clear that these methods are essential in drug design campaigns. Finally, this chapter will explore approaches used in drug design, from the past to the present, from classical methods such as bioisosterism, molecular simplification, and hybridization, to computational methods such as docking, molecular dynamics (MD) simulations, and virtual screenings, and how these methods have been vital to the identification and design of promising drugs or compounds. Finally, we hope that this chapter guides researchers worldwide in rational drug design methods in which readers will learn about approaches and choose the one that best fits their research.

Keywords: CADD, Computational methods, Drug design, Drug discovery, Drug Development, Docking, FBDD, LBDD, QSAR, Rational Design, SBDD.

* **Corresponding author Igor José dos Santos Nascimento:** Pharmacy Department, Estácio of Alagoas College, Maceió-AL, Brazil; Cesmac University Center, Pharmacy Department, Maceió-AL, Brazil; and Programa de Pós-Graduação em Ciências Farmacêuticas (PPGCF), Departamento de Farmácia, Universidade Estadual da Paraíba, Campina Grande-PB, Brazil; Tel.: (+55)8299933-5457; E-mail: igorjsn@hotmail.com & igor.nascimento@cesmac.edu.br

INTRODUCTION

The process of designing and developing new drugs is challenging and has evolved constantly in recent years, from empirical approaches related to natural products to the current phase with the use of computers and artificial intelligence [1 - 3]. One of the most significant advances in this area has been high-throughput screening (HTS), in which thousands of compounds can be screened in a few hours. In addition, the growth of genomics, proteomics, metabolomics, and molecular modeling promoted substantial advances in the knowledge of critical biochemical pathways for the P&D of drugs [4 - 6]. Associated with this, the synthetic approach exploring combinatorial chemistry could masterfully explore the available chemical space, supporting the discovery of new molecules [5]. However, the high financial cost and time-related to these approaches have driven researchers to adopt *in silico* methods [7, 8]. In this way, Computer-Aided Drug Design (CADD) emerged and perfected itself, indispensable in any new drug design discovery program [7, 9].

Traditionally, the discovery of a new drug can take between 10 and 15 years, with an investment of approximately US$800 million to US$1.8 billion [10, 11]. In this context, developing new drug design tools has become a constant quest to overcome old paradigms and speed up the discovery process at a lower financial investment [10, 12]. Over time, the scientific community accepted the new paradigm in the rational design of new drugs through CADD [13, 14]. The main reason is constant failures in the clinical evolution of prototypes identified and designed through classical techniques [13]. Thus, this paradigm shift facilitated the identification of new drugs, designing drugs with optimal physicochemical properties, and evaluating their potential *in silico* before they were synthesized [13]. With this, virtual screenings (VS) are increasingly explored, finding drug candidates in libraries of thousands of compounds. In addition, this method can be used in scaffolds identification as a starting point in molecular modeling studies, further confirming the *in silico* methods and rational design in the new era of P&D of drugs [15].

CADD can usually be divided into Structure-Based Drug Design (SBDD), and Ligand-Based Drug Design (LBDD) approaches. The researcher's choice between these approaches is related to the availability of key information about the clinical condition or known compounds against the same [16, 17]. For an SBDD protocol, the main requirement is the knowledge and availability of the target related to the explored clinical condition, in which the ligands are designed to interact with the target in question [18, 19]. On the other hand, in LBDD, there is no information about the target, but there are ligands of known activity against the clinical condition in question, and new molecules can be designed based on the production of pharmacophoric models or Quantitative Structure-Activity Relationship studies (QSAR) [20]. Traditionally, SBDD is preferred by the scientific community mainly due to the easy access to software and the wide availability of experimental structures of biological targets [21 - 23].

Currently, CADD methods using SBDD or LBDD approaches are vital in discovering new molecules and identifying critical information in drug design. Thus, this chapter will present a historical perspective on the evolution of drug design methods to CADD approaches. We hope that this chapter will guide drug developers in deciding on the type of strategy in their studies, increasingly promoting scientific advances in rational drug design.

DRUG DESIGN AND DISCOVERY: PAST AND TODAY METHODS AND OTHER APPROACHES

Strategies used in drug design and discovery have improved over the years [24]. In a historical context, each strategy was responsible for numerous discoveries. However, the improvement of methods made the process faster and more effective in the search for innovative molecules until the arrival of computational methods [25]. The following topics will address the evolution of the methods and their historical importance.

Natural Compounds (NC)

Before any study of rational drug design, Natural Compounds (NC) were the primary sources of drugs explored. During the last five decades, NCs were the target of isolation or total syntheses, as they presented high biological potential and challenging structural complexity. The discovery of numerous NCs against threatening diseases like cancer and infectious diseases has increased the interest in discovering new revolutionary NCs [26]. Indeed, most drugs introduced into the pharmaceutical market since 1994 are NCs or modified synthetic analogs, highlighting their potential for many years [27].

Traditionally, drug discovery by NCs starts with testing the extract of interest in vitro or in vivo assays. After demonstrating the pharmacological effect, the responsible compounds are then isolated [27, 28]. These compounds can be modified from then on to improve their pharmacological effect [27]. In a more current approach, drug repurposing using known NCs is used to find new potentials for available structures [29, 30]. Examples of natural compounds include Artemisinin (**1**), Atropine (**2**), Metformin (**3**), and Quinine (**4**) (Fig. **1**). It is essential to highlight that these molecules were useful as molecular scaffolds that led to important clinical discoveries, which highlights the role of NCs in the

R&D process [31]. However, synthetic difficulties, low quantities isolation, and challenges in optimizing these NCs led to the disuse of this approach. Thus, pharmaceutical companies and academics opted for rational strategies that promoted faster results with less financial costs and improved drug design techniques [27, 32].

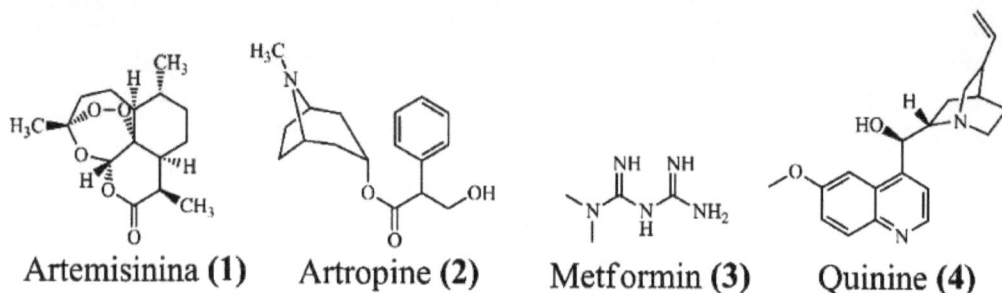

Artemisinina (1)   Artropine (2)   Metformin (3)   Quinine (4)

Fig. (1). Chemical structure of any drugs discovered by NCs.

Synthetic Drugs: Classical Approaches

The passing of the years revealed a decrease in the search for new drugs based on natural products. As previously mentioned, the low yield of isolated products, difficulties in purification processes, and toxicity related to the "wild" structural framework has increasingly instigated the synthetic approach to obtain new drugs. Furthermore, rational molecular modification strategies were proposed and improved over time, leading to the rational design era [33, 34]. These approaches will be presented in the following topics.

Bioisosterism

After identifying an NC with promising activity but an inadequate pharmaco-kinetic and/or toxicological profile, the need arose for strategies that could be coupled to organic synthesis aimed at a rational modification of this chemical scaffold [35]. In this sense, bioisosterism emerges as one of the most effective strategies available to medicinal chemists [36]. The requirements for the application of this strategy are *i)* knowledge about pharmacophoric groups; *ii)* mechanism of action; *iii)* metabolic inactivation pathways; *iv)* physical-chemical properties that determine its bioavailability and side effects [37, 38]. Pharmaceutical companies frequently use this strategy for sales competitiveness (me-too drugs) [38]. This strategy suggests changes that aim to modulate its pharmacological potency, physicochemical properties, and pharmacokinetics [39]. The success of this strategy can be exemplified by several marketed drugs, such as Piroxicam **(5)**, Celecoxib **(6)**, Zidovudine **(7)**, and Cimetidine **(8)** (Fig. 2) through different bioisosteric replacement approaches [38].

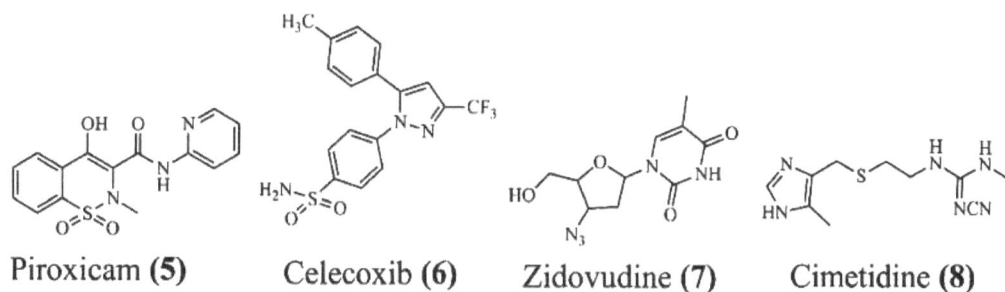

Fig. (2). Drugs designed using bioisosterism approach.

Molecular Simplification

In fact, part of the NCs present high structural complexity, which some authors call "molecular obesity". This is one factor that makes using these structures unfeasible as drugs, as they have an inappropriate molecular pattern and are related to several side effects. For example, molecular simplification was applied to the Morphine **(9)** to obtain the Tramadole **(12)** (Fig. **3**), showing less side effects and highlighting the promising potential if the molecular simplification approach [40]. In this context, the molecular simplification strategy arises, in which molecular groups not essential for the compound's activity are removed, representing one of the most commonly used strategies in *lead* optimization [41, 42]. One of the critical factors in this approach is the elimination of redundant chiral centers, reduction in the number of rings, and scaffold hopping. Thus, structural units are rationally removed to determine their biological importance, and based on structure-activity relationships (SAR) and pharmacophore groups, the essential units are maintained [41]. Furthermore, the reduction in molecular weight is considered a critical factor in improving the pharmacodynamic and pharmacokinetic profile of the simplified analog. Thus, simplifying the molecular structure while maintaining its pharmacological activity is also related to better synthetic accessibility of the analog and accelerates the discovery process. Classic examples of molecular simplification include Morphine **(9)** analogs such as Butorphanol **(10)**, Pentazocine **(11)**, and Tramadol **(12)** (Fig. **3**) [40].

Molecular Hybridization

Another crucial molecular modification strategy performed through organic synthesis is molecular hybridization [43]. Through this, new ligands are designed based on the molecular recognition of two or more pharmacophoric subunits of compounds with known activity. Thus, the fusion of these subunits leads to a hybrid molecule that maintains the biological properties of the original templates, such as physical-chemical, pharmacological, and toxicological properties [44, 45]. Among the advantages of applying this strategy and producing multiple ligands, including *i)* the same molecule interacting with different targets, improving the therapeutic potential, and *ii)* improving bioavailability, and can be used in the production of prodrugs [46]. This strategy is exciting and constantly applied in developing drugs against clinical conditions involving multiple pathways, such cancer and Alzheimer Disease, in which a multitarget inhibitor would be desired [47 - 49]. Based on this strategy, Abourehab *et al.* performed the synthesis of ibuprofen and ketoprofen with pyrrolizine/indolizine aiming at anticancer activity. Thus compounds **(13)**, **(14)**, and **(15)** (Fig. 4) showed promising results against MCF-7 cells (IC^{50} of 7.61, 1.07, and 3.16 µM, respectively). This study exemplifies the potential of this strategy, especially against multifactorial diseases such as cancer [50].

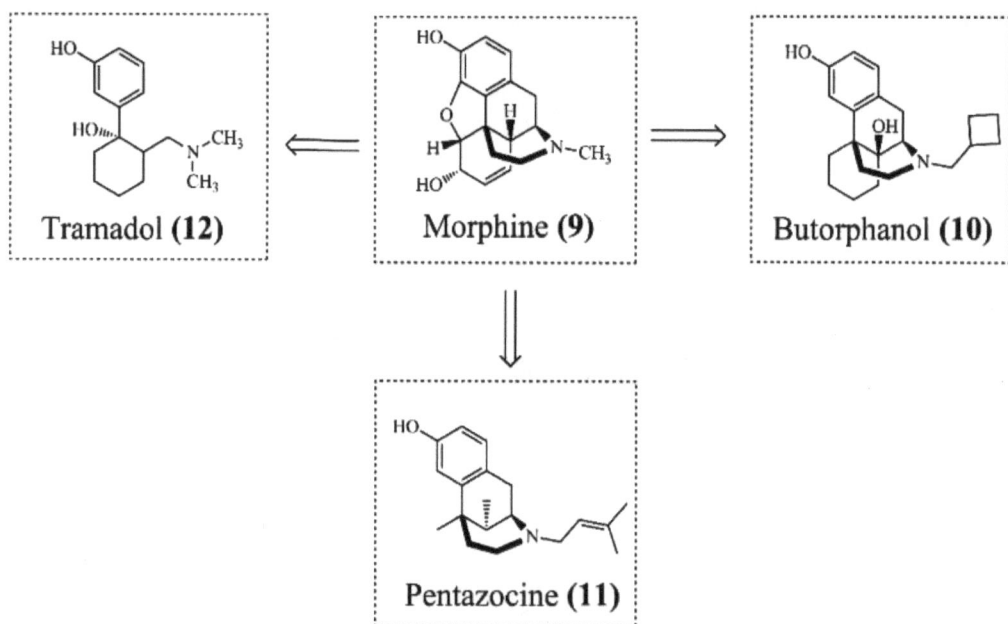

Fig. (3). Morphine and their analogs discovered through molecular simplification.

Combinatorial Chemistry

The improvement of organic synthesis methods linked to the need to find new promising molecules resulted from combinatorial chemistry. One of its objectives is to carry out the serial synthesis of a large number of molecules in less time and financial cost and consequent biological evaluation of this structural diversity [51]. In addition to a synthetic approach, combinatorial chemistry can also be applied to NCs, where ample chemical space is explored to find promising *hits* and *leads* [52].

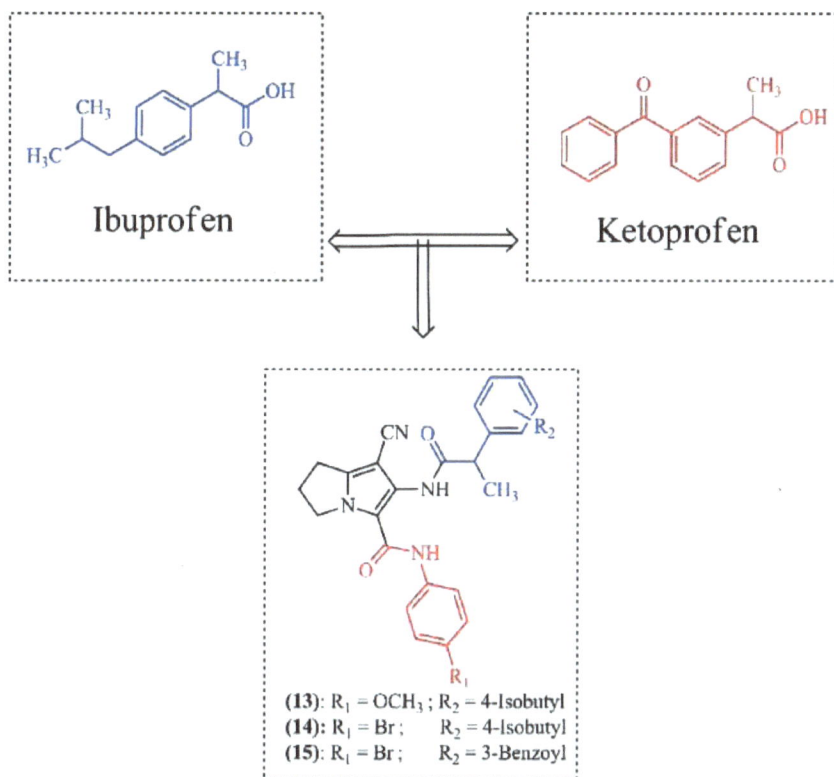

Fig. (4). Drugs designed using a molecular hybridization approach.

Basically, in the combinatorial chemistry approach, a library of structurally diverse compounds is built through repetitive, systematic, or covalent linkages of building blocks (chemical groups). After their synthesis, they are tested against the biological targets of interest. The first studies with the application of combinatorial chemistry were carried out around the 1980s, in which several combinations of peptides synthesized through the solid phase were evaluated. Over time, this method was used for *lead* compound discovery and optimization within a combinatorial library [53].

Despite being an exciting and well-explored strategy, around the 2000s, academics and pharmaceutical companies stopped using this strategy mainly due to the delay in obtaining large libraries and failures in the clinical evolution of the compounds. It became clear then that synthesizing 1,000 to 10,000 compounds would not increase the probability of finding a promising candidate for pre-clinical trials, as discovering a drug is planning, not just numerical combinations. In addition, it is challenging to obtain 10,000 different varieties of a single chemical scaffold, as many times it could present a small structural diversity [54].

High Throughput Screening (HTS)

With the popularization of combinatorial chemistry, the need arose for technologies that could evaluate these large libraries of compounds quickly. Furthermore, the availability of multiple molecular targets has raised the need for technologies that can evaluate libraries against a wide range of targets. In this way, the High Throughput Screening (HTS) technology emerged, in which thousands of compounds can be evaluated in enzymatic or cellular assays, resulting in the discovery of useful leads and hits in drug design [55, 56]. Through HTS, around 10,000 to 100,000 compounds are screened daily, whereas over 100,000 can be performed by ultra-high throughput screening (uHTS) [56].

The most common readings on HTS are fluorescence and bioluminescence. Several fluorescence methods have been developed for HTS, the main ones being fluorescence resonance energy transfer (FRET) and fluorescence quenching energy transfer (QFRET). For example, the b-lactamase gene reporter assay is a widely used FRET-based assay. In addition, luminescence assays can be used, which present a lower signal than fluorescence, but with a more excellent range. The most used luminescence technology is luciferase reporter genes. Furthermore, detection in HTS can also be performed using atomic absorption spectroscopy, high throughput electrophysiology, absorbance, and scintillation proximity assays [57].

However, this approach fell into disuse mainly due to its high financial cost and the low quality of the data generated. The greater the number of compounds tested, the lower the quality and credibility of the results. Another problem is that the compounds are tested at a concentration, which can generate several false negatives [58, 59]. In addition, detection technologies are susceptible to false positives and negatives due to the physical properties of the compounds, which greatly limited and made researchers opt for more effective assay methods, driving the growth of pharmacological approaches in drug discovery [57, 60].

Target-Based Drug Discovery (TBDD)

The Target-Based Drug Discovery (TBDD) approach is one of the most traditional in drug discovery. This is an experimental approach in which compounds are screened against a specific biochemical target of the studied disease. A target can be conceptualized as a gene, gene product, or molecular mechanism identified through biological observations. A genetic target is a gene product or a gene that carries mutations that increase the predisposition to develop the disease (Ex. Alzheimer's disease, schizophrenia, or depression). On the other hand, a mechanistic target is an enzyme or receptor identified through biological observations or mechanism of action of known drugs [61, 62]. In this way, the assay against the specific target can be used to discover or design new compounds, identified as *hits* and *leads,* that can be optimized to improve their pharmacological potency [63, 64]. For many years, this strategy was one of the most used in drug discovery, and it resulted in the discovery of commercial drugs, such as Gefitinib **(16)**, Imatinib **(17)**, Raltegravir **(18)**, and Zanamivir **(19)** (Fig. **5**) [65]. However, using target-based approaches, which do not reflect the reality of the system, is a significant limitation of this technique. Furthermore, the target-based strategy based on a single target cannot simulate entire organisms, which can lead to later failures. Finally, the wrong choice of target can also lead to the identification of molecules that are not useful against the given clinical condition, which the Phenotypic-Based Drug Discovery (PBDD) approach tries to overcome this challenge [66].

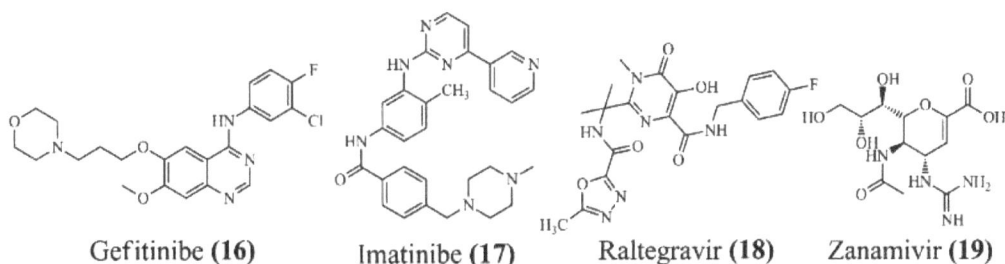

Gefitinibe **(16)**   Imatinibe **(17)**   Raltegravir **(18)**   Zanamivir **(19)**

Fig. (5). Chemical structure of some drugs identified using TBDD approach.

Phenotypic-Based Drug Discovery (PBDD)

Unlike TBDD, in the Phenotypic-Based Drug Discovery (PBDD) approach, drugs are screened against a biological system, usually, cellular assays, in which the evaluation is based on phenotypic modification [67]. Also known as classical pharmacology or direct pharmacology, PBDD evaluates the phenotypic change free of any target hypothesis, in which several biochemical pathways are evaluated and may be involved in the activity of the compounds [68]. For example, inflammation or cancer involves multiples events, and in phenotypic screenings, several drugs targets can be involved in the compounds activity. This approach is historically more promising and related to several successful cases in drug discovery, such as Ezetimibe **(20)**, Vorinostat **(21)**, and Linezolid **(22)** (Fig. **6**). However, not long ago, due to the higher cost and difficulties in optimizing the compounds due to the lack of knowledge of the molecular target, decreased its prevalence in research, and researchers began to opt for TBDD [65, 69]. But because it is a complete approach and uses more complete organisms to screen molecules, academics and pharmaceutical companies are returning to using PBDD as the main approach in their drug discovery programs [67, 70].

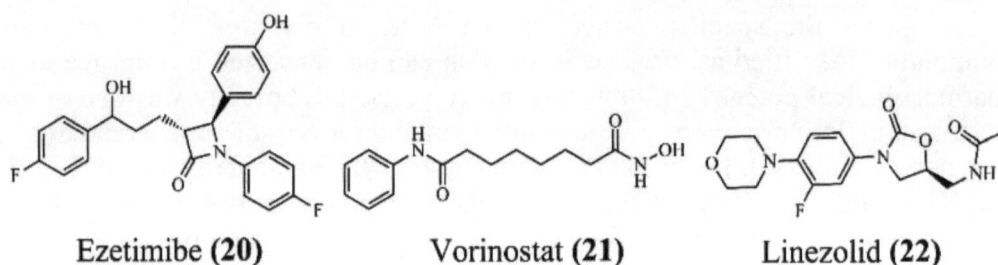

Ezetimibe **(20)**   Vorinostat **(21)**   Linezolid **(22)**

Fig. (6). Chemical structure of some drugs identified using PBDD approach.

Multitarget Drug Design (MDD)

As shown in this chapter, drug discovery has been influenced for many years by many biological targets available. However, despite numerous advances in drug discovery, the number of successful drugs has been falling each year [71, 72]. It is increasingly evident that one of the reasons is the focus on just one target in the search for new molecules. Currently, a new concept of how drugs interact in the body is emerging, increasingly leaving the lock and critical model aside. This is mainly due to polypharmacology, side effects, and drug repositioning. Thus, the concept that a drug interacts with multiple targets grows more and more, in which academics are trying to find a drug and its various biological targets [73, 74].

In the mid-2000s, the general principles of Multitarget Drug Design (MDD) were proposed. Thus, a lead compound that interacts with multiple targets can be a single molecule or composed by the fusion of several structural nuclei in which each one interacts with a different target. One of the requirements is that these fused cores are smaller than the molecule that binds directly to the target, and their separate core binding efficiency is lower than that of clinically used ligands. After identifying the potential drug, it can become an optimized *lead* to improve activity, selectivity, and physicochemical and pharmacokinetic properties. Finally, multitarget drug design constantly uses *in silico* methods [75]. Advances in this technique are evidenced by Entrectinib **(23)** (Fig. **7**), a multitarget inhibitor that acts against the MAPK, PLC-γ, and PI3K pathways in the treatment of nonsmall cell lung cancer with a positive ROS1 fusion gene. Another example is Imeglimin **(24)** (Fig. **7**), an antidiabetic that acts inhibition of the hepatic production of glucose and amplification glucose-stimulated insulin secretion (GSIS) [49].

Computer-Aided Drug Design (CADD)

As shown so far, designing and discovering new drugs is time-consuming and demands a high financial cost. In this context, strategies that could predict the potential of molecules before synthesis and biological evaluation would be ideal since they would make the process agile with less investment [76]. Thus, *in silico* methods of Computer-Aided Drug Design (CADD) appear, in which computer simulations are used to predict the pharmacological potential of a given ligand, which is extremely useful in the initial processes of any drug development campaign [76]. Technological advancement and the "Big Data" era benefited academics and pharmaceutical companies from these methods. These include the availability of large databases of biological targets and ligands and broad access to high-performance computing and high-throughput software, making essential CADD methods in searching for new drugs [77, 78].

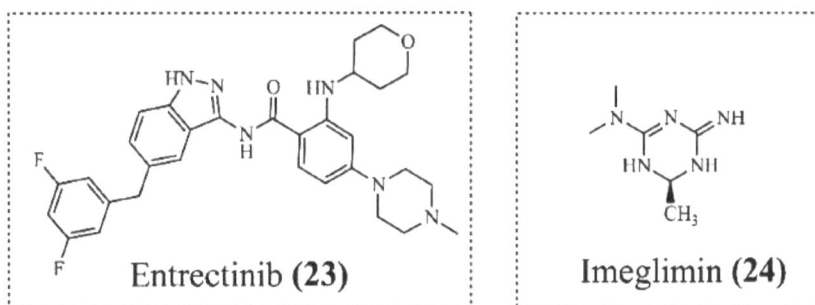

Fig. (7). Approved drug with Multitarget mechanism.

CADD is a multidisciplinary tool used as a shortcut to discover, analyze and develop new drugs. This tool can identify targets and ligands or promote rational changes in lead compounds through two main approaches: SBDD and LBDD (discussed in the next topic). Furthermore, this tool is constantly used to optimize the pharmacokinetic properties of ligands related to absorption, distribution, metabolism, excretion, and toxicity (ADMET). Furthermore, essential requirements for using these tools are practiced in a computational environment to speed up the lead identification/optimization process. As an advantage, it can remove molecules with unwanted properties, choosing only the most promising ones [79]. The success of this technique is highlighted mainly in the discovery of the Human Immunodeficiency Virus (HIV) protease inhibitors Saquinavir (**25**), Indinavir (**26**), Ritonavir (**27**), Amprenavir (**28**) (Fig. **8**), and others that were revolutionaries in the treatment against this virus and continue to be explored in the search for more potent inhibitors [80].

SBDD AND LBDD METHODS IN DRUG DESIGN

Structure-Based Drug Design (SBDD)

A classical protocol of SBDD is applied when there is experimental information about the drug target, with their 3D co-crystallized structure complexed with a ligand. In this way, millions of molecules can be screened and chosen based on the best affinity and complementarity with the target binding site [20, 81]. Molecular docking is a powerful tool used in these protocols, generating information about the interactions between ligands and macromolecules and the best conformation to binding in the active or binding site. In fact, this tool used with pharmacophore models can lead to a more specific screening of ligands [81]. On the other hand, if the target is known and the experimental 3D structure is unavailable, SBDD protocols can be applied using homology modeling. In this way, the target structure is built based on a homologous structure [81]. The following topics will show the main approaches used in SBDD campaigns.

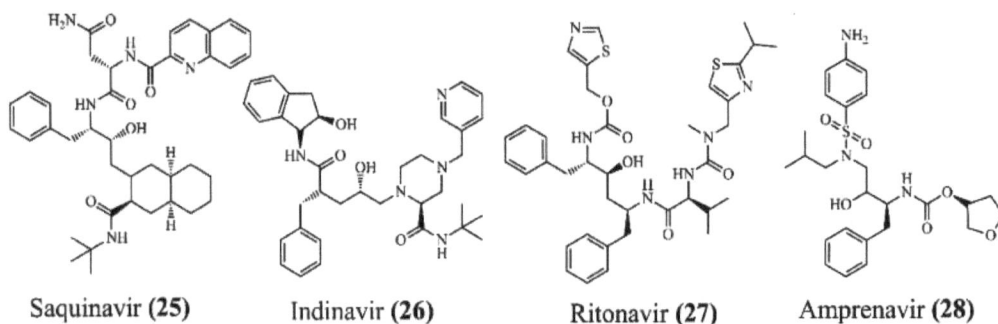

Saquinavir (**25**)   Indinavir (**26**)   Ritonavir (**27**)   Amprenavir (**28**)

Fig. (8). Chemical structure of some HIV protease inhibitors discovered by CADD methods.

Homology Modeling

For success in an SBDD protocol, knowledge about the structure and function of the biological target is essential [82]. Thousands of experimental structures of these drug targets can be easily found in the Research Collaboratory for Structural Bioinformatics Protein Data Bank (RCSB PDB). In addition, thousands of protein and amino acid sequences are available in the National Center for Biotechnology Information (NCBI) database, which can be used for alignment and homology verification or even model building. In fact, the number of sequences is greater than that of experimental 3D structures, which makes homology modeling a powerful tool in drug design by SBDD [83]. As the number of sequences is greater than the number of experimental structures, computational methods are constantly used for the structure prediction of these sequences. Through homology modeling, a 3D model of a target protein is constructed using an amino acid sequence (usually in FASTA format), generating a 3D structure for the intended target, which has not been characterized experimentally [84, 85].

A similarity above 30% between the experimental structure and the sequence is necessary for the targets to be considered homologous. Basically, a procedure for building a model by homology can be applied as follows: *i)* identification of the 3D structure homologates the sequence; *ii)* Alignment of sequences of the 3D structure and the homologous sequence; *iii)* construction of the model from the alignment; *iv)* validation of the constructed structure. This procedure can be repeated with other sequences if the model is not adequately validated [86]. A powerful tool used in this type of protocol is the SWISS-MODEL web server, which performs all these steps and is widely used by the scientific community [87].

Homology modeling was helpful during the COVID-19 pandemic due to the pathological agent SARS-CoV-2 since it was necessary to identify a drug that could prevent this disease from spreading faster and faster worldwide. As an example, one can cite the study by Zhang and Zhou (2020) [88] in the search for a promising compound against RNA-dependent RNA polymerase (RdRp) that could be used against this disease. Thus, RdRP was constructed through homology modeling, using the SARS-CoV RdRp (Identity of 95.8%) as a template. Then, molecular dynamics simulations were performed to obtain the native structure of the protein. Finally, the authors identified Remdesivir **(29)** (Fig. **9**) as a potential inhibitor of RdRp. It should be noted that this drug was approved for hospital use in patients infected with SARS-CoV. This study highlights the importance of homology modeling in the drug discovery process.

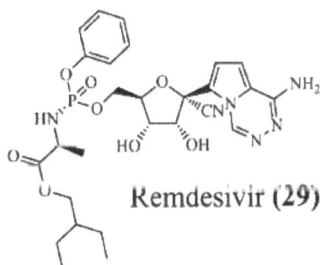

Fig. (9). Chemical structure of Remdesivir.

Molecular Docking and Molecular Dynamics Simulations

Molecular docking and Molecular Dynamics (MD) simulations are essential tools for any drug discovery and development campaign. Through these tools, it is possible to rationally design new compounds and model biochemical processes to propose mechanisms of action and interaction involved in the target inhibition process. Thus, target conformation and ligands favorable to the inhibition process are revealed, helping researchers to search for innovative molecules. This is an *in silico* identification, which can be performed on large ligands libraries even before the compound synthesis [89, 90].

Molecular docking aims to identify the best ligand conformation in its biological receptor (binding or active site). In this way, several conformations, known as binding modes or poses, are generated, and those with greater affinity for the region are analyzed. Clearly, the target 3D structure is essential for this approach. If it is unavailable, homology modeling can be used for structure construction (as explained in the previous topic). Based on this information, this technique is based on: *i)* Generating the binding poses and *ii)* Ranking these conformations and choosing the one with the most excellent affinity for the interaction site [91].

One of the problems of molecular docking is that current programs do not consider target flexibility, which can lead to a mistake in identifying promising molecules. These problems can be solved using MD simulations, assuming the flexibility of proteins and ligands. In addition, more modern methods such as Molecular Mechanics/Poisson-Boltzmann and Surface Area (MM-PBSA) are constantly used to estimate the binding energies of a ligand for its target with values closer to the real, increasing the possibilities of identifying active molecules [92].

With advances in drug design, it is observed that the key-lock model that explains how a drug interacts with its biological receptor has fallen into disuse. This occurs mainly because this model does not consider the proteins' flexibility in interacting with the ligand. Thus, several studies of molecular docking highlight that it is essential to predict the conformation of the receptor during the interaction process. This decreases the probability of identifying false positives in drug discovery campaigns. Thus, a couple of MD simulations and Molecular Docking are constantly used. MD simulation is used to find the best match of a target before the docking protocol or even the use of docking to obtain the complex, followed by an MD simulation to identify the most appropriate target and ligand conformation [93].

The combination of MD simulation protocols with molecular docking is responsible for identifying several commercial drugs. One of the most current is Vaborbactam **(30)** (Fig. **10**), launched on the market in 2017. It is one of the newest β-lactamase inhibitors, useful as an adjuvant in antimicrobial therapy. Its genesis was based on the flexible docking strategy through the ICMdocking software [94]. Other marketed drugs discovered through docking and MD simulations include the protease inhibitors Amprenavir **(28)** (Fig. **8**), approved for clinical use in 1999, in which MD simulations proposed a weak hydrogen bond of an amide group, replaced by *N,N*-dialkyl sulfonamide, improving affinity for the enzyme and resulting in the highlighted compound. However, Amprenavir **(28)** was discontinued in 2004 and replaced by its prodrug, Fosamprenavir **(31)** (Fig. **10**) [94]. Another drug discovered by coupling docking with MD simulations is Raltegravir **(18)** (Fig. **5**) [80].

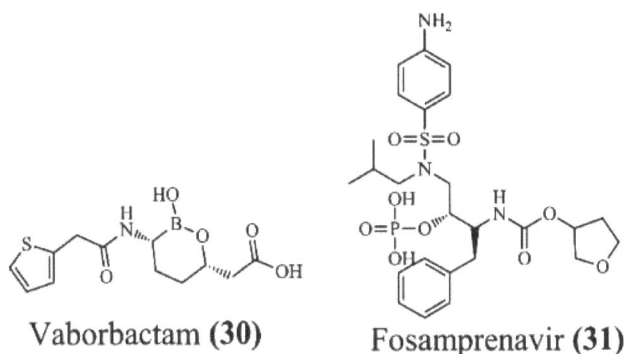

Vaborbactam (**30**)   Fosamprenavir (**31**)

Fig. (10). Chemical structure of the drugs discovered using MD simulations and molecular docking.

Fragment-Based Drug Design (FBDD) or *de novo* Drug Design

Another strategy widely explored in drug discovery campaigns is the Fragment-Based Drug Discovery (FBDD) strategy, which aims to find active molecular fragments and optimize them to improve activity against a specific target [95, 96]. One of the first authors who presented this strategy was Fesik and collaborators (Abbott Laboratories) in 1996, with its conceptualization carried out by Hol and colleagues in the 90s [95]. Since then, FBDD has been associated with several successful cases of discovering new drugs where other methods failed to achieve the same objectives [95, 96].

The first step in an FBDD protocol is constructing a library of low molecular weight molecular fragments designed against a specific target [97]. These fragments must have sufficient size for the interaction site and low chemical complexity so that the interaction occurs most favorably, avoiding unfavorable interactions [96, 97]. Thus, the initial focus is the identification of attractive fragments, followed by their optimization. A fragment must have a weak binding affinity in the range of mM and µM [96]. Thus, the optimization of the power of this fragment is obtained through the fusion, linking, and growing strategies, better-known approaches in FBDD campaigns [96, 97].

The growing approach is preferred by researchers using FBDD [95]. In this approach, a fragment is placed at the target interaction site to perform significant interactions, and this fragment grows to complement this binding site [95]. This protocol is carried out on an experimental structure of the target, obtained by X-rays or NMR, with the placement of the fragments through molecular docking [95, 96]. During the process, it is essential to maintain the interactions of the initial fragment in the optimized compost to obtain a molecule with superior activity [95].

Although less used than the growth approach, the ligation approach is quite versatile in designing promising molecules and optimizing fragments [95]. Thus, the basis of this approach is the addition of fragments in strategic locations of the target binding site, followed by the addition of linkers to unite them and form a single compound with optimized properties [95, 98]. Although simple, this approach is more challenging, as it requires optimal linkage between fragments [98].

On the other hand, the *de novo* drug discovery strategy, similar to previous approaches, is based on the 3D structure of the receptor and pharmacophore groups that constitute the ligand [99]. This approach uses optimizable *lead* compounds from zero to design new chemical structures adapted to the selected target binding site [100]. There are two main ways of using this approach: *i)* outside-in, also known as linking, in which fragments are added to interact at the main sites of the binding site, followed by linking these fragments to obtain a single compound; and *ii)* inside-out, known as growing, in which a fragment is added at the binding site, and it grows in such a way as to present complementarity and essential interactions with the interaction site resulting in the formation of an optimized compound [100].

It is possible to observe that the concepts of FBDD and *de novo* drug discovery are similar. Interestingly, the differences between the two approaches are little explored in the literature, and in most cases, the concepts are confused. Some authors point to the building blocks as one of the differences between the two approaches. In *de novo* drug design starts with smaller fragments. However, the fragment size is not precise, and thus the definition between *de novo* or FBDD is defined by the study authors due to this literature gap [101].

As shown in FBDD and other techniques presented in this chapter, new drug design is explored through computer-aided drug design. Caveat and SPROUT are software exploited using this technique, which places fragments using an outside-in method. On the other hand, the inside-out approach can be applied through the LUDI software [100]. In addition, the LigBuilder V3 program has become a fundamental tool in drug design through the *de novo* approach [102].

The use of this technique resulted in the discovery of Dacomitinib **(32)** (Fig. **11**) by researchers from Pfizer performed. Thus, *in vitro* screening of molecular fragments was performed. The compound was built using different combinations to synthesize new analogs against the epidermal growth factor receptor (EGFR) in the search for new anticancer drugs. Finally, Dacomitinib **(32)** was chosen because it had better pharmacokinetic properties [103]. On the other hand, using virtual fragment screening in a collaboration between academics and pharmaceutical companies (Janssen), Erdafitinib (**33**) (Fig. **10**) was discovered, an inhibitor of Fibroblast Growth Factor Receptors (FGFR) in cancer cell lines (IC_{50} between 0.1 to 130 nM) and selectivity against the inhibition of Vascular Endothelial Growth Factor Receptor 2 (VEGFR2). Its potency and excellent pharmacokinetic properties resulted in its approval in clinical trials and later approval by the FDA Food and Drug Administration [104]. Finally, Zanubrutinib (**34**) (Fig. **11**) was discovered through FBDD as a Novel Covalent Inhibitor of Bruton's Tyrosine Kinase, useful as an anticancer drug [105].

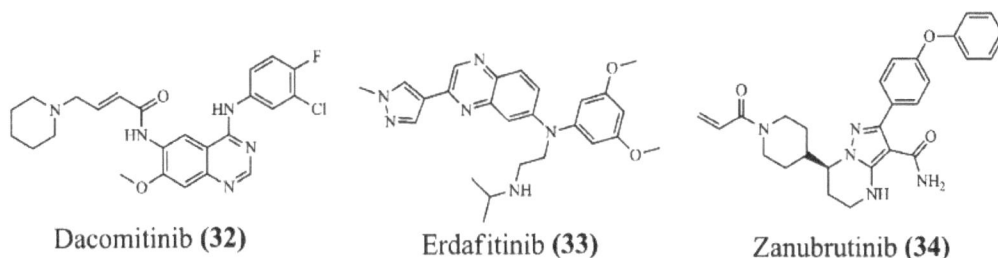

Dacomitinib (**32**)   Erdafitinib (**33**)   Zanubrutinib (**34**)

Fig. (11). Chemical structure of the drugs discovered through FBDD or *de novo* drug design.

Density Function Theory (DFT)

As shown so far, SBDD is based on identifying promising molecules based on their binding energies. However, more accurate methods are still needed to calculate these energies and the interactions of a ligand with the macromolecule. In this context, there is a growing tendency to use quantum mechanics (QM) through CADD approaches. This is due to advances in computing and trends in QM providing more accurate data, making it a powerful tool for medicinal chemists [106, 107]. One of the most explored methods in QM is the Density Functional Theory (DFT), in which ground state energy and other molecular properties are determined through their electronic densities and how they are related to the pharmacological activity of a molecule [108, 109].

Currently, DFT is widely used in the design of drugs in active sites of enzymes, inhibiting their catalytic activity. Despite molecular docking not being a handy tool in these studies, DFT can calculate affinity energy values more accurately, increasing the predictive ability to find new drugs [110]. It is used to predict the most significant interaction forces, yielding critical information that can be used to design new drugs [108]. One of the most explored applications is predicting covalent bonds between an inhibitor and an enzyme. Through this method, it is possible to predict the binding energy for the formation of this type of bond and predict whether the drug can act covalently, an approach that has been widely explored in the design of anticancer, antimicrobial, and antiparasitic drugs [108, 111].

QM also provides *ab initio* methods, widely used in drug design, which require less computational power when compared to DFT. Similar to DFT, they use the electronic Schrödinger equation to calculate binding energies, electron densities, atomic and nuclear coordinates, and other parameters. Thus, using this approach, it was possible to discover Dorzolamide (35) (Fig. 12), a carbonic anhydrase (CA) II inhibitor used in the treatment of glaucoma. Basically, ab initio modeling was crucial in its discovery since its *S*-enantiomer had higher activity than the *R*-enantiomer. In this way, ab initio calculations could estimate the energy related to the conformational state and suggest the most appropriate binding mode. Thus, Dorzolamide (35) was identified as more promising, approved by the FDA, and launched on the market by Merck in 1994 [94].

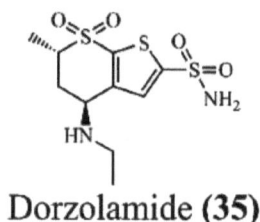

Dorzolamide (35)

Fig. (12). The compound's chemical structure was discovered using the quantum chemistry approach in *ab initio* studies.

Ligand-Based Drug Design (LBDD)

As previously highlighted, LBDD is widely used by drug developers worldwide. Its core concept is based on evaluating a set of ligands' chemical and biological properties, mainly when the target structure is not solved experimentally or is unknown [112]. Thus, an alignment of known ligands with a similar biological effect is carried out, in which their molecular descriptors and the most appropriate mathematical method are identified [113, 114]. The most used are Pharmacophore Modeling and Quantitative structure-activity relationship (QSAR), in which "fingerprints" of molecular attributes are generated in banks of molecules that are used in the design of new compounds with optimized properties [115, 116].

Quantitative Structure-Activity Relationship (QSAR)

Similar to other approaches, QSAR plays a vital role in discovering and developing new drugs as an efficient, economic, and fast alternative to select molecules for pharmacological assays or to optimize them [117, 118]. Both QSAR and QSPR (quantitative structure-property relationship) can predict molecular activity based on the properties of their substituents [119, 120]. A typical QSAR protocol is computational, and the result is mathematical equations that characterize molecular descriptors that relate the ligands and their biological activities [121, 122]. In this way, the process is based on the preparation of the model, analysis, and validation of the results. Thus, molecular descriptors relating chemical-quantum, physical-chemical, geometric and topological properties are needed. With this, the most relevant descriptors are selected and mapped to predict pharmacological activity [123, 124]. It is important to emphasize that a well-validated model depends on the quality of the biological data, and the statistical validation of the results is a crucial factor for the method's reliability [125, 126].

One of the first applications of QSAR resulted in the discovery of the drug Norfloxacin (**36**) (Fig. **13**), a fluoroquinolone with antibacterial activity produced by the company Kyorin Pharmaceutical (Japan) in the 1980s. By activity, and with the Hansch equation demonstrating that disubstituted analogs were more promising than monosubstituted ones, the molecule was modulated until Norfloxacin (**36**) was obtained, approved in 1986 by the FDA and distributed by Merck [94]. Further, combining concepts of LBDD with SBDD, Captopril (**37**) (Fig. **13**) was discovered, demonstrating once again the predictive capacity of LBDD methods and that they can be improved in combination with SBDD methods [80].

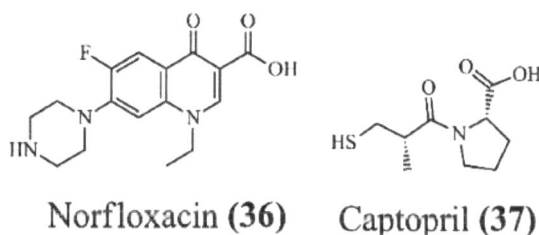

Norfloxacin (**36**)   Captopril (**37**)

Fig. (13). Chemical compounds of the drugs identified using QSAR.

Pharmacophore Modeling

A chemical group essential for the biological activity of a specific molecule is known as a pharmacophore [127, 128]. Such chemical groups have gained prominence in recent years in drug discovery campaigns, useful in virtual screenings based on pharmacophores to identify promising *hits* and *leads* [127]. Although knowledge about the pharmacophore concept is older than the use of computers in drug design, it has become essential in CADD campaigns, mainly through the molecular docking technique [127, 129]. In this approach, each atom has its molecular recognition properties, such as hydrophobic, H-bond donors or acceptors, aromatics, cations, anions, or a combination of these interactions that aid molecular recognition [127, 130].

The construction of a pharmacophore model can be either structure-based or ligand-based. In the first, interaction points are designed into the target structure, and ligands are designed to interact with these regions. On the other hand, in the second approach, design is done based on the essential chemical groups of a set of superimposed active ligands [128]. In this way, pharmacophoric modeling provides an overview of the initial structure of a ligand to improve its interaction with the receptor and its biological activity [131].

A successful application of pharmacophoric modeling in the search for new drugs was the discovery of Zolmitriptan (**38**) (Fig. **14**), a 5-hydroxytryptamine 1 (5-HT$_1$) receptor agonist indicated for the treatment of acute migraine. This drug was discovered at the time by Wellcome Research Laboratories (today Glaxo Wellcome) but licensed and released by AstraZeneca. Thus, through a pharmacophoric model of known ligands with activity against 5HT$_1$ and molecular mechanics calculations, a pharmacophore hypothesis was generated (protonated amine, aromatic, hydrophobic interaction, and two H-bonds), and a pharmacokinetic study based on LogP resulted in the design of Zolmitriptan (**38**), approved by the FDA in 2003 [94]. Further, Tirofiban (**39**) (Fig. **14**), an antiplatelet agent that inhibits Integrin (GP) IIb/IIIa and Fibrinogen receptor, was also identified through pharmacophore-based screens. In addition, Saroglitazar (**40**) (Fig. **14**), an antidiabetic drug that inhibits PPAR, was discovered through screenings based on pharmacophoric models and SBDD protocols, highlighting the predictive capacity of drug identification with these methods [80].

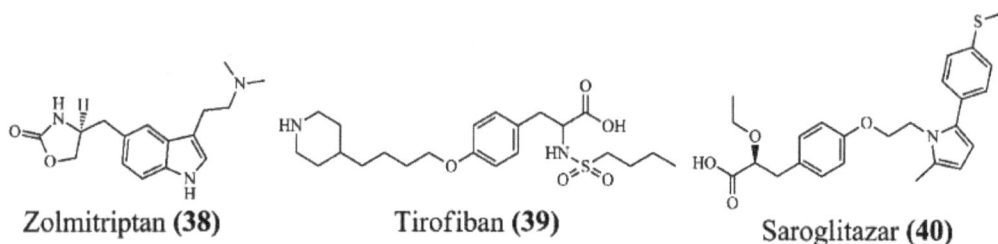

Zolmitriptan (**38**)   Tirofiban (**39**)   Saroglitazar (**40**)

Fig. (14). Chemical structure of the drugs identified by Pharmacophore Modeling in an LBDD approach.

Machine and Deep Learning and Artificial Methods

In recent years, the computational revolution and the vast amount of data made available, known as the "Big Data" *era*, together with high computational capacity using Graphics Processing Units (GPU), has created a new phase in drug design [132]. This is because the large volume of accumulated chemical and biological data provided a path for using methods based on Artificial Intelligence (AI) [132, 133]. Thus, it is known that physicochemical properties and other descriptors are related to a molecule's pharmacological activity and pharmacokinetic properties, and that AI results can help design new drugs [132]. Typically, AI methods comprise three steps: *i)* Selection of specific descriptors capable of predicting the essential properties in the bioactivity of the molecule; *ii)*

A metric or score assigned to the molecules to compare them; and *iii)* a Machine Learning (ML) algorithm is used to qualitatively and quantitatively distinguish active and inactive composites [134].

In this context, the Machine Learning (ML) technique arises, in which the best known are Artificial Neural Networks (ANN), Partial Least Squares (PLS), and k-Nearest Neighbors (kNN). Within AI and ML, there is Deep Learning (DL), which focuses on making computers learn from the data obtained, in which the developed protocols can identify new drugs. It is worth highlighting the ANN, most used in the research of new drugs. In DL, the term "deep" is related to the deepest layers of the network, expanding the generation of essential data in drug design [132]. DL are ML algorithms that use ANNs in several layers of processing units in a non-linear way, modeling high-level abstractions of data [135]. Like neural networks in the human brain, ANNs are organized in such a way that each neuron is a processing unit that interconnects with several others, showing the incredible power of computing in solving complex operations [135].

The importance of methods in DL is demonstrated in the study by Liu and colleagues (2021) [136] in a drug repurposing study to find drugs against coronary artery disease. In this way, the authors generated a high-performance protocol with a correlation of treatment effect biases, maintaining the ability to interpret and estimate critical confounding factors. Drugs such as Amlodipine **(41)**, Diltiazem **(42)**, and Rosuvastatin **(43)** (Fig. **15**) were identified as promising, demonstrating that their protocol is superior to others available in the literature.

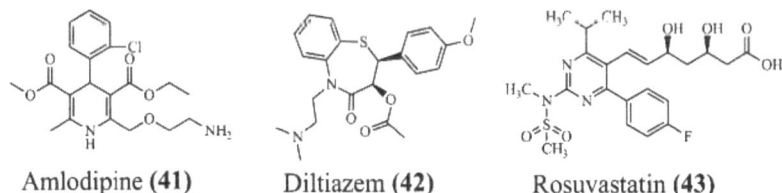

Amlodipine **(41)**   Diltiazem **(42)**   Rosuvastatin **(43)**

Fig. (15). Chemical compounds identified using DL methods.

CHALLENGES AND OPPORTUNITIES IN LBDD AND SBDD APPROACHES TO DESIGN AND DISCOVER NEW DRUGS

Notably, there have been significant advances in CADD, especially in the SBDD and LBDD approaches. However, some challenges still need to be overcome, and new opportunities are generated for drug developers. One of them is the limited processing capacity in academic environments. Despite advances, more significant investment in high-quality hardware software is still needed to sustain research and provide new opportunities in the P&D process in academic settings [137].

Despite advances in the development of molecular docking software in SBDD approaches, there is still great difficulty in incorporating target flexibility through this tool. It is scientifically widespread that the use of rigid structures in this type of protocol is one of the main factors responsible for errors in virtual drug screenings. Although MD simulations try to solve this flexibility problem, there is still a time-related difficulty, as it is a time-consuming and computationally expensive technique. Thus, there is a need to develop docking programs that produce flexibility in the ligand and the target, reinforcing new opportunities in this area [22, 24, 138]. Furthermore, one of the challenges to be overcome in LBDD is the lack of databases of low-energy ligand conformations. The lack of these conformations limits the probability of finding and designing a promising molecule, mainly based on molecules with multiple rotating bonds [78, 138].

Furthermore, validation of results and fully automated processes are necessary for CADD. For example, the secret of success in a QSAR-based screening is protocol validation, which is considered a critical step that needs further improvement. Furthermore, the *de novo* approach (FBDD) becomes tedious and time-consuming due to the manual process. Thus, there is still a need for fully automated protocols that can overcome these difficulties and increase the probability of finding promising molecules [100, 139].

Finally, the choice of target in SBDD protocols is the most relevant factor for the study's success. Choosing the wrong target can make the whole research unfeasible since its intervention may not generate an effective response in the studied clinical condition [137]. Thus, advances in genomics, proteomics, and metabolomics are crucial and need to improve more and more to avoid this type of problem, showing new opportunities in these research fields [140].

CONCLUSION

It is clear that discovering a new drug with unique and innovative properties is a process that demands time, financial cost, and integration between different areas of research. Despite the remarkable success of the classic rational design and drug identification approaches, CADD methods are increasingly widespread and used by academics and pharmaceutical companies, essential in any drug discovery program. The SBDD and LBDD approaches are constantly explored and have resulted in promising drugs. Increasingly, molecular docking, MD simulations, pharmacophoric models, FBDD, and QSAR are being used and improved, with protocols close to accurate, leading to the identification of clinically useful analogs. In addition, methods based on machine learning, such as deep learning and artificial intelligence algorithms, are emerging that increasingly drive CADD methods in the search for an innovative molecule. Improving these methods is the next step in the drug development process. Finally, we hope that this chapter provides insights and serves as a guide for researchers worldwide in searching for an innovative molecule capable of improving the health and quality of life of the world's population.

ACKNOWLEDGMENTS

The authors thank the Coordenação de Aperfeiçoamento de Pessoal de Nível Superior – Brazil (CAPES) and the National Council for Scientific and Technological Development (CNPq) – Brazil for their support to the Brazilian Post-Graduate Programs.

REFERENCES

[1]   Lima AN, Philot EA, Trossini GHG, Scott LPB, Maltarollo VG, Honorio KM. Use of machine learning approaches for novel drug discovery. Expert Opin Drug Discov 2016; 11(3): 225-39.
[http://dx.doi.org/10.1517/17460441.2016.1146250] [PMID: 26814169]

[2]   Guido RVC, Oliva G, Andricopulo AD. Modern drug discovery technologies: Opportunities and challenges in lead discovery. Comb Chem High Throughput Screen 2011; 14(10): 830-9.
[http://dx.doi.org/10.2174/138620711797537067] [PMID: 21843147]

[3]   Ban TA. The role of serendipity in drug discovery. Dialogues Clin Neurosci 2006; 8(3): 335-44.
[http://dx.doi.org/10.31887/DCNS.2006.8.3/tban] [PMID: 17117615]

[4]   Drews J. Drug discovery: A historical perspective Science 2000; 287(5460): 1960-4.
[http://dx.doi.org/10.1126/science.287.5460.1960]

[5]   Ferreira LLG, Andricopulo AD. Editorial: Chemoinformatics approaches to structure- and ligand-based drug design. Front Pharmacol 2018; 9: 1416.
[http://dx.doi.org/10.3389/fphar.2018.01416] [PMID: 30564124]

[6]   Mandal S, Moudgil M, Mandal SK. Rational drug design. Eur J Pharmacol 2009; 625(1-3): 90-100.
[http://dx.doi.org/10.1016/j.ejphar.2009.06.065] [PMID: 19835861]

[7]   Nascimento IJS, de Aquino TM, da Silva-Júnior EF. The new *era* of drug discovery: The power of computer-aided drug design (CADD). Lett Drug Des Discov 2022; 19(11): 951-5.
[http://dx.doi.org/10.2174/1570180819666220405225817]

[8]   Nascimento IJS, da Silva Santos-Júnior PF, de Araújo-Júnior JX, da Silva-Júnior EF. Strategies in medicinal chemistry to discover new hit compounds against ebola virus: Challenges and perspectives in drug discovery. Mini Rev Med Chem 2022; 22(22): 2896-924.
[http://dx.doi.org/10.2174/1389557522666220404085858] [PMID: 35379146]

[9]   Nascimento IJS, de Aquino TM, da Silva-Júnior EF. Drug repurposing: A strategy for discovering inhibitors against emerging viral infections. Curr Med Chem 2021; 28(15): 2887-942.
[http://dx.doi.org/10.2174/1875533XMTA5rMDYp5] [PMID: 32787752]

[10]   Macalino SJY, Gosu V, Hong S, Choi S. Role of computer-aided drug design in modern drug discovery. Arch Pharm Res 2015; 38(9): 1686-701.
[http://dx.doi.org/10.1007/s12272-015-0640-5] [PMID: 26208641]

[11]   Huang HJ, Yu HW, Chen CY, *et al.* Current developments of computer-aided drug design. J Taiwan Inst Chem Eng 2010; 41(6): 623-35.
[http://dx.doi.org/10.1016/j.jtice.2010.03.017]

[12]   Taft CA, da Silva VB, da Silva CHTP. Current topics in computer-aided drug design. J Pharm Sci 2008; 97(3): 1089-98.

[http://dx.doi.org/10.1002/jps.21293] [PMID: 18214973]

[13]   Song CM, Lim SJ, Tong JC. Recent advances in computer-aided drug design. Brief Bioinform 2009; 10(5): 579-91.
[http://dx.doi.org/10.1093/bib/bbp023] [PMID: 19433475]

[14]   Yu W, MacKerell AD. Computer-Aided Drug Design Methods. In: Sass, P. (eds) Antibiotics. Methods in Molecular Biology, 2017, vol 1520. Humana Press, New York, NY.
[http://dx.doi.org/10.1007/978-1-4939-6634-9]

[15]   Talevi A. Computer-aided drug design: An overview. Gore M, Jagtap U. Computational Drug Discovery and Design. Methods in Molecular BiologyNew York, NY: Humana Press 2018; 1762: pp. 1-19.
[http://dx.doi.org/10.1007/978-1-4939-7756-7_1]

[16]   Grinter S, Zou X. Challenges, applications, and recent advances of protein-ligand docking in structure-based drug design. Molecules 2014; 19(7): 10150-76.
[http://dx.doi.org/10.3390/molecules190710150] [PMID: 25019558]

[17]   He H, Liu B, Luo H, Zhang T, Jiang J. Big data and artificial intelligence discover novel drugs targeting proteins without 3D structure and overcome the undruggable targets. Stroke Vasc Neurol 2020; 5(4): 381-7.
[http://dx.doi.org/10.1136/svn-2019-000323] [PMID: 33376199]

[18]   Ferreira L, dos Santos R, Oliva G, Andricopulo A. Molecular docking and structure-based drug design strategies. Molecules 2015; 20(7): 13384-421.
[http://dx.doi.org/10.3390/molecules200713384] [PMID: 26205061]

[19]   Garofalo M, Grazioso G, Cavalli A, Sgrignani J. How computational chemistry and drug delivery techniques can support the development of new anticancer drugs. Molecules 2020; 25(7): 1756.
[http://dx.doi.org/10.3390/molecules25071756] [PMID: 32290224]

[20]   Surabhi S, Singh BK. Computer aided drug design: An overview. J Drug Deliv Ther 2018; 8(5): 504-9.
[http://dx.doi.org/10.22270/jddt.v8i5.1894]

[21]   Nascimento IJS, da Silva-Júnior EF, de Aquino TM. Molecular modeling targeting transmembrane serine protease 2 (TMPRSS2) as an alternative drug target against coronaviruses. Curr Drug Targets 2022; 23(3): 240-59.
[http://dx.doi.org/10.2174/1389450122666210809090909] [PMID: 34370633]

[22]   Nascimento IJS, de Aquino TM, da Silva Júnior EF. Computer-aided drug design of anti-inflammatory agents targeting microsomal prostaglandin E2 synthase-1 (mPGES-1). Curr Med Chem 2022; 29(33): 5397-419.
[http://dx.doi.org/10.2174/0929867329666220317122948] [PMID: 35301943]

[23]   Nascimento IJS, da Silva Júnior EF, de Aquino TM. Repurposing FDA-approved drugs targeting SARS-CoV2 3CLpro: A study by applying virtual screening, molecular dynamics, MM-PBSA calculations and covalent docking. Lett Drug Des Discov 2022; 19(7): 637-53.
[http://dx.doi.org/10.2174/1570180819666220106110133]

[24]   Nascimento IJS, de Aquino TM, and da Silva-Júnior EF. Molecular Dynamics Applied to Discover Antiviral Agents. In: Ul-Haq Z, K. Wilson A. Frontiers in Computational Chemistry. Bentham Science Publishers 2022, pp. 62–131.
[http://dx.doi.org/10.2174/9789815036848122060005]

[25]   Nascimento IJS, de Aquino TM, da Silva Santos-Júnior PF, de Araújo-Júnior JX, da Silva-Júnior EF. Molecular modeling applied to design of cysteine protease inhibitors. A powerful tool for the identification of hit compounds against neglected tropical diseases. Front Computa Chem 2020; 5: 63-110.
[http://dx.doi.org/10.2174/9789811457791120050004]

[26] Beghyn T, Deprez-Poulain R, Willand N, Folleas B, Deprez B. Natural compounds: Leads or ideas? Bioinspired molecules for drug discovery. Chem Biol Drug Des 2008; 72(1): 3-15.
[http://dx.doi.org/10.1111/j.1747-0285.2008.00673.x] [PMID: 18554253]

[27] Harvey AL, Clark RL, Mackay SP, Johnston BF. Current strategies for drug discovery through natural products. Expert Opin Drug Discov 2010; 5(6): 559-68.
[http://dx.doi.org/10.1517/17460441.2010.488263] [PMID: 22823167]

[28] Hao H, Zheng X, Wang G. Insights into drug discovery from natural medicines using reverse pharmacokinetics. Trends Pharmacol Sci 2014; 35(4): 168-77.
[http://dx.doi.org/10.1016/j.tips.2014.02.001] [PMID: 24582872]

[29] Rastelli G, Pellati F, Pinzi L, Gamberini MC. Repositioning natural products in drug discovery. Molecules 2020; 25(5): 1154.
[http://dx.doi.org/10.3390/molecules25051154] [PMID: 32143476]

[30] da Silva-Júnior EF, Nascimento IJS. TNF-α inhibitors from natural compounds: An overview, CADD approaches, and their exploration for anti-inflammatory agents. Comb Chem High Throughput Screen 2022; 25(14): 2317-40.
[http://dx.doi.org/10.2174/1386207324666210715165943] [PMID: 34269666]

[31] Zhang L, Song J, Kong L, *et al.* The strategies and techniques of drug discovery from natural products. Pharmacol Ther 2020; 216: 107686.
[http://dx.doi.org/10.1016/j.pharmthera.2020.107686] [PMID: 32961262]

[32] Nascimento IJS, de Aquino TM, da Silva-Júnior EF. Molecular docking and dynamics simulation studies of a dataset of NLRP3 inflammasome inhibitors. Rec Adv Inflam All Drug Disc 2022; 15(2): 80-6.
[http://dx.doi.org/10.2174/2772270816666220126103909]

[33] Karimi A, Majlesi M, Rafieian-Kopaei M. Herbal versus synthetic drugs; beliefs and facts. J nephropharmacology 2015; 4(1): 27-30. Available from: http://www.ncbi.nlm.nih.gov/pubmed/28197471

[34] Bade R, Chan HF, Reynisson J. Characteristics of known drug space. Natural products, their derivatives and synthetic drugs. Eur J Med Chem 2010; 45(12): 5646-52.
[http://dx.doi.org/10.1016/j.ejmech.2010.09.018] [PMID: 20888084]

[35] Patani GA, LaVoie EJ. Bioisosterism: A rational approach in drug design. Chem Rev 1996; 96(8): 3147-76.
[http://dx.doi.org/10.1021/cr950066q] [PMID: 11848856]

[36] Brown N. Bioisosteres and scaffold hopping in medicinal chemistry. Mol Inform 2014; 33(6-7): 458-62.
[http://dx.doi.org/10.1002/minf.201400037] [PMID: 27485983]

[37] Jayashree BS, Nikhil PS, Paul S. Bioisosterism in drug discovery and development- An overview. Med Chem 2022; 18(9): 915-25.
[http://dx.doi.org/10.2174/1573406418666220127124228] [PMID: 35086456]

[38] Lima L, Barreiro E. Bioisosterism: A useful strategy for molecular modification and drug design. Curr Med Chem 2005; 12(1): 23-49.
[http://dx.doi.org/10.2174/0929867053363540] [PMID: 15638729]

[39] Papadatos G, Brown N. *In silico* applications of bioisosterism in contemporary medicinal chemistry practice. Wiley Interdiscip Rev Comput Mol Sci 2013; 3(4): 339-54.
[http://dx.doi.org/10.1002/wcms.1148]

[40] Wang S, Dong G, Sheng C. Structural simplification: An efficient strategy in lead optimization. Acta Pharm Sin B 2019; 9(5): 880-901.
[http://dx.doi.org/10.1016/j.apsb.2019.05.004] [PMID: 31649841]

[41] Wang S, Dong G, Sheng C. Structural simplification of natural products. Chem Rev 2019; 119(6): 4180-220.
[http://dx.doi.org/10.1021/acs.chemrev.8b00504] [PMID: 30730700]

[42] Pinacho Crisóstomo FR, Carrillo R, León LG, Martín T, Padrón JM, Martín VS. Molecular simplification in bioactive molecules: Formal synthesis of (+)-muconin. J Org Chem 2006; 71(6): 2339-45.
[http://dx.doi.org/10.1021/jo0524674] [PMID: 16526782]

[43] Nepali K, Sharma S, Sharma M, Bedi PMS, Dhar KL. Rational approaches, design strategies, structure activity relationship and mechanistic insights for anticancer hybrids. Eur J Med Chem 2014; 77: 422-87.
[http://dx.doi.org/10.1016/j.ejmech.2014.03.018] [PMID: 24685980]

[44] Fraga CAM. Drug hybridization strategies: Before or after lead identification? Expert Opin Drug Discov 2009; 4(6): 605-9.
[http://dx.doi.org/10.1517/17460440902956636] [PMID: 23489153]

[45] Viegas-Junior C, Danuello A, da Silva Bolzani V, Barreiro EJ, Fraga CAM. Molecular hybridization: A useful tool in the design of new drug prototypes. Curr Med Chem 2007; 14(17): 1829-52.
[http://dx.doi.org/10.2174/092986707781058805] [PMID: 17627520]

[46] Bosquesi PL, Melo TRF, Vizioli EO, Santos JL, Chung MC. Anti-inflammatory drug design using a molecular hybridization approach. Pharmaceuticals 2011; 4(11): 1450-74.
[http://dx.doi.org/10.3390/ph4111450] [PMID: 27721332]

[47] Szumilak M, Wiktorowska-Owczarek A, Stanczak A. Hybrid drugs- A strategy for overcoming anticancer drug resistance? Molecules 2021; 26(9): 2601.
[http://dx.doi.org/10.3390/molecules26092601] [PMID: 33946916]

[48] Lage-Rupprecht V, Schultz B, Dick J, et al. A hybrid approach unveils drug repurposing candidates targeting an Alzheimer pathophysiology mechanism. Patterns 2022; 3(3): 100433.
[http://dx.doi.org/10.1016/j.patter.2021.100433] [PMID: 35510183]

[49] Nascimento IJS, de Moura RO. Would the development of a multitarget inhibitor of 3CLpro and TMPRSS2 be promising in the fight against SARS-CoV-2? Med Chem 2022; 18.
[http://dx.doi.org/10.2174/1573406418666221011093439] [PMID: 36221875]

[50] Abourehab MAS, Alqahtani AM, Almalki FA, et al. Pyrrolizine/Indolizine-NSAID hybrids: Design, synthesis, biological evaluation, and molecular docking studies. Molecules 2021; 26(21): 6582.
[http://dx.doi.org/10.3390/molecules26216582] [PMID: 34770990]

[51] Patel D, Gordon E. Applications of small-molecule combinatorial chemistry to drug discovery. Drug Discov Today 1996; 1(4): 134-44.
[http://dx.doi.org/10.1016/1359-6446(96)89062-3]

[52] Ortholand JY, Ganesan A. Natural products and combinatorial chemistry: Back to the future. Curr Opin Chem Biol 2004; 8(3): 271-80.
[http://dx.doi.org/10.1016/j.cbpa.2004.04.011] [PMID: 15183325]

[53] Liu R, Li X, Lam KS. Combinatorial chemistry in drug discovery. Curr Opin Chem Biol 2017; 38: 117-26.
[http://dx.doi.org/10.1016/j.cbpa.2017.03.017] [PMID: 28494316]

[54] Kennedy JP, Williams L, Bridges TM, Daniels RN, Weaver D, Lindsley CW. Application of combinatorial chemistry science on modern drug discovery. J Comb Chem 2008; 10(3): 345-54.
[http://dx.doi.org/10.1021/cc700187t] [PMID: 18220367]

[55] Carnero A. High throughput screening in drug discovery. Clin Transl Oncol 2006; 8(7): 482-90.
[http://dx.doi.org/10.1007/s12094-006-0048-2] [PMID: 16870538]

[56] Mayr LM, Bojanic D. Novel trends in high-throughput screening. Curr Opin Pharmacol 2009; 9(5):

580-8.
[http://dx.doi.org/10.1016/j.coph.2009.08.004] [PMID: 19775937]

[57]   Attene-Ramos MS, Austin CP, Xia M. High throughput screening. Encyclopedia of Toxicology Elsevier 2014; 916-7.
[http://dx.doi.org/10.1016/B978-0-12-386454-3.00209-8]

[58]   Mayr LM, Fuerst P. The future of high-throughput screening. SLAS Discov 2008; 13(6): 443-8.
[http://dx.doi.org/10.1177/1087057108319644] [PMID: 18660458]

[59]   Macarron R, Banks MN, Bojanic D, *et al.* Impact of high-throughput screening in biomedical research. Nat Rev Drug Discov 2011; 10(3): 188-95.
[http://dx.doi.org/10.1038/nrd3368] [PMID: 21358738]

[60]   Inglese J, Johnson RL, Simeonov A, *et al.* High-throughput screening assays for the identification of chemical probes. Nat Chem Biol 2007; 3(8): 466-79.
[http://dx.doi.org/10.1038/nchembio.2007.17] [PMID: 17637779]

[61]   Vasaikar S, Bhatia P, Bhatia PG, Chu Yaiw K. Complementary approaches to existing target based drug discovery for identifying novel drug targets. Biomedicines 2016; 4(4): 27.
[http://dx.doi.org/10.3390/biomedicines4040027] [PMID: 28536394]

[62]   Samsdodd F. Target-based drug discovery: Is something wrong? Drug Discov Today 2005; 10(2): 139-47.
[http://dx.doi.org/10.1016/S1359-6446(04)03316-1] [PMID: 15718163]

[63]   Brown D. Unfinished business: Target-based drug discovery. Drug Discov Today 2007; 12(23-24): 1007-12.
[http://dx.doi.org/10.1016/j.drudis.2007.10.017] [PMID: 18061878]

[64]   Kana BD, Karakousis PC, Parish T, Dick T. Future target-based drug discovery for tuberculosis? Tuberculosis 2014; 94(6): 551-6.
[http://dx.doi.org/10.1016/j.tube.2014.10.003] [PMID: 25458615]

[65]   Swinney DC. Phenotypic vs. target-based drug discovery for first-in-class medicines. Clin Pharmacol Ther 2013; 93(4): 299-301.
[http://dx.doi.org/10.1038/clpt.2012.236] [PMID: 23511784]

[66]   Croston GE. The utility of target-based discovery. Expert Opin Drug Discov 2017; 12(5): 427-9.
[http://dx.doi.org/10.1080/17460441.2017.1308351] [PMID: 28306350]

[67]   Berg EL. The future of phenotypic drug discovery. Cell Chem Biol 2021; 28(3): 424-30.
[http://dx.doi.org/10.1016/j.chembiol.2021.01.010] [PMID: 33529582]

[68]   Zheng W, Thorne N, McKew JC. Phenotypic screens as a renewed approach for drug discovery. Drug Discov Today 2013; 18(21-22): 1067-73.
[http://dx.doi.org/10.1016/j.drudis.2013.07.001] [PMID: 23850704]

[69]   Moffat JG, Vincent F, Lee JA, Eder J, Prunotto M. Opportunities and challenges in phenotypic drug discovery: An industry perspective. Nat Rev Drug Discov 2017; 16(8): 531-43.
[http://dx.doi.org/10.1038/nrd.2017.111] [PMID: 28685762]

[70]   Childers WE, Elokely KM, Abou-Gharbia M. The resurrection of phenotypic drug discovery. ACS Med Chem Lett 2020; 11(10): 1820-8.
[http://dx.doi.org/10.1021/acsmedchemlett.0c00006] [PMID: 33062159]

[71]   Ramsay RR, Popovic-Nikolic MR, Nikolic K, Uliassi E, Bolognesi ML. A perspective on multi-target drug discovery and design for complex diseases. Clin Transl Med 2018; 7(1): 3.
[http://dx.doi.org/10.1186/s40169-017-0181-2] [PMID: 29340951]

[72]   Csermely P, Agoston V, Pongor S. The efficiency of multi-target drugs: The network approach might help drug design. Trends Pharmacol Sci 2005; 26(4): 178-82.
[http://dx.doi.org/10.1016/j.tips.2005.02.007] [PMID: 15808341]

[73]   Espinoza-Fonseca LM. The benefits of the multi-target approach in drug design and discovery. Bioorg Med Chem 2006; 14(4): 896-7.
[http://dx.doi.org/10.1016/j.bmc.2005.09.011] [PMID: 16203151]

[74]   Medina-Franco JL, Giulianotti MA, Welmaker GS, Houghten RA. Shifting from the single to the multitarget paradigm in drug discovery. Drug Discov Today 2013; 18(9-10): 495-501.
[http://dx.doi.org/10.1016/j.drudis.2013.01.008] [PMID: 23340113]

[75]   Zhang W, Pei J, Lai L. Computational multitarget drug design. J Chem Inf Model 2017; 57(3): 403-12.
[http://dx.doi.org/10.1021/acs.jcim.6b00491] [PMID: 28166637]

[76]   Llorach-Pares L, Nonell-Canals A, Avila C, Sanchez-Martinez M. Computer-aided drug design (CADD) to de-orphanize marine molecules: Finding potential therapeutic agents for neurodegenerative and cardiovascular diseases. Mar Drugs 2022; 20(1): 53.
[http://dx.doi.org/10.3390/md20010053] [PMID: 35049908]

[77]   Cole DJ, Horton JT, Nelson L, Kurdekar V. The future of force fields in computer-aided drug design. Future Med Chem 2019; 11(18): 2359-63.
[http://dx.doi.org/10.4155/fmc-2019-0196] [PMID: 31544529]

[78]   Zhao L, Ciallella HL, Aleksunes LM, Zhu H. Advancing computer-aided drug discovery (CADD) by big data and data-driven machine learning modeling. Drug Discov Today 2020; 25(9): 1624-38.
[http://dx.doi.org/10.1016/j.drudis.2020.07.005] [PMID: 32663517]

[79]   Rajkishan T, Rachana A, Shruti S, Bhumi P, Patel D. Computer-aided drug designing.Advances in Bioinformatics. Singapore: Springer Singapore 2021; pp. 151-82.
[http://dx.doi.org/10.1007/978-981-33-6191-1_9]

[80]   Sabe VT, Ntombela T, Jhamba LA, *et al.* Current trends in computer aided drug design and a highlight of drugs discovered *via* computational techniques: A review. Eur J Med Chem 2021; 224: 113705.
[http://dx.doi.org/10.1016/j.ejmech.2021.113705] [PMID: 34303871]

[81]   Njogu PM, Guantai EM, Pavadai E, Chibale K. Computer-aided drug discovery approaches against the tropical infectious diseases malaria, tuberculosis, trypanosomiasis, and leishmaniasis. ACS Infect Dis 2016; 2(1): 8-31.
[http://dx.doi.org/10.1021/acsinfecdis.5b00093] [PMID: 27622945]

[82]   Batool M, Ahmad B, Choi S. A structure-based drug discovery paradigm. Int J Mol Sci 2019; 20(11): 2783.
[http://dx.doi.org/10.3390/ijms20112783] [PMID: 31174387]

[83]   Muhammed MT, Aki-Yalcin E. Homology modeling in drug discovery: Overview, current applications, and future perspectives. Chem Biol Drug Des 2019; 93(1): 12-20.
[http://dx.doi.org/10.1111/cbdd.13388] [PMID: 30187647]

[84]   França TCC. Homology modeling: An important tool for the drug discovery. J Biomol Struct Dyn 2015; 33(8): 1780-93.
[http://dx.doi.org/10.1080/07391102.2014.971429] [PMID: 25266493]

[85]   Munsamy G, Soliman MES. Homology modeling in drug discovery- An update on the last decade. Lett Drug Des Discov 2017; 14(9): 1099-111.
[http://dx.doi.org/10.2174/1570180814666170110122027]

[86]   Cavasotto CN, Phatak SS. Homology modeling in drug discovery: Current trends and applications. Drug Discov Today 2009; 14(13-14): 676-83.
[http://dx.doi.org/10.1016/j.drudis.2009.04.006] [PMID: 19422931]

[87]   Sliwoski G, Kothiwale S, Meiler J, Lowe EW Jr. Computational methods in drug discovery. Pharmacol Rev 2014; 66(1): 334-95.
[http://dx.doi.org/10.1124/pr.112.007336] [PMID: 24381236]

[88]   Zhang L, Zhou R. Structural basis of the potential binding mechanism of remdesivir to SARS-CoV-2

RNA-dependent RNA polymerase. J Phys Chem B 2020; 124(32): 6955-62.
[http://dx.doi.org/10.1021/acs.jpcb.0c04198] [PMID: 32521159]

[89]   Stefaniu A. Introductory chapter: Molecular docking and molecular dynamics techniques to achieve rational drug design. Molecular Docking and Molecular Dynamics. IntechOpen 2019.
[http://dx.doi.org/10.5772/intechopen.84200]

[90]   Pak Y, Wang S. Application of a molecular dynamics simulation method with a generalized effective potential to the flexible molecular docking problems. J Phys Chem B 2000; 104(2): 354-9.
[http://dx.doi.org/10.1021/jp993073h]

[91]   Salmaso V, Moro S. Bridging molecular docking to molecular dynamics in exploring ligand-protein recognition process: An overview. Front Pharmacol 2018; 9: 923.
[http://dx.doi.org/10.3389/fphar.2018.00923] [PMID: 30186166]

[92]   Okimoto N, Futatsugi N, Fuji H, *et al.* High-performance drug discovery: Computational screening by combining docking and molecular dynamics simulations. PLOS Comput Biol 2009; 5(10): e1000528.
[http://dx.doi.org/10.1371/journal.pcbi.1000528] [PMID: 19816553]

[93]   Kothandan G, Ganapathy J. A short review on the application of combining molecular docking and molecular dynamics simulations in field of drug discovery. J Chos Nat Sci 2014; 7(2): 75-8.
[http://dx.doi.org/10.13160/ricns.2014.7.2.75]

[94]   Athanasiou C, Cournia Z. From computers to bedside: Computational chemistry contributing to FDA approval. Biomol Simul Struc Bas Drug Disc. Wiley Online Library 2018; pp. 168-203.
[http://dx.doi.org/10.1002/9783527806836.ch7]

[95]   Kumar A, Voet A, Zhang KYJ. Fragment based drug design: From experimental to computational approaches. Curr Med Chem 2012; 19(30): 5128-47.
[http://dx.doi.org/10.2174/092986712803530467] [PMID: 22934764]

[96]   Erlanson DA, Davis BJ, Jahnke W. Fragment-based drug discovery: Advancing fragments in the absence of crystal structures. Cell Chem Biol 2019; 26(1): 9-15.
[http://dx.doi.org/10.1016/j.chembiol.2018.10.001] [PMID: 30482678]

[97]   Kirsch P, Hartman A M, Hirsch A K H, Empting M. Concepts and core principles of fragment-based drug design. Molecules 2019; 24(23): 4309.

[98]   Murray CW, Rees DC. The rise of fragment-based drug discovery. Nat Chem 2009; 1(3): 187-92.
[http://dx.doi.org/10.1038/nchem.217] [PMID: 21378847]

[99]   Hartenfeller M, Schneider G. *De novo* drug design. Methods Mol Biol 2010; 672: 299-323.
[http://dx.doi.org/10.1007/978-1-60761-839-3_12] [PMID: 20838974]

[100]   Kalyaanamoorthy S, Chen YPP. Structure-based drug design to augment hit discovery. Drug Discov Today 2011; 16(17-18): 831-9.
[http://dx.doi.org/10.1016/j.drudis.2011.07.006] [PMID: 21810482]

[101]   Loving K, Alberts I, Sherman W. Computational approaches for fragment-based and *de novo* design. Curr Top Med Chem 2010; 10(1): 14-32.
[http://dx.doi.org/10.2174/156802610790232305] [PMID: 19929832]

[102]   Yuan Y, Pei J, Lai L. LigBuilder V3: A multi-target *de novo* drug design approach. Front Chem 2020; 8(February): 142.
[http://dx.doi.org/10.3389/fchem.2020.00142] [PMID: 32181242]

[103]   Reed JE, Smaill JB. The discovery of dacomitinib, a potent irreversible EGFR inhibitor. ACS Symposium Series. ACS Publications 2016; 1: pp. 207-33.
[http://dx.doi.org/10.1021/bk-2016-1239.ch008]

[104]   Murray CW, Newell DR, Angibaud P. A successful collaboration between academia, biotech and pharma led to discovery of erdafitinib, a selective FGFR inhibitor recently approved by the FDA. MedChemComm 2019; 10(9): 1509-11.

[http://dx.doi.org/10.1039/C9MD90044F]

[105] Guo Y, Liu Y, Hu N, *et al.* Discovery of zanubrutinib (BGB-3111), a novel, potent, and selective covalent inhibitor of bruton's tyrosine kinase. J Med Chem 2019; 62(17): 7923-40.
[http://dx.doi.org/10.1021/acs.jmedchem.9b00687] [PMID: 31381333]

[106] Manathunga M, Götz AW, Merz KM Jr. Computer-aided drug design, quantum-mechanical methods for biological problems. Curr Opin Struct Biol 2022; 75: 102417.
[http://dx.doi.org/10.1016/j.sbi.2022.102417] [PMID: 35779437]

[107] Zhou T, Huang D, Caflisch A. Quantum mechanical methods for drug design. Curr Top Med Chem 2010; 10(1): 33-45.
[http://dx.doi.org/10.2174/156802610790232242] [PMID: 19929831]

[108] LaPointe S, Weaver D. A review of density functional theory quantum mechanics as applied to pharmaceutically relevant systems. Curr Computeraided Drug Des 2007; 3(4): 290-6.
[http://dx.doi.org/10.2174/157340907782799390]

[109] Sulpizi M, Folkers G, Rothlisberger U, Carloni P, Scapozza L. Applications of density functional theory-based methods in medicinal chemistry. Quant Struct-Act Relationsh 2002; 21(2): 173-81.
[http://dx.doi.org/10.1002/1521-3838(200207)21:2<173::AID-QSAR173>3.0.CO;2-B]

[110] Rozhenko AB. Density functional theory calculations of enzyme–inhibitor interactions in medicinal chemistry and drug design. Application of computational techniques in pharmacy and medicine. challenges and advances in computational chemistry and physics dordrecht: Springer 2014; 17: 207-40.
[http://dx.doi.org/10.1007/978-94-017-9257-8_7]

[111] Nascimento IJS, de Aquino TM, da Silva-Júnior EF. Cruzain and rhodesain inhibitors: Last decade of advances in seeking for new compounds against american and african trypanosomiases. Curr Top Med Chem 2021; 21(21): 1871-99.
[http://dx.doi.org/10.2174/18734294MTE10MTEoz] [PMID: 33797369]

[112] Wilson GL, Lill MA. Integrating structure-based and ligand-based approaches for computational drug design. Future Med Chem 2011; 3(6): 735-50.
[http://dx.doi.org/10.4155/fmc.11.18] [PMID: 21554079]

[113] Moro S, Bacilieri M, Deflorian F. Combining ligand-based and structure-based drug design in the virtual screening arena. Expert Opin Drug Discov 2007; 2(1): 37-49.
[http://dx.doi.org/10.1517/17460441.2.1.37] [PMID: 23496036]

[114] Acharya C, Coop A, Polli JE, Mackerell AD Jr. Recent advances in ligand-based drug design: Relevance and utility of the conformationally sampled pharmacophore approach. Curr Comput Aided Drug Des 2011; 7(1): 10-22.
[http://dx.doi.org/10.2174/157340911793743547] [PMID: 20807187]

[115] Lee CH, Huang HC, Juan HF. Reviewing ligand-based rational drug design: The search for an ATP synthase inhibitor. Int J Mol Sci 2011; 12(8): 5304-18.
[http://dx.doi.org/10.3390/ijms12085304] [PMID: 21954360]

[116] Baskin II. The power of deep learning to ligand-based novel drug discovery. Expert Opin Drug Discov 2020; 15(7): 755-64.
[http://dx.doi.org/10.1080/17460441.2020.1745183] [PMID: 32228116]

[117] Patel HM, Noolvi MN, Sharma P, *et al.* Quantitative structure–activity relationship (QSAR) studies as strategic approach in drug discovery. Med Chem Res 2014; 23(12): 4991-5007.
[http://dx.doi.org/10.1007/s00044-014-1072-3]

[118] Wang T, Wu MB, Lin JP, Yang LR. Quantitative structure–activity relationship: promising advances in drug discovery platforms. Expert Opin Drug Discov 2015; 10(12): 1283-300.
[http://dx.doi.org/10.1517/17460441.2015.1083006] [PMID: 26358617]

[119] Nantasenamat C, Isarankura-Na-Ayudhya C, Naenna T, Prachayasittikul V. A practical overview of

quantitative structure-activity relationship. EXCLI J 2009; 8: 74-88.
[http://dx.doi.org/10.17877/DE290R-690]

[120] Du QS, Huang RB, Wei YT, Du LQ, Chou KC. Multiple field three dimensional quantitative structure–activity relationship (MF-3D-QSAR). J Comput Chem 2008; 29(2): 211-9.
[http://dx.doi.org/10.1002/jcc.20776] [PMID: 17559075]

[121] Muhammad U, Uzairu A, Arthur DE. Quantitative structure activity relationship (QSAR) modeling. Int J Adv Acad Res 2018; 4(5): 1-9.

[122] Perkins R, Fang H, Tong W, Welsh WJ. Quantitative structure-activity relationship methods: Perspectives on drug discovery and toxicology. Environ Toxicol Chem 2003; 22(8): 1666-79.
[http://dx.doi.org/10.1897/01-171] [PMID: 12924569]

[123] Tropsha A. Best practices for QSAR model development, validation, and exploitation. Mol Inform 2010; 29(6-7): 476-88.
[http://dx.doi.org/10.1002/minf.201000061] [PMID: 27463326]

[124] Dudek A, Arodz T, Gálvez J. Computational methods in developing quantitative structure-activity relationships (QSAR): A review. Comb Chem High Throughput Screen 2006; 9(3): 213-28.
[http://dx.doi.org/10.2174/138620706776055539] [PMID: 16533155]

[125] Roy K. On some aspects of validation of predictive quantitative structure–activity relationship models. Expert Opin Drug Discov 2007; 2(12): 1567-77.
[http://dx.doi.org/10.1517/17460441.2.12.1567] [PMID: 23488901]

[126] Shahlaei M. Descriptor selection methods in quantitative structure-activity relationship studies: A review study. Chem Rev 2013; 113(10): 8093-103.
[http://dx.doi.org/10.1021/cr3004339] [PMID: 23822589]

[127] Voet A, Qing X, Lee XY, *et al.* Pharmacophore modeling: Advances, limitations, and current utility in drug discovery. J Recep Lig Chan Res 2014; 7: 81-92.
[http://dx.doi.org/10.2147/JRLCR.S46843]

[128] Yang SY. Pharmacophore modeling and applications in drug discovery: Challenges and recent advances. Drug Discov Today 2010; 15(11-12): 444-50.
[http://dx.doi.org/10.1016/j.drudis.2010.03.013] [PMID: 20362693]

[129] Akram M, Waratchareeyakul W, Haupenthal J, Hartmann RW, Schuster D. Pharmacophore modeling and *in silico/in vitro* screening for human cytochrome P450 11B1 and cytochrome P450 11B2 inhibitors. Front Chem 2017; 5(DEC): 104.
[http://dx.doi.org/10.3389/fchem.2017.00104] [PMID: 29312923]

[130] Schaller D, Šribar D, Noonan T, *et al.* Next generation 3D pharmacophore modeling. Wiley Interdiscip Rev Comput Mol Sci 2020; 10(4): 1-20.
[http://dx.doi.org/10.1002/wcms.1468]

[131] Khedkar S, Malde A, Coutinho E, Srivastava S. Pharmacophore modeling in drug discovery and development: An overview. Med Chem 2007; 3(2): 187-97.
[http://dx.doi.org/10.2174/157340607780059521] [PMID: 17348856]

[132] Lipinski CF, Maltarollo VG, Oliveira PR, da Silva ABF, Honorio KM. Advances and perspectives in applying deep learning for drug design and discovery. Front Robot AI 2019; 6: 108.
[http://dx.doi.org/10.3389/frobt.2019.00108] [PMID: 33501123]

[133] Klambauer G, Hochreiter S, Rarey M. Machine learning in drug discovery. J Chem Inf Model 2019; 59(3): 945-6.
[http://dx.doi.org/10.1021/acs.jcim.9b00136] [PMID: 30905159]

[134] Gawehn E, Hiss JA, Schneider G. Deep learning in drug discovery. Mol Inform 2016; 35(1): 3-14.
[http://dx.doi.org/10.1002/minf.201501008] [PMID: 27491648]

[135] Lavecchia A. Deep learning in drug discovery: Opportunities, challenges and future prospects. Drug

Discov Today 2019; 24(10): 2017-32.
[http://dx.doi.org/10.1016/j.drudis.2019.07.006] [PMID: 31377227]

[136] Liu R, Wei L, Zhang P. A deep learning framework for drug repurposing *via* emulating clinical trials on real-world patient data. Nat Mach Intell 2021; 3(1): 68-75.
[http://dx.doi.org/10.1038/s42256-020-00276-w] [PMID: 35603127]

[137] Bajorath J, Chávez-Hernández AL, Duran-Frigola M, *et al.* Chemoinformatics and artificial intelligence colloquium: Progress and challenges in developing bioactive compounds. J Cheminform 2022; 14(1): 82.
[http://dx.doi.org/10.1186/s13321-022-00661-0] [PMID: 36461094]

[138] Hassan Baig M, Ahmad K, Roy S, *et al.* Computer aided drug design: Success and limitations. Curr Pharm Des 2016; 22(5): 572-81.
[http://dx.doi.org/10.2174/1381612822666151125000550] [PMID: 26601966]

[139] Nascimento IJS, da Silva Rodrigues ÉE, da Silva MF, de Araújo-Júnior JX, de Moura RO. Advances in computational methods to discover new NS2B-NS3 inhibitors useful against dengue and zika viruses. Curr Top Med Chem 2022; 22(29): 2435-62.
[http://dx.doi.org/10.2174/1568026623666221122121330] [PMID: 36415099]

[140] Russell C, Rahman A, Mohammed AR. Application of genomics, proteomics and metabolomics in drug discovery, development and clinic. Ther Deliv 2013; 4(3): 395-413.
[http://dx.doi.org/10.4155/tde.13.4] [PMID: 23442083]

CHAPTER 2

Quantitative Structure-activity Relationship (QSAR) in Studying the Biologically Active Molecules

Serap ÇETINKAYA[1], Burak TÜZÜN[2,*] and Emin SARIPINAR[3]

[1] *Department of Molecular Biology and Genetics, Science Faculty, Sivas Cumhuriyet University, Sivas, Turkey*

[2] *Plant and Animal Production Department, Technical Sciences Vocational School of Sivas, Sivas Cumhuriyet University, Sivas, Turkey*

[3] *Department of Chemistry, Faculty of Science, Erciyes University, Melikgazi, 38039, Kayseri, Turkey*

Abstract: Recently, many new methods have been used in the research and development of a new drug. In this article, QSAR, which is one of the usable areas of artificial intelligence during molecule research, and the analysis and formulation studies related to the suitability of this area are discussed. It is explained how a model to be created is prepared and calculation formulas for how to verify this model are shown. Examples of the most recent 4D-QSAR calculations are given.

Keywords: Molecular Modelling, Pharmacophore, QSAR, Quantitative Structure-activity Relationship, Validation.

INTRODUCTION

Quantitative structure-activity relationship (QSAR) analysis uses the molecular structure of a compound or ligand to predict its biological activity. It presupposes that similar biological activities are retained in similar molecular structures [1]. It also uses known biological activity data to predict unknown activities. This approach has been adapted to diverse but related scientific disciplines [2-5], including the design of new chemical entities (NCEs) [5, 6] with high biological potentials.

* **Corresponding author Burak TÜZÜN:** Plant and Animal Production Department, Technical Sciences Vocational School of Sivas, Sivas Cumhuriyet University, Sivas, Turkey; E-mail: theburaktuzun@yahoo.com

Igor José dos Santos Nascimento (Ed.)

QSAR is a systematic multi-step process (Fig. **1**), made up of dataset preparation, selection, and generation of molecular descriptors; derivation of mathematical or statistical models; model training and validation using a training dataset; and model testing on a test dataset [7 - 10].

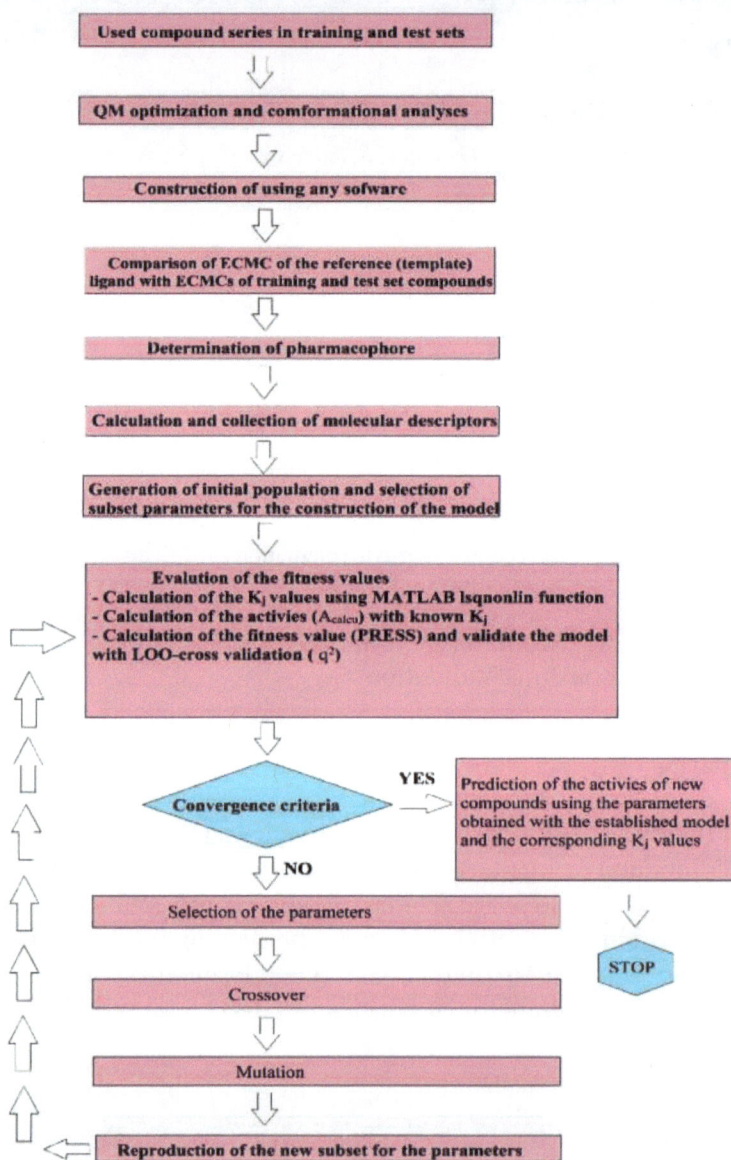

Fig. (1). Outlines of a QSAR model development.

In order to create a consistent QSAR model, it is central to utilize high-quality data that have been derived from bioassays, and to use an adequate number of compounds. Biological data are preferred to have been produced in a single laboratory [5, 11].

Selection and generation of molecular identifiers form the second step. Here the selection of appropriate descriptors, describing structural variations, is important. Various methods, such as machine learning techniques (*e.g.*, forward selection) and evolutionary algorithms (*e.g.*, genetic algorithm [11]), are utilized for descriptor/variable/feature selection.

A suitable mathematical or statistical model must be chosen to define the correlation between relevant descriptors and biological activities. The model can be linear partial least squares (PLS) [12], multiple linear regression (MLR) [13] or nonlinear. The selected model is then trained on a randomly chosen dataset, and the rest is used as test compounds. Model training often involves validation procedures, for example, exclusion cross validation (LOOCV) [14]. The training process is reiterated in order to reach an acceptable performance. The final step involves the testing process [11].

The concept of QSAR was first envisioned by Free, Wilson, Hansch and Fujita in 1964 [15, 16]. Subsequently, a new 3D-QSAR method, named comparative molecular field analysis (CoMFA) [17], has been worked out to overwhelm general 3D-QSAR problems. It has provided the basis for the development of multidimensional (nD) QSARs.

QSAR's Use

QSAR should not be seen as an academic tool that allows for the subsequent rationalization of data. It aims to derive molecular structure relationships between biology and chemistry for a valid reason. Models can be developed from these relationships and are thought to be predictive with common sense, luck, and expertise. A QSAR model can have many practical commitments [18, 19]:

- Rational estimation of biological activity and physicochemical properties.
- Understand and rationalize the action mechanisms of a wide variety of chemicals.
- Cost-effective product development.
- Minimization of the production time.
- Elimination of the ethical concerns.
- Spurring of "green" chemistry.

The ability to predict biological activity is of great importance in industry. Some QSARs have many applications in industry, academia and government (regulatory bodies). Some potential application areas can be seen below [18]:

- Rational compound identification with significant pharmacological, bactericidal or insecticidal activity.
- Optimization of the available biological activities.
- Design of versatile products to be used in diverse fields.
- Determination of toxicity in general and side effects in pharmaceuticals.
- Determination of compound stability.
- Rationalization and estimation of the combinatorial effects.

QSAR Model Development

A prerequisite in the creation of a QSAR model is the accuracy and reliability of the experimental data. If the experimental data are not accurate and precise, it would not be possible to obtain and develop a meaningful model. To develop an appropriate QSAR model, these data must include a wide range of chemically similar compounds [20]. Various biologically active compounds should be determined experimentally, and predetermined chemical structures should be selected [21].

While classical Hansch-like applications predominantly use the properties of substituents, atomic properties can also be taken into consideration, including charge, logP, orbital energy, and molar refractivity. The diversity of experimentally viable molecular variables contributes most to the strength of a given model. Software for any chemical structure determination variables are commercially available [22, 23], and these consider geometric, structural, topological, electrostatic, thermodynamic, and quantum mechanical properties of the products. In CODESSA, generally GAUSSIAN [24] and AMPAC [25] output files are used for the production of variables. DRAGON program on the other hand utilizes Molfiles, Multiple SD, Sybyl, SMILES, HyperChem, and MacroModel output files [26]. These programs compute compound parameters at the lowest energy fitness and exclude complex components. Therefore, the programs CODESSA and DRAGON only formulate those parameters related to one conformer of a compound. Spartan software [27] can calculate the Monte Carlo structure of compounds and calculate their relative energies, providing an overwhelming solution for the study of complex compounds. Another program, EMRE, uses the output files of SPARTAN 02, SPARTAN 04, SPARTAN 06, SPARTAN 08, SPARTAN 10, SPARTAN 14. An advantage of EMRE is that after determining both the matrix and the pharmacophore group, it also provides information on the perpendicular distance of the farthest atom in the molecule to the pharmacophore plane, the angle between the pharmacophore plane and the two atoms, the angle between atoms, the torsion angle, and the van der Waals radius [28 - 30].

As mentioned above, a statistically robust model relies on the quality of the selection of variables that can help explain the evolution of the structure-activity relationship. The QSAR choice, on the other hand, is crucial to develop an appropriate structure-activity correlation. Today, least squares (PLS) and multiple linear regression (MLR) are used in QSAR modelling. The commonest problem is chanced upon in the biological activity calculations if the number of variables is larger than that of the compounds. It is important in statistical calculations that the number of parameters produced at the end of the calculation should not exceed the number of compounds (Table **1**) [21].

Table 1. QSAR techniques and used parameters.

QSAR techniques	Used Parameters
1D-QSAR	Physicochemical parameters such as pKa, logP
2D-QSAR	1D-QSAR + Structural, geometric, electrostatic, thermodynamic parameters
3D-QSAR	2D-QSAR + Electrostatic, steric, hydrophobic parameters
4D-QSAR	3D-QSAR + Parameters related to conformations, protonation and stereoisomers
5D-QSAR	4D-QSAR + Parameters related to conformational changes in ligand-protein binding
6D-QSAR	5D-QSAR + Parameters related to dissolution

2D-QSAR Analysis

The 2D-QSAR analysis uses linear regression to attain quantitative relationships. Statistical data for most 2D-QSAR techniques are questionable [31]. It focuses on the initial interaction between ligand and receptor. Thus, 2D QSAR analysis is not as precise as 3D-QSAR. It typically performs structure-based drug identification and focuses on reaction information, structural characterization predictions, and chemical information [32].

There are generally three types of molecular parameters in 2D-QSAR:

• Two-dimensional molecular coupling parameters
• Three-dimensional molecular surface area parameters
• LogP

2D-QSAR involves structural-, geometric-, topological-, quantum-, electrostatic-, and thermodynamic parameters. In its simplest definition, a structure parameter reflects a molecular structure, excluding its geometric or electronic structure. Topology parameters contain the bonding information between two pairs of atoms in a molecule. Geometric parameters consider molecular space, density, and volume. Quantum chemical parameters are descriptive symbols such as HOMO and LUMO that provide information about the electronic properties of a molecule. Thermodynamic parameters cover heat formation, hydrophobicity, and molar refractivity [33, 34].

Fragment-Based 2D-QSAR Methods

As time and technology progressed, modelling calculations of many ligands or compounds can be performed, including 2D methods, 3D crystal receptors or where target structures are not available [35, 36].

Hologram-QSAR (HQSAR)

The hologram-QSAR method is fragment-based HQSAR (Hologram QSAR), originally developed by Tripos [37, 38]. The first step in the HQSAR method is to create molecular holograms that contain structural, geometric, electrostatic, and thermodynamic parameters of a large number of molecules and can be correlated with 2D molecular structures. Calculations are continued by deriving a mathematical regression equation to correlate these prepared hologram box values or components with the corresponding biological activities.

Fragment-Based QSAR (FB-QSAR)

In 2009, Du *et al.* [39] showed that a 2D-QSAR method based on molecular fragments uses two equations, the first being the Hansch-Fujita [40] linear free energy equation and the second the Free-Wilson [41] equation.

Fragment-Similarity Based QSAR (FS-QSAR)

A fragment similarity based QSAR (FS-QSAR) method [42] was developed using the original Free-Wilson method by introducing the concept of fragment similarity in the linear regression equation in 2010. It was applied for the first time to develop the traditional Free-Wilson equation instead of using physicochemical properties that often produce non-unique solutions. In this approach, part similarity calculations are made on the basis of similarity, using the lowest or highest eigenvalues calculated from the BCUT matrices [43, 44]. The matrices contain the partial charges of the individual atoms as well as the atomic connectivity information contained in each individual part.

3D-QSAR

Even when computational studies did not exist, chemists were able to predict structure-activity relationships in a wide variety of molecules, and earlier QSAR approaches have often been effectively used to elucidate drug-receptor interactions. The most important limitation in their use is the total absence of numerical parameters which are essential in the production of substituents.

The main 3D-QSAR approaches, emerged in the 1980s, are the Active Analog Approach [35], Comparative Molecular Field Analysis [45, 46], and Molecular Shape Analysis [38]. Many others were developed in the 1990s, and have evolved considerably over the past decade [47]. Because the three-dimensional structure determines the biological activity of a compound, 3D-QSAR helps understand how structural modifications change biological properties. The 3D-QSAR uses informatics to reveal structure-activity relationships [33].

The general rules in the 3D-QSAR modelling are as follows [48]:

1. Conformational data are to be used in the selection of minimum energy conformers.

2. Distance between two atoms, optimization of the ring system, donor-acceptor hydrogen bond or charged functional groups should be determined.

3. Active Cluster Analysis is to be used in the selection of the linker.

4. Atomic properties such as dipole moment, polarity, point charges, molecular electrostatic potential, HOMO and LUMO orbitals, shape and volume that affect biological activity are to be determined.

5. Pharmacophore properties are to be selected [49].

Thanks to the use of the correlation between biological activity and the compound structure, 3D-QSAR has become the most popular approach in recent years. Parameters describing the chemical structure are used for the activity calculation. The 3D-QSAR method is divided into two basic classes according to the parameter type. One involves volume-based parameters while the other employs surface-based parameters. As a volume-based method, CoMFA is the most preferred 3D-QSAR method [34].

4D-QSAR

Hopfinger added another dimension to 3D-QSAR, in 1997, which is the mean of the matching conformers, and invented the term 4D-QSAR. Later on, Vedani has developed the 4D-QSAR approach. The latter used the concepts of 4D-QSAR by combining the effects of multidirectional conformations. This method is particularly useful in estimating the ligand's free binding energy to its receptor when the receptor structure is unknown [50]. The most active conformer of a given compound is defined at the minimum energy state (2 Kcal/mol). In this method, the active conformation is not a minimum energy conformer and instead it uses multi-temperature molecular dynamics (MDS). Hence, it is useful to estimate the potential energy of all low-energy conformers on a 3-D surface. Best results are produced when low energy conformers are calculated individually. The molecular coincidence problem is overwhelmed by similar sampling and computation. The conformers of each compound are placed on a predetermined rectangular surface to find the best agreement [51, 52]. The main difference of this method from CoMFA is that it combines both molecular conformers and overlapping molecules within a set of known molecular structures [33].

5D- and 6D-QSARs

Vedani [53] has added another dimension to 4D-QSAR, giving rise to 5D-QSAR. The latter examines the conformational changes inflicted on a protein by the binding of its ligand. The 6D-QSAR, also developed by Vedani, is able to map various dissolution patterns formed between solvent and solute molecules [54, 55].

Molecular Modelling and QSAR

In simple terms, molecular modelling can be thought of as a computer technique that can either predict molecular and biological properties or analyze molecules and molecular systems, based on experimental knowledge and theoretical chemical methods. Available techniques help describe properties that are responsible for biological activity, such as hydrophobicity, electronic properties of atoms and molecules, and geometry of the molecule. All these features are important in understanding the structure-activity relationship in drug design. SYBYL, AMBER, DOCK, MODELER and RasMol are the most used programs in molecular modelling [56].

Molecular modelling employs computer chemistry and graphical design techniques to explore structural properties of a molecule [57, 58]. Here three of the outstanding approaches are molecular dynamics, molecular mechanics, and quantum mechanics [59]. Models are generated by QM and MM calculations. The models first allow the energy regulation of molecules and atoms in a calculated system, and they help understand how alterations in the position of atoms and molecules change the energy of a system. Three main steps in a molecular modelling can be stated as follows [60]:

1. Geometric work involves the construction of bond angle and bond lengths.

2. Fragments with appropriate geometry are demonstrated.

3. Structure is constructed using the experimental information, produced by X-ray crystallography, neutron diffraction or NMR.

A fourth step involves Monte Carlo Simulation or molecular dynamics. Finally, the calculations are analyzed.

Importance of the Validation of QSAR Models

QSAR has largely been matured, but it needs further improvement. Estimating the precision of calculations is for instance a serious problem [61]. Thus, the need for the proof of QSAR models has been brought into consideration [62]. Four common tools have so far been set for the assessment of the proof of QSAR models [63]: (i) cross validation, (ii) bootstrapping, (iii) randomization of response data, and (iv) external validation.

Some proposals have also been made for evaluating the rationality of QSAR models at an International workshop held in Setubal (Portugal) that have been revised later on at the OECD QSAR Work Program in 2004 [64, 65].

Means of Proof for QSAR Models

The proof is needed both to confirm the prediction strength of a model and to conclude the intricacy of an equation (Fig. **2**).

Least squares fit (R^2), cross-validation (Q^2) [66, 67], adjusted R^2 (R^2adj), chi-square test (χ^2) [68], root mean square error (RMSE) [69], bootstrapping, and shuffling (Y-Randomization) [70, 71] are internal methods for validating a model. However, external statistical methods serve best to make sure that the models created are robust and impartial [64].

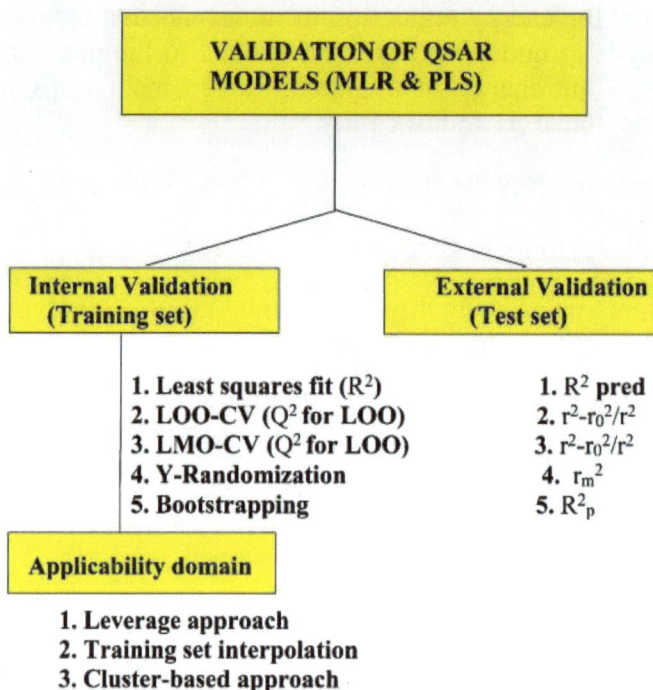

Fig. (2). A diagram of QSAR model validation.

Internal Validation

Least Squares Fit

It is an internal validation method, similar to linear regression. A solid straight line fit is produced to compute R^2. A substitute of this method excludes compounds in the training set (outliers) to validate the QSAR model. A difference between R^2 and R^2adj value of less than 0.3 shows that the number of descriptors included in the QSAR model is satisfactory [64].

$$R^2 = \left[\frac{N\Sigma XY - (\Sigma x)(\Sigma Y)}{\sqrt{([N\Sigma X^2 - (\Sigma X)^2][N\Sigma Y^2 - (\Sigma Y)^2])}} \right]$$ (1)

Fit of the Model

The prediction power of a QSAR model, reflected in R^2, can also be validated by chi-square (χ^2) and mean square error (RMSE) tests. *The chi-square* value is the difference between experimental and theoretical bioactivity scores. RMSE on the other hand finds the error between the mean of the experimental- and the predicted figures [64].

$$\chi^2 = \sum_{i=1}^{n} \left(\frac{(y_i - \hat{y}_i)^2}{\hat{y}_i} \right) \tag{2}$$

$$RMSE = Sqrt \left(\sum_{i=1}^{n} \frac{(\hat{y}_i - y_m)^2}{n-1} \right) \tag{3}$$

y and $\hat{y}$, the experimental and predicted bioactivity for a single compound in the training set; ym, the mean of the experimental bioactivities; and n, the number of molecules in the studied data set.

High *chi*-square or RMSE scores (≥ 0.5 and 1.0, respectively) imply that the model built is of poor quality despite its high R^2 scores (≥ 0.7). An appropriate model should have *chi*- and RMSE values of 0.5 and <0.3, respectively [72] but they are insufficient for the complete validity test [21].

Cross-validation

Cross-validation is another internal validating test (CV, Q^2, q^2 or jack-knifing). The CV process reiterates the regression on the data subsets. It is often used to determine the suitability of a model to a given data set. In the test, the molecules are sequentially scrutinized and the missing molecule is used in the computation of R.

Cross-validation is exploited to understand the predictive capability of a model and to find out whether the model overfits, that is, a predictive model can only define the relationship between predictors and response in the existing compounds. This phenomenon comes to the fore when the difference between R^2 and Q^2 is higher than 0.3) [21, 66, 67].

Cross-validation test can be cross-checked using majority-out (LMO). This method is mostly handy when the training set is small (≤ 20 composites) or does not exist.

Equation for CV:

$$Q^2 = 1 - \frac{PRESS}{\sum_{i=1}^{N}(y_i - y_m)^2} \tag{4}$$

$$PRESS = \sum_{i=1}^{N}(y_{pred,i} - y_i)^2 \tag{5}$$

y_i, the data values not used to build the CV model; PRESS [65], the estimated remaining sum of squares.

Bootstrapping

Bootstrapping [73, 74] is another internal validation test which analyzes randomized subsamples rather than analyzing subsets. Here groups are indicated with K and object numbers with n. The test performed with randomly selected n objects is employed to assess target features for excepted samples. A high average Q^2 indicates the robustness of a models.

Randomization Test (Scrambling Model)

In this test, activity values are casually reallocated and the whole procedure is reiterated. Scrambled Model controls the identifiers to warrant that the chosen descriptors are proper. It is created using the original identifiers. For this test R^2 and Q^2 values are also computed. Lower scores established by repetitive testing indicate the robustness of the model built and there should not be a strong correlation between R^2 ($R^2>0.50$) [58] and Q^2 [75, 76].

External Validation

It has sometimes been argued that the best way to test the prediction capacity of a QSAR model is to compare the predicted and observed activities of compounds which are not included in the model creation process [77, 78]. The steps of the test have been featured as follows: computing the correlation coefficient, R and R^2 [79]; and defining the k and k' regression slopes passing through the origin.

$$R^2_{pred} > 0.6 \tag{6}$$

$$r^2 - r_0^2/r^2 < 0.1, \qquad r^2 - r_0^2/r^2 < 0.1 \tag{7}$$

$$0.85 \leq k \leq 1.15 \; or \; 0.85 \leq k' \leq 1.15 \tag{8}$$

The prediction power of the selected model should also be verified by the external R^2 predicted R^2 values. R2 >0.6 is a sufficient indicator of proper external predictability.

$$R^2_{Pred} = 1 - \frac{\sum_{i=1}^{test}(y_{exp} - y_{pred})^2}{\sum_{i=1}^{test}(y_{exp} - \bar{y}_{tr})^2} \tag{9}$$

y_{tr}, the mean value of the dependent variable for the training set.

The final tool available to check the external predictability of the selected model is r^2_m, which has been put forward by Roy and Paul [80]:

$$r^2_m = r^2(1 - \sqrt{r^2 - r^2_0} \tag{10}$$

r^2, the square of the correlation coefficient between the experimental and predicted values; and r^2_0, the square of the correlation coefficient between the experimental and predicted values with the intercept set to zero. An r^2_m value greater than 0.5 can be taken as an indicator of good external predictability [81].

The r^2_m (overall) statistic can be beneficial when the test set size is quite small and the regression-based external validation [82, 83] parameter is less dependent and highly reliant on test set observations. The r^2_m (overall) statistic can be utilized to choose the best analytical models among comparable models, because some models produce better internal validation parameters while others are good at generating external validation parameters.

Another proof parameter, R^2_p, checks the fitness of the selected model [82]. It penalizes the R^2 model (Eq. 11).

$$R^2_p = R^2 * \left(1 - \sqrt{|R^2 - R^2_r|}\right) \tag{11}$$

The r^2_m (overall) value defines whether the range of activity values predicted for the entire dataset is actually close to the experimental results. Therefore, a QSAR model can be satisfactory if the r^2_m (total) and R^2_p values are equal to or greater than 0.5 [52].

QSAR calculations are an effective theoretical method used to design more efficient and more active molecules (Fig. **3**). These calculations can be used to design new high activity molecules [84, 85].

Easily Reproducible QSAR Protocol

Spartan 10's software [23] is typically used for optimizations, conformational analysis of related molecules, and the creation of molecular structures at the chosen basis set level. Water was deemed to be the best solvent for cancer in general. When extremely sophisticated approaches like DFT (Density Functional Theory) are combined with enormous basis sets, it's a crucial technique for getting more precise findings. However, creating high basis sets usually takes long time, since molecular conformers are also taken into account. A relatively simple method and a larger basis set should be used to produce trustworthy results within acceptable time schedules.

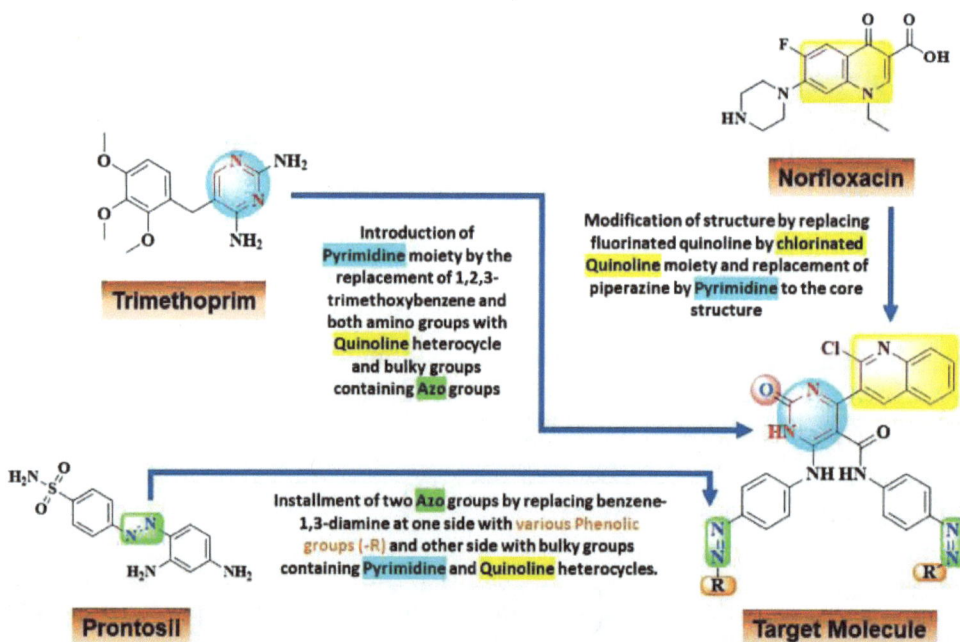

Fig. (3). Molecule design.

Low-energy conformers are known to be densely positioned at room temperature because conformers are ordered according to Boltzmann weights, which rely on the energies of the conformers relative to the energy of the conformer. Then, using the Monte Carlo random search approach, conformational analysis is carried out to look for these conformers. When the conformational search is complete, conformers with lower energies are retained and those with a Boltzmann distribution of less than 1/10000 are removed. Mulliken charges in diagonal elements, and bond configurations/atomic distances in non-diagonal elements were then carried out using EMRE software, and computations of electron conformational conformity matrices (ECMC) with residual conformers [2, 22, 24]. The conformers of each molecule were used to generate the ECMC based on the findings of the quantum chemical calculations. We defined a template compound that generates a predictive model for use in the comparison process of ECMCs [86]. Typically, the template was chosen from the atoms in the backbone of the lowest energy molecule. The molecules in the examined series were ranked according to their activities in order to distinguish between active and inactive compounds. The molecules were split into two after this sorting, by taking into account the order of the activities. The row's top individuals were labeled as active molecules. The lowest group was labeled as having inactive molecules. The pharmacophore, also known as the electron conformational activity sub-matrix (ECSA), was then defined by comparing the model compound's ECMC with all other ECMCs that fell within the specified tolerance range. Then, using the MATLAB application and a variety of produced quantum chemical parameters, QSAR calculations were performed. These computations were used to build a model.

QSAR studies are important in the designing of new and more effective molecules, based on the existing group of molecules. Activity estimations are made using many different programs. These programs are made by some researchers using programs developed by them and paid advanced programs. There are many studies conducted in the light of this information. Some of them are as follows:

In a study by Sahin and co-workers [87], a theoretical comparison of the activities of isatin derivatives was made against the BCL-2 inhibitor with the EMRE program using the Electron-Conformational Genetic Algorithm (EC-GA) hybrid method. In their calculations, it is seen that the molecules are prepared in a total of 801 parameters under four different categories ((i) electronic (ii) quantum-chemical (iii) geometrical and (iv) thermodynamic). It is seen that a good model is obtained by using the best seven parameters among these 801 parameters (Table **2**).

Table 2. κj values and description of the optimum parameters.

$a_{ni}^{(j)}$	Molecular Descriptors	κj values
1	C2 Nucleophilic Atom Boundary Electron Density (eV)	−15.43
2	H5 Electrophilic Atom Boundary Electron Density(eV)	−22.80
3	C7 Nucleophilic Atom Boundary Electron Density (eV)	−4.11
4	C4 Fukui Atomic Electrophilic Reactivity Index (eV)	152.31
5	log P	−0.03
6	PSA (u)	0.00
7	max.el. pot.(u)	0.00

In the model, it was seen that activity calculations were made for 27 molecules whose activity was unknown. It can be seen that very high results are obtained for the model when more than one internal and external validation is made. If these are the results of the validation; q^2_{ext1}=0.79, q^2_{ext2}=0.79, q^2_{ext3}=0.83, CCC_1=0.97, CCC_2=0.90 (Table **2**).

Another study has identified 80 methanone derivatives and biological comparison of molecules [52]. Electron Conformational Compatibility Matrices were obtained with EMRE software by using the geometric, thermodynamic and topological properties of 80 molecules studied. A model was created using 804 parameters. (Table **3**).

Table 3. κj values and description of the optimum parameters.

$a_{ni}^{(j)}$	Molecular Descriptors	κj values
1	The distance of O2–F1	0.027
2	Bond degree of C10–N2 (Lowdin)	−3.720
3	Lowdin values of the N2 atom (e−)	1.295
4	Orthogonal distance of the C11 atom to the N3 C20 O1 plane	−0.129
5	Bond degree of N2–C16 (Lowdin)	1.801
6	Orthogonal distance of the C11 atom to the C12 N1 O1 plane	−0.069
7	Bond degree of C6–C1 (Lowdin)	−0.435
8	H6 C6 C1 angle	−0.752

The numerical values of the model were were $R^2_{training}$ =0.834, q^2 =0.768 and $SE_{training}$ =0.075, q^2_{ext1} =0.875, q^2_{ext2} =0.839, q^2_{ext3} =0.764, ccc_{tr} =0.908, ccc_{test} =0.929 and ccc_{all} =0.920.

A chemical property-based pharmacophore model was developed for tetrahydrodibenzazosine derivatives with the EMRE package program [88]. All QSAR models were built with 40 compounds (training set), and then a test set was created with an additional nine compounds (test set) to create a consistent model (Table **4**).

Table 4. κj values and description of the optimum parameters.

$a_{ni}^{(j)}$	Molecular Descriptors	κj values
1	Orthogonal distance from the C2 atoms to C23–C14–C10 plane	−0.202
2	Orthogonal distance from the C2 atoms to C23–N1–C10 plane + van der Waals radius of C2 atom	−0.255
3	Orthogonal distance from the C12 atoms to C14–N1–C10 plane + van der Waals radius of C12 atom	0.135
4	The angle (degree) between C15–N1–C4atoms	−0.130
5	Fukui Lumo of the C5 atom	4.386
6	Fukui Lumo of the C13 atom	−5.912
7	Mulliken charge of C9 atom	−5.434
8	E (kcal/Mol)	0.002
9	Rel Eaq (kcal/Mol)	0.003
10	PSA (u)	0.008

In the end, a statistically valid 4D-QSAR (R^2_{training} =0:856, R^2_{test} =0:851 and $q^2 -$ 0.650) model was obtained with a good good external set estimation

The activity comparison of 86 alkynylphenoxyacetic acid derivatives has been made for the CRTh2 receptor [85]. A model was then created using the genetic algorithm and nonlinear least squares regression methods (Table **5**).

Table 5. κj values and description of the optimum parameters.

$a_{ni}^{(j)}$	Molecular Descriptors	κj values
1	C22-C18 distance (Å)	0.0216
2	C22-C18 distance (Å)+ vdW radius	0.0564
3	Orthogonal distance from the C22 atom to the C1-O1-O3 plane (Å)	0.0243
4	Orthogonal distance from the O2 atom to the C1-H6-C12 plane (Å) + vdW radius of O2 atom	−0.0139
5	Orthogonal distance from the C6 atom to the C1-O1-C12 plane (Å) + vdW radius of C6 atom	−0.0713
6	The angle of between C11-O3-C18 atoms	0.0108

(Table 5) cont.....

$a_{ni}^{(j)}$	Molecular Descriptors	κj values
7	The angle between the O3-H6-C12 plane and the O2-O4 lines	0.2155
8	The angle between the C6-O3-H6 plane and the Cl1-C18 lines	0.1040
9	The dihedral angle between the O2-C5-C22-C18 plane	−0.0109

A theoretical comparison of the anti-HIV-1 activity of 52 tetrahy- droimidazo [4,5,1-jk] 1,4] benzodiazepinone derivatives has been reported (Table **6**) [89].

Table 6. κj values and description of the optimum parameters.

$a_{ni}^{(j)}$	Molecular Descriptors	κj values
1	Distance between S1- the farthest atom bonded to N3 (C14)	0.127
2	Angle between line of N3-M* atoms and H25-S1-C9 plane	−0.003
3	Angle between S1-C1-N1 atoms	−0.466
4	Dihedral angle between H25-N1-C12- the farthest atom bonded to C12	−0.024
5	Angle (radian) between S1-C1-C12- the farthest atom bonded to C12 atoms	−95.346
6	Distance between N1 and C9+ van der Waals radius of C9 atom	5.124
7	Distance between N2 and C11+ van der Waals radius of the farthest atom bonded to C11	−0.118

CONCLUSION

QSAR calculations are an important method used to determine the parameter that contributes to the activities of molecules by comparing the numerical values of these properties by examining many properties of the molecules made before the experimental processes in order to design more effective and more active molecules.

REFERENCES

[1]   Esposito EX, Hopfinger AJ, Madura JD. Methods for applying the quantitative structure-activity relationship paradigm. In Chemoinformatics Humana Press 2014; pp. 131-213.

[2]   Bradbury SP. Quantitative structure-activity relationships and ecological risk assessment: An overview of predictive aquatic toxicology research. Toxicol Lett 1995; 79(1-3): 229-37.
[http://dx.doi.org/10.1016/0378-4274(95)03374-T] [PMID: 7570660]

[3]   Chen JZ, Han XW, Liu Q, Makriyannis A, Wang J, Xie XQ. 3D-QSAR studies of arylpyrazole antagonists of cannabinoid receptor subtypes CB1 and CB2. A combined NMR and CoMFA approach. J Med Chem 2006; 49(2): 625-36.
[http://dx.doi.org/10.1021/jm050655g] [PMID: 16420048]

[4]   Hansen C, Telzer BR, Zhang L. Comparative QSAR in toxicology: Examples from teratology and cancer chemotherapy of aniline mustards. Crit Rev Toxicol 1995; 25(1): 67-89.
[http://dx.doi.org/10.3109/10408449509089887] [PMID: 7734060]

[5]   Perkins R, Fang H, Tong W, Welsh WJ. Quantitative structure-activity relationship methods:

perspectives on drug discovery and toxicology. Environ Toxicol Chem 2003; 22(8): 1666-79.
[http://dx.doi.org/10.1897/01-171] [PMID: 12924569]

[6]   Salum LB, Andricopulo AD. Fragment-based QSAR: Perspectives in drug design. Mol Divers 2009; 13(3): 277-85.
[http://dx.doi.org/10.1007/s11030-009-9112-5] [PMID: 19184499]

[7]   Myint KZ. QSAR methods development, virtual and experimental screening for cannabinoid ligand discovery. Dissertation 2012.

[8]   Akyüz L, Sarıpınar E. Conformation depends on 4D-QSAR analysis using EC–GA method: Pharmacophore identification and bioactivity prediction of TIBOs as non-nucleoside reverse transcriptase inhibitors. J Enzyme Inhib Med Chem 2013; 28(4): 776-91.
[http://dx.doi.org/10.3109/14756366.2012.684051] [PMID: 22591319]

[9]   Sahin K, Saripinar E. A novel hybrid method named electron conformational genetic algorithm as a 4D QSAR investigation to calculate the biological activity of the tetrahydrodibenzazosines. J Comput Chem 2020; 41(11): 1091-104.
[http://dx.doi.org/10.1002/jcc.26154] [PMID: 32058616]

[10]   Akyüz L, Sarıpınar E, Kaya E, Yanmaz E. 4D-QSAR study of HEPT derivatives by electron conformational–genetic algorithm method. SAR QSAR Environ Res 2012; 23(5-6): 409-33.
[http://dx.doi.org/10.1080/1062936X.2012.665082] [PMID: 22452710]

[11]   Keskin H. 4D-QSAR analysis of benzotienopyrimidin compounds by electron conformation-genetic algorithm (EC-GA) method. Master thesis. 2019.

[12]   Köprü S. 4D-QSAR analysis of phenylpyrazine, alkinylpenoxyacetic acid, phthalazine derivatives by electron conformational-genetic algorithm (EC-GA) method. Master thesis. 2018.

[13]   Yao XJ, Panaye A, Doucet JP, et al. Comparative study of QSAR/QSPR correlations using support vector machines, radial basis function neural networks, and multiple linear regression. J Chem Inf Comput Sci 2004; 44(4): 1257-66.
[http://dx.doi.org/10.1021/ci049965i] [PMID: 15272833]

[14]   Bahar R. Pharmacophore modelling and 4D-QSAR analysis of phenylpyrazole glutamic acid piperazine by electron conformational-genetic algorithm method. Master thesis. 2018.

[15]   Free SM Jr, Wilson JW. A mathematical contribution to structure-activity studies. J Med Chem 1964; 7(4): 395-9.
[http://dx.doi.org/10.1021/jm00334a001] [PMID: 14221113]

[16]   Hansch C, Fujita T. ρ−σ−π Analysis. A method for the correlation of biological activity and chemical structure. J Am Chem Soc 1964; 86(8): 1616-26.
[http://dx.doi.org/10.1021/ja01062a035]

[17]   Kulak L. QSAR studies with CoMCET method on 2-amino-6-arylsulfonyl benzonitrile compounds as HIV-1 inhibitor. Master thesis. 2018.

[18]   Puzyn T, Leszczynski J, Cronin MT, Eds. Recent advances in QSAR studies: methods and applications. Springer 2010.
[http://dx.doi.org/10.1007/978-1-4020-9783-6]

[19]   Aydın S. Qsar studies with electron conformation-genetic algorithm (EC-GA) method on 3-(1, 1-dioxo-2h-(1,2,4)-benzothiadiazine-3-yl)-4-hydroxy-2(1h)-quinoline compounds as hepatitis c inhibitor. Master thesis. 2018.

[20]   Ravi M, Hopfinger AJ, Hormann RE, Dinan L. 4D-QSAR analysis of a set of ecdysteroids and a comparison to CoMFA modeling. J Chem Inf Comput Sci 2001; 41(6): 1587-604.
[http://dx.doi.org/10.1021/ci010076u] [PMID: 11749586]

[21]   Tüzün B. 4D-QSAR study of some pyridine carboxylic acid, oxadiazole, pyrimidine and oxazol derivatives by electron conformational-genetic algorithm method. Phd thesis. 2018.

[22]   Karelson M, Maran U, Wang Y, Katritzky AR. QSPR and QSAR models derived using large molecular descriptor spaces. A review of CODESSA applications. Collect Czech Chem Commun 1999; 64(10): 1551-71.
[http://dx.doi.org/10.1135/cccc19991551]

[23]   Polishchuk PG, Kuz'min VE, Artemenko AG, Muratov EN. Universal approach for structural interpretation of QSAR/QSPR models. Mole Informa 2013; 32(9-10): 843-53.
[http://dx.doi.org/10.1002/minf.201300029]

[24]   Frisch MJ, Trucks GW, Schlegel HB, *et al.* revision D01. Wallingford: Gaussian Inc 2009.

[25]   Holder AJ, Ye L, Yourtee DM, Agarwal A, Eick JD, Chappelow CC. An application of the QM-QSAR method to predict and rationalize lipophilicity of simple monomers. Dent Mater 2005; 21(7): 591-8.
[http://dx.doi.org/10.1016/j.dental.2004.08.004] [PMID: 15978267]

[26]   Altaf R, Nadeem H, Iqbal MN, *et al.* Synthesis, biological evaluation, 2D-QSAR, and molecular simulation studies of dihydropyrimidinone derivatives as alkaline phosphatase inhibitors. ACS Omega 2022; 7(8): 7139-54.
[http://dx.doi.org/10.1021/acsomega.1c06833] [PMID: 35252705]

[27]   Wavefunction inc. 18401 Von Karman Avenue. USA: Suite 370, Irvine 2006.

[28]   Maria de Assis T, Ramalho TC, Ferreira da Cunha EF. 4D-QSAR models applied to the study of TGF-β1 receptor inhibitors. Curr Top Med Chem 2021; 21(13): 1157-66.
[http://dx.doi.org/10.2174/1568026621666210727161431] [PMID: 34315368]

[29]   Liu J, Pan D, Tseng Y, Hopfinger AJ. 4D-QSAR analysis of a series of antifungal p450 inhibitors and 3D-pharmacophore comparisons as a function of alignment. J Chem Inf Comput Sci 2003; 43(6): 2170-9.
[http://dx.doi.org/10.1021/ci034142z] [PMID: 14632469]

[30]   de Souza APM, Costa MCA, de Aguiar AR, *et al.* Leishmanicidal and cytotoxic activities and 4D-QSAR of 2-arylidene indan-1,3-diones. Arch Pharm 2021; 354(10): 2100081.
[http://dx.doi.org/10.1002/ardp.202100081] [PMID: 34323311]

[31]   Altaf R, Nadeem H, Iqbal MN, *et al.* "Synthesis, biological evaluation, 2D-QSAR, and molecular simulation studies of dihydropyrimidinone derivatives as alkaline phosphatase Inhibitors". ACS Omega 2022; 7(8): 7139-54.
[http://dx.doi.org/10.1021/acsomega.1c06833] [PMID: 35252705]

[32]   Yu X. Prediction of inhibitory constants of compounds against SARS-CoV 3CLpro enzyme with 2D-QSAR model. J Saudi Chem Soc 2021; 25(7): 101262.
[http://dx.doi.org/10.1016/j.jscs.2021.101262]

[33]   Kocakaplan I. Investigation of Q2/R2 rating according to the training and test sets cutting and molecular distribution for 4d QSAR model of acetyl cholinesterase inhibitor activity Master thesis 2021.

[34]   Kaushal T, Khan S, Fatima K, Luqman S, Khan F, Negi AS. Synthesis, molecular docking, and 2D-QSAR modeling of quinoxaline derivatives as potent anticancer agents against triple-negative breast cancer. Curr Top Med Chem 2022; 22(10): 855-67.
[http://dx.doi.org/10.2174/1568026622666220324151808] [PMID: 35331094]

[35]   Myint KZ, Xie XQ. Recent advances in fragment-based QSAR and multi-dimensional QSAR methods. Int J Mol Sci 2010; 11(10): 3846-66.
[http://dx.doi.org/10.3390/ijms11103846] [PMID: 21152304]

[36]   SYBYL80. Discovery Software for Computational Chemistry and Molecular Modeling. St. Louis, MO, USA: Tripos 2008.

[37]   Salum LB, Andricopulo AD. Fragment-based QSAR: Perspectives in drug design. Mol Divers 2009;

13(3): 277-85.
[http://dx.doi.org/10.1007/s11030-009-9112-5] [PMID: 19184499]

[38]   Lowis D. HQSAR: A new, highly predictive QSAR technique, tripos technique notes. St. Louis, MO, USA: Tripos 1997.

[39]   Du QS, Huang RB, Wei YT, Pang ZW, Du LQ, Chou KC. Fragment-based quantitative structure-activity relationship (FB-QSAR) for fragment-based drug design. J Comput Chem 2009; 30(2): 295-304.
[http://dx.doi.org/10.1002/jcc.21056] [PMID: 18613071]

[40]   Hansch C, Fujita T. ρ–σ–π Analysis. A method for the correlation of biological activity and chemical structure. J Am Chem Soc 1964; 86(8): 1616-26.
[http://dx.doi.org/10.1021/ja01062a035]

[41]   Free SM Jr, Wilson JW. A mathematical contribution to structure-activity studies. J Med Chem 1964; 7(4): 395-9.
[http://dx.doi.org/10.1021/jm00334a001] [PMID: 14221113]

[42]   Myint KZ, Ma C, Wang L, Xie XQ. The Fragment-similarity-based QSAR (FS-QSAR): A Novel 2D-QSAR method to predict biological activities of triaryl Bis-sulfone and COX2 analogs. 2010.

[43]   Burden FR. Molecular identification number for substructure searches. J Chem Inf Comput Sci 1989; 29(3): 225-7.
[http://dx.doi.org/10.1021/ci00063a011]

[44]   Xie XQ, Chen JZ. Data mining a small molecule drug screening representative subset from NIH PubChem database. J Chem Inf Model 2008; 48(3): 465-75.
[http://dx.doi.org/10.1021/ci700193u] [PMID: 18302356]

[45]   DePriest SA, Mayer D, Naylor CB, Marshall GR. 3D-QSAR of angiotensin-converting enzyme and thermolysin inhibitors: A comparison of CoMFA models based on deduced and experimentally determined active site geometries. J Am Chem Soc 1993; 115(13): 5372-84.
[http://dx.doi.org/10.1021/ja00066a004]

[46]   Cramer RD, Depriest SA, Patterson DE, Hecht P (1993) The developing practice of comparative molecular field analysis. In: Kubinyi H, editor. 3D QSAR in drug design: theory, method and applications, pp 443-85. ESCOM, Leiden.

[47]   Kubinyi H. Comparative molecular field analysis (CoMFA). The encyclopedia of computational chemistry, 1998; 1: 448-60.

[48]   Mezey PG. The degree of similarity of three-dimensional bodies: Application to molecular shape analysis. J Math Chem 1991; 7(1): 39-49.
[http://dx.doi.org/10.1007/BF01200814]

[49]   Kubinyi H, Folkers G, Martin YC. 3D QSAR in Drug Design: Recent Advances. Springer Dordrecht 1998.

[50]   Kubinyi H. QSAR and 3D QSAR in drug design Part 1: Methodology. Drug Discov Today 1997; 2(11): 457-67.
[http://dx.doi.org/10.1016/S1359-6446(97)01079-9]

[51]   Kieser KJ, Kim DW, Carlson KE, Katzenellenbogen BS, Katzenellenbogen JA. Characterization of the pharmacophore properties of novel selective estrogen receptor downregulators (SERDs). J Med Chem 2010; 53(8): 3320-9.
[http://dx.doi.org/10.1021/jm100047k] [PMID: 20334372]

[52]   Debnath AK. Quantitative structure-activity relationship (QSAR) paradigm-Hansch era to new millennium. Mini Rev Med Chem 2001; 1(2): 187-95.
[http://dx.doi.org/10.2174/1389557013407061] [PMID: 12369983]

[53]   Çatalkaya S, Sabancı N, Yavuz SÇ, Sarıpınar E. The effect of stereoisomerism on the 4D-QSAR study

of some dipeptidyl boron derivatives. Comput Biol Chem 2020; 84: 107190.
[http://dx.doi.org/10.1016/j.compbiolchem.2019.107190] [PMID: 31918171]

[54]   Tüzün B, Saripinar E. Molecular docking and 4D-QSAR model of methanone derivatives by electron conformational-genetic algorithm method. J Iranian Chem Soc 2020; 17(5): 985-1000.

[55]   Vedani A, Briem H, Dobler M, Dollinger H, McMasters DR. Multiple-conformation and protonation-state representation in 4D-QSAR: The neurokinin-1 receptor system. J Med Chem 2000; 43(23): 4416-27.
[http://dx.doi.org/10.1021/jm000986n] [PMID: 11087566]

[56]   Qin D, Zeng X, Zhao T, Cai B, Yang B, Tu G. 5D-QSAR studies of 1H-pyrazole derivatives as EGFR inhibitors. J Mol Model 2022; 28(12): 379.
[http://dx.doi.org/10.1007/s00894-022-05370-x] [PMID: 36342554]

[57]   Yanmaz E, Sarıpınar E, Şahin K, Geçen N, Çopur F. 4D-QSAR analysis and pharmacophore modeling: Electron conformational-genetic algorithm approach for penicillins. Bioorg Med Chem 2011; 19(7): 2199-210.
[http://dx.doi.org/10.1016/j.bmc.2011.02.035] [PMID: 21419636]

[58]   Gund T. Molecular modelling of small molecules. In: Guideb on Molec Model Drug Desi, Cohen 1996; 55-92.

[59]   Hopfinger AJ, Wang S, Tokarski JS, et al. Construction of 3D-QSAR models using the 4D-QSAR analysis formalism. J Am Chem Soc 1997; 119(43): 10509-24.
[http://dx.doi.org/10.1021/ja9718937]

[60]   Cai Z, Zafferani M, Akande OM, Hargrove AE. Quantitative structure–activity relationship (QSAR) study predicts small-molecule binding to RNA structure. J Med Chem 2022; 65(10): 7262-77.
[http://dx.doi.org/10.1021/acs.jmedchem.2c00254] [PMID: 35522972]

[61]   Smith C. Molecular modeling -seeing the whole picture with modeling software packages. Scientist 1998; 12: 17-31.

[62]   Esmel A. Insight into binding mode of nitrile inhibitors of Plasmodium falciparum Falcipain-3, QSAR and Pharmacophore models, virtual design of new analogues with favorable pharmacokinetic profiles. SDRP J Comput Chem Mole Model 2017; 2(1): 1-22.

[63]   Tong W, Hong H, Xie Q, Shi L, Fang H, Perkins R. Assessing QSAR limitations-A regulatory perspective. Curr Computer-Aided Drug Des 2005; 1(2): 195-205.
[http://dx.doi.org/10.2174/1573409053585663]

[64]   Worth AP, Van Leeuwen CJ, Hartung T. The prospects for using (Q)SARs in a changing political environment-high expectations and a key role for the european commission's joint research centre. SAR QSAR Environ Res 2004; 15(5-6): 331-43.
[http://dx.doi.org/10.1080/10629360412331297371] [PMID: 15669693]

[65]   Worth AP, Hartung T, Van Leeuwen CJ. The role of the European centre for the validation of alternative methods (ECVAM) in the validation of (Q)SARs. SAR QSAR Environ Res 2004; 15(5-6): 345-58.
[http://dx.doi.org/10.1080/10629360412331297362] [PMID: 15669694]

[66]   Veerasamy R, Rajak H, Jain A, Sivadasan S, Varghese CP, Agrawal RK. Validation of QSAR models-strategies and importance. Int J Drug Des Dis 2011; 3: 511-9.

[67]   Von der Ohe PC, Kühne R, Ebert RU, Schüürmann G. Comment on "Discriminating toxicant classes by mode of action: 3. Substructure indicators" (M. Nendza and M. Müller, SAR QSAR Environ. Res. 18 155 (2007)). SAR QSAR Environ Res 2007; 18(7-8): 621-4.
[http://dx.doi.org/10.1080/10629360701698571] [PMID: 18038362]

[68]   Leach AR. Molecular modeling: Principles and applications. Harlow, England: Pearson Education Ltd. 2001.

[69] Shao J. Linear model selection by cross-validation. J Am Stat Assoc 1993; 88(422): 486-94.
[http://dx.doi.org/10.1080/01621459.1993.10476299]

[70] Tallarida RJ, Murray RB. Chi-square test. Manual of pharmacologic calculations. New York, NY: Springer 1987; pp. 140-2.
[http://dx.doi.org/10.1007/978-1-4612-4974-0_43]

[71] Chai T, Draxler RR. Root mean square error *(RMSE)* or mean absolute error (MAE). Geosci Model Dev Discuss 2014; 7(1): 1525-34.

[72] Wold S, Ericksson L. Partial least squares projections to latent structures (PLS) in chemistry. In: Encyclopedia of computationalchemistry. Ragu , Schleyer P. Chichester: John Wiley & Sons 1998; 3: pp. 2006-21.

[73] Yasri A, Hartsough D. Toward an optimal procedure for variable selection and QSAR model building. J Chem Inf Comput Sci 2001; 41(5): 1218-27.
[http://dx.doi.org/10.1021/ci010291a] [PMID: 11604021]

[74] Schneider G, Wrede P. Artificial neural networks for computer-based molecular design. Prog Biophys Mol Biol 1998; 70(3): 175-222.
[http://dx.doi.org/10.1016/S0079-6107(98)00026-1] [PMID: 9830312]

[75] Thakur M, Thakur A, Khadikar PV. QSAR studies on psychotomimetic phenylalkylamines. Bioorg Med Chem 2004; 12(4): 825-31.
[http://dx.doi.org/10.1016/j.bmc.2003.10.027] [PMID: 14759743]

[76] Zhai X, Chen M, Lu W. Predicting the toxicities of metal oxide nanoparticles based on support vector regression with a residual bootstrapping method. Toxicol Mech Methods 2018; 28(6): 440-9.
[http://dx.doi.org/10.1080/15376516.2018.1449278] [PMID: 29644916]

[77] Choudhary G, Karthikeyan C, Hari Narayana Moorthy NS, Sharma SK, Trivedi P. QSAR analysis of some cytotoxic thiadiazinoacridines. Inter Elect J Mol Des 2005; 4: 793-802.

[78] Litang QIN, Shushen LIU, Qianfen XIAO, Qingsheng WU. Internal and external validtions of QSAR model. Environ Chem 2013; 32(7): 1205-11.

[79] Rücker C, Rücker G, Meringer M. y-Randomization and its variants in QSPR/QSAR. J Chem Inf Model 2007; 47(6): 2345-57.
[http://dx.doi.org/10.1021/ci700157b] [PMID: 17880194]

[80] Golbraikh A, Tropsha A. Beware of q2!. J Mol Graph Model 2002; 20(4): 269-76.
[http://dx.doi.org/10.1016/S1093-3263(01)00123-1] [PMID: 11858635]

[81] Guha R, Jurs PC. Determining the validity of a QSAR model-a classification approach. J Chem Inf Model 2005; 45(1): 65-73.
[http://dx.doi.org/10.1021/ci0497511] [PMID: 15667130]

[82] Sachs L. Applied Statistics: A Handbook of Techniques. Berlin, Germany: Springer-Verlag 1984.
[http://dx.doi.org/10.1007/978-1-4612-5246-7]

[83] Roy K, Paul S. Exploring 2D and 3D QSARs of 2, 4-diphenyl-1, 3-oxazolines for ovicidal activity against Tetranychus urticae. QSAR Comb Sci 2009; 28(4): 406-25.
[http://dx.doi.org/10.1002/qsar.200810130]

[84] Pratim Roy P, Paul S, Mitra I, Roy K. On two novel parameters for validation of predictive QSAR models. Molecules 2009; 14(5): 1660-701.
[http://dx.doi.org/10.3390/molecules14051660] [PMID: 19471190]

[85] Roy K. On some aspects of validation of predictive quantitative structure–activity relationship models. Expert Opin Drug Discov 2007; 2(12): 1567-77.
[http://dx.doi.org/10.1517/17460441.2.12.1567] [PMID: 23488901]

[86] Özalp A, Yavuz SÇ, Sabancı N, Çopur F, Kökbudak Z, Sarıpınar E. 4D-QSAR investigation and

pharmacophore identification of pyrrolo[2,1-c][1,4]benzodiazepines using electron conformational–genetic algorithm method. SAR QSAR Environ Res 2016; 27(4): 317-42.
[http://dx.doi.org/10.1080/1062936X.2016.1174152] [PMID: 27121415]

[87] Sahin K, Saripinar E, Durdagi S. Combined 4D-QSAR and target-based approaches for the determination of bioactive Isatin derivatives. SAR QSAR Environ Res 2021; 32(10): 769-92.
[http://dx.doi.org/10.1080/1062936X.2021.1971760] [PMID: 34530651]

[88] Köprü S, Saripinar E. 4D-QSAR analysis and pharmacophore modeling for alkynylphenoxyacetic acids as CRTh2 (DP2) receptor antagonists. Turk J Chem 2018; 42(6): 1577-97.
[http://dx.doi.org/10.3906/kim-1801-86]

[89] Akyüz L, Sarıpınar E. Conformation depends on 4D-QSAR analysis using EC–GA method: pharmacophore identification and bioactivity prediction of TIBOs as non-nucleoside reverse transcriptase inhibitors. J Enzyme Inhib Med Chem 2013; 28(4): 776-91.
[http://dx.doi.org/10.3109/14756366.2012.684051] [PMID: 22591319]

CHAPTER 3

Pharmacophore Mapping: An Important Tool in Modern Drug Design and Discovery

Dharmraj V. Pathak[1], Abha Vyas[1], Sneha R. Sagar[1], Hardik G. Bhatt[2] and Paresh K. Patel[1,*]

[1] *Department of Pharmaceutical Chemistry, L.J. Institute of Pharmacy, L.J. University, Ahmedabad 382 210, India*

[2] *Department of Pharmaceutical Chemistry, Institute of Pharmacy, Nirma University, Ahmedabad 382 481, India*

Abstract: Computer-Aided Drug Design (CADD) has become an integral part of drug discovery and development efforts in the pharmaceutical and biotechnology industry. Since the 1980s, structure-based design technology has evolved, and today, these techniques are being widely employed and credited for the discovery and design of most of the recent drug products in the market. Pharmacophore-based drug design provides fundamental approach strategies for both structure-based and ligand-based pharmacophore approaches. The different programs and methodologies enable the implementation of more accurate and sophisticated pharmacophore model generation and application in drug discovery. Commonly used programmes are GALAHAD, GASP, PHASE, HYPOGEN, ligand scout *etc*. In modern computational chemistry, pharmacophores are used to define the essential features of one or more molecules with the same biological activity. A database of diverse chemical compounds can then be searched for more molecules which share the same features located at a similar distance apart from each other. Pharmacophore requires knowledge of either active ligands and/or the active site of the target receptor. There are a number of ways to build a pharmacophore. It can be done by common feature analysis to find the chemical features shared by a set of active compounds that seem commonly important for receptor interaction. Alternately, diverse chemical structures for certain numbers of training set molecules, along with the corresponding IC_{50} or *Ki* values, can be used to correlate the three-dimensional arrangement of their chemical features with the biological activities of training set molecules. There are many advantages in pharmacophore based virtual screening as well as pharmacophore based QSAR, which exemplify the detailed application workflow. Pharmacophore based drug design process includes pharmacophore modelling and validation, pharmacophore based virtual screening, virtual hits profiling, and lead identification. The current chapter on pharmacophores also describes case studies and applications of pharmacophore mapping in finding new drug molecules of specific targets.

---

* **Corresponding author Paresh K. Patel:** Department of Pharmaceutical Chemistry, L.J. Institute of Pharmacy, L.J. University, Ahmedabad 382 210, India; Tel: +91 9712151531; E-mail: pareshpharmacist@gmail.com

Igor José dos Santos Nascimento (Ed.)

Keywords: Features, Ligand Based, Pharmacophore Query, Pharmacophore, Structure Based, Virtual Screening.

INTRODUCTION

Currently engaged in creating a new medicine, drug design and development is a costly and time-consuming process [1]. From foundational research to commercial products, a new medicine requires 10 to 14 years of research and billions of dollars *via* several preclinical and clinical phases [2]. With the amazing advancement of computational resources, computer aided drug design (CADD) and discovery technologies are highly valued all over the world. Designing small lead and drug-like molecules with expected multitarget actions increasingly employs both ligand and structure-based methods. CADD has advanced significantly in recent years, boosting the comprehension of multiple and complicated biological processes, allowing for the fast development of novel pharmacologically active drugs [3]. One such CADD tool employed in drug design and discovery is pharmacophore mapping or pharmacophore modeling. In the late 19th century, *Paul Ehrlich* was the first who propose that certain groups inside a molecule (phoros) are responsible for a molecule's biological activity (pharmacon), giving rise to the idea of "pharmacophores" [4, 5]. The pharmacophore theory postulates that a collection of shared properties that engage a group of contrasting locations on a biological target can explain how a class of chemicals recognizes that target on a molecular level [6]. In the contemporary drug discovery process, the pharmacophore approach serves as a helpful bridge between medicinal chemistry and computational chemistry, both in virtual screening (VS) and library design for effective hit finding and in the optimization of lead compounds to final therapeutic candidates.

Definitions of Pharmacophore

As per the IUPAC definition, "A pharmacophore is the ensemble of steric and electronic features that is necessary to ensure the optimal supramolecular interactions with a specific biological target structure and to trigger (or to block) its biological response."

Apart from the official IUPAC definition, other similar definitions have also been given in the literature. "A pharmacophore does not represent a real molecule or a real association of functional groups, but a purely abstract concept that accounts for the common molecular interaction capacities of a group of compounds with their target structure."

"A pharmacophore is defined by pharmacophoric descriptors, including H-bonding, hydrophobic, and electrostatic interaction sites, defined by atoms, ring centers, and virtual points.

"A pharmacophore can be considered the largest common denominator shared by a set of active molecules". This definition discards a misuse often found in the medicinal chemistry literature, which consists of naming as pharmacophores simple chemical functionalities such as guanidine, sulphonamides, or dihydroimidazoles (formerly imidazolines), or typical structural skeletons such as flavones, phenothiazines, prostaglandins, or steroids [5, 7].

To describe unique functional groups or chemical classes with biological activity, scientists frequently use the terms "pharmacophore" or "pharmacophoric group. In this context, the word "pharmacophore" is used in conjunction with the concept of "privileged structures," which refers to the alternative idea of structure and function. The chemical scaffolds and retroactive examination of medicinal molecules' chemical structures allowed for the identification of a few structural motifs that are frequently linked to bioactive compounds. Evans *et al.* referred to these patterns as "privileged structures" to describe substructures that bestow activity against a number of different targets [8]. Dihydropyridines, arylethylamines, *N*-arylpiperazines, diphenylmethane derivatives, biphenyls, pyridazines, sulphonamides and benzodiazepines are a few well-known instances of the advantaged structures [7 - 10].

Pharmacophore: History

The pharmacophore was first envisioned by Paul Ehrlich, the pioneer of chemotherapy, and that idea has remained unchanged for the past 100 years [11]. Langley, who coined the phrase "receptive substance," first proposed the notion that bioactive compounds interact with receptors in 1878 [12]. Paul Ehrlich, meanwhile, coined the word "receptor" a few years down the line [13], as well as introduced the term "pharmacophore". In conjunction with Emil Fischer's lock-and-key concept, it tends to be evident but not the properties of a molecule, the "key", are equally significant aimed at biological action [14]. Biological activity can be dramatically altered by small changes in some parts of a molecule, while minor changes in others can do the same. Modern drug discovery and development is based on Langley, Ehrlich, and Fischer's concepts. As soon as they were confirmed according to the earliest protein-ligand complex crystal structures half a decade later, they established a new paradigm [15]. Before the development of computers and modelling software, simple pharmacophores were documented in the literature and recognised as tools for the discovery of novel compounds. Modest 2D models remained first proposed in the 1940s based on the first structure-activity relationship assumptions [16]. According to the research by Woods and Fildes, p-aminobenzoic acid (PABA) and p-amino benzene sulphonamide target the PABA, which bind with equal efficiency and then prevent tetrahydrofolic acid production, because they have similar critical distances. One of the earliest 2D pharmacophore models was this one [17]. The notion of showing and modifying 3D structures became viable with the development of computer and modeling software. In the instance of (R)-(−)-adrenaline [= (R)-(−)-epinephrine], Easson, Stedman, and Beckett's three-point contact model was a pioneering 3D pharmacophoric strategy [18, 19]. The pharmacophore idea and its use in structure-activity connections were invented by Kier and Marshall [20, 21]. Corresponding to this, a different three-dimensional method for assessing the effect of clonidine on the central norepinephrine receptor was discovered by researchers in the early 1970s. It has been found that there are three major interactions by which the natural ligand norepinephrine interacts towards its target: an ionic bond between the binding pocket's anion (carboxylate, phosphate) and the protonated $-NH_2$ functional group, a hydrogen bond (HB) between the binding site's NH-CO group and the secondary alcoholic hydroxyl, and a π-stacking in between phenyl ring and protonated imidazole of a histidine residue [22]. In order to successfully explain the pharmacophoric similarity between clonidine and norepinephrine, Pullmann *et al.* computed the violent intramolecular ranges for the aforementioned key interactions in their 3D pharmacophore model of the norepinephrine receptor. This, in turn, helps clonidine to produce the same type of interactions as norepinephrine [23]. Peter Gund developed the first *in silico* screening technique in the 1970s, using software to check a drug library for pharmacophoric patterns [24]. One of the earliest automated technologies for pharmacophore production was the active analogue method created by Garland Marshall's team. The foundation for several succeeding pharmacophore modeling initiatives in that field was Marshall's method. Since then, several software development firms and academic institutions have created computerized pharmacophore discovery algorithms [25, 26].

Until recently, pharmacophore models were primarily created manually with the use of easy-to-use interactive molecular graphics visualization applications. Eventually, highly developed computational programs were needed for the identification and application of pharmacophore models due to the increasing complexity of molecular structures. Although computational chemistry has advanced an essential understanding of a pharmacophore model, to become a straightforward geometrical representation of the important chemical interface has not altered. There are many automated techniques for modelling and using pharmacophores which have arisen as a result of the developments in computational chemistry during the past 20 years. Since the techniques were introduced for the pharmacophore, a sizable number of investigations have been conducted [27]. Lead optimization, *de novo* design, VS, and multitarget drug design all make extensive use of pharmacophore techniques [28].

Pharmacophoric Features

A pharmacophore model, according to the definition, reflects the binding interaction of bioactive compounds with a receptor or proteins by the use of receptor binding pockets through a distinctive 3D arrangement of intangible interaction characteristics taken into consideration by various kinds of non-covalent interactions, columbic interactions, hydrogen bonds (HB) formation, metal interactions, hydrophobic contacts, charge transfer or aromatic stacking interactions are a few examples of these interaction types. In general, a pharmacophore model describes the typical way that several ligands bind to a particular target. In pharmacophore modeling, a molecules are first divided into a group of structures, each of these represents a reliable kind of interaction with the residues in the binding position. Then, for least-squares fitting, *i.e.,* superimposition of molecules with one another, the specific feature is represented by points which are as follows [29].

- Hydrophobic groups (H)
- Hydrogen bond donor (HBD)
- Aromatic Rings (R)
- Hydrogen bond acceptor (HBA)
- Negatively charged groups (N)
- Positively charged groups (P)

Hydrogen bond donor (HBD): Hydrogen bond donor groups play an important role in the interaction of bioactive molecules with electronegative atoms or residues at the receptor sites. Donors are often identified as hydroxyl group (-OH), hydrogens linked to acetylenic CH groups, nitrogen and thiols (SH). The -SH and -CH groups are considered lesser donors. Other kinds of -CH, such as those found in heterocycles containing a nitrogen atoms of several kinase inhibitors, are occasionally regarded as donors in addition to acetylenes. Basic amines like $RCH_2N(Me)_2$ are taken into account as donors when considering protonation. Tautomeric and ionized states have a significant impact on how pharmacophore features are defined since they can change a feature's characteristics. So, all potential protonation/ionization states of molecules ought to supply to the pharmacophore clarification programmes.

Hydrogen bond acceptor (HBA): In general, electronegative atoms like N, O, and S that have accessible lone pairs of electrons are regarded as acceptors. But, a few programmes do not take into consideration the oxygen atoms that are present in furan/oxazole rings since crystallographic and theoretical data show that they are extremely flimsy acceptors. Fixing the positions of the complementary, including aspects which will overlap in the eventual pharmacophore, is essential in addition to specifying the HB characteristics. Because of this, pharmacophore modeling programmes are associated with the acceptor`s and donor`s characteristics with analogous ligand atoms as well as the ascribed positions of the harmonized atom`s receptor, which is engaged in the interaction.

Aromatic Rings (R): phenyl and heterocyclic aromatic rings contribute hydrophobicity in bioactive molecules and play a crucial role in the interaction with aromatic amino acids through π- π interaction. According to the geometry and protonation state, the relative potency of anion and cation interactions was examined [30]. It was shown that while cation-π contacts increase due to the electrostatic contributions, edge-to-face interactions and π-π stacking are required for the favourable electron correlation. Aromatic rings are taken into consideration as a specific type of hydrophobic characteristic represented by vectors rather than dots to imitate the orientation of interactions such as cation- π and π-π -stacking.

Hydrophobic features (H): The selection of atoms or groupings which ought to consider hydrophobic is neither simple nor obvious. The Greene *et al.* algorithm is the most frequently employed [31]. Assign each atom a hydrophobicity score based on a set of empirical guidelines developed since the opinions of pharmaceutical chemists, and then cluster atoms with disproportionately high hydrophobicity scores. The centroid of any such cluster is then designated as a hydrophobic feature point. Alkyl chains are followed by groups like $-CF_3$ and then ring/ring atoms in terms of hydrophobicity score [32]. Some simple methods represent the form of molecules by treating all non-acceptors, non-donors, and non-charged atoms as structural groups (the analogue of aquaphobic groups).

Positive and negative features (P and N): If they are not a component of a dipole, atoms with formal charges are regarded as (+) positive or (-) negative characteristics in molecules. A net formal charge is also seen as either beneficial or unbeneficial aspects of groups. The location where the positively and negatively charged features are often located is the centroid of the heteroatoms in a group. On occasion, the ionizability of the characteristics is used to accentuate both their good and negative aspects. For instance, $R-NH_3^+$ is assessed as a positively ionizable characteristic, but $R-N(Me)_3^+$ is not because of the considerably different synergy established through these two groups [29]. Pharmacophoric features and examples of chemical substituents are enlisted in Table **1**. A 2D representation of phramacophoric features is also provided in Fig. (**1**).

Table 1. Pharmacophoric features and their examples.

Pharmacophoric features	Examples
Hydrogen bond donors	• Nitrogen donors (aliphatic) • Nitrogen donors(aromatic) • Hydroxyl • Thiols
Hydrogen bond acceptors	• Carboxylic • Carbonyl • S/P oxygens • Hydroxyl/phenol • Ether • Ester • Nitro • Nitrogen acceptors
Positively ionizable centers	• Amidine • Protonable amine • Guanidinium • 2/4-amino pyridine • Charged nitrogen atoms
Negatively ionizable centers	• Carboxylic acid • Tetrazole • Acid sulphonamides • Hydroxamic acid • S/P acids
Hydrophobic site	• Six/five-member rings • Tert-butyl • Hydrophobic moiety • Halogens

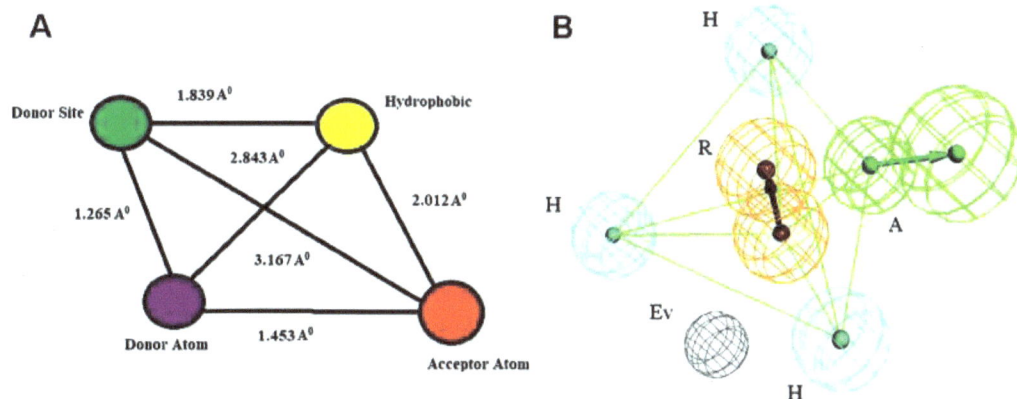

Fig. (1). Typical pharmacophore features representation using **A**) SYBYL-X-2.0 **B**) Discovery studio.

Additional steric restriction characteristics are defined by the most modern modeling software for pharmacophore modelling. Exclusion volumes are what they are, and they represent the arrangements of atom impact on the binding pocket [33, 34]. The molecules can be aligned quickly and logically to create features, which is also useful for scoring. The level of overlay is quantitatively described by the RMSD (root mean square deviation) amongst matched features, which is frequently employed as a fitness score [35]. As a result, the positioning of feature points must be precise, and one must exercise caution when determining whether to take into account all feasible characteristics or to pick a small number that would provide sufficient information on the molecular arrangement of a group in space. For instance, the alignment may be biased and create a model with a high score when there are far more hydrophobic properties than other features, but the model won't be useful because it lacks specificity [29].

Models of pharmacophores can be generated by structure-based and ligand-based approaches. Ligand based approach uses several active molecules from which common chemical features responsible for biological activity can be extracted. In the structure-based approach, possible interaction points of ligands with macromolecules can be identified. The workflow for the generation of the pharmacophore model through structure-based pharmacophores and ligand-based pharmacophores is highlighted in Fig. (**2**).

LIGAND BASED PHARMACOPHORE

When a target or receptor's 3D structure is unavailable, but the chemical structure of the group of powerful active small molecules of the same target is known or available, then ligand-based pharmacophore modeling is utilized as a great tool for compound database screening [36, 37]. Training molecules refer to this group of known active small compounds that are active against the target receptor or enzyme [37, 38]. To identify other compounds that could be active against the targets, training molecules are arranged together to identify the common chemical properties/functionalities that indicate crucial interactions between ligands and specific protein targets [36, 37]. Pharmacophore generation from multiple ligands (also known as training set compounds) entails two steps: initially, each ligand in the training set is given its own conformational space to represent its structural diversity by rotating it to obtain an improved version of the same ligand, which is then retained in the database of ligand. Step two attempts to impose the various training sets of the ligand and identify its crucial chemical characteristics to build pharmacophore models [28, 36, 37, 39]. To accommodate for a smaller molecule's flexibility in the process, several conformation forms of the same ligand are developed. The resulting model, a 4–7 featured pharmacophore, lists the attributes that a brand-new small molecule must possess in order to successfully interact with the target. To find novel active compounds, this pharmacophore can be used to search against another ligand database. In ligand-based pharmacophore modeling, managing ligand conformational flexibility and conducting molecular alignment constitute the core methodologies and major challenges. Several techniques, including genetic algorithms and Monte Carlo methods, are used to produce the conformation of ligands. Another significant drawback is the possibility that the conformation chosen for the pharmacophore is not the ligand's active form due to free energy. Each ligand-based pharmacophore has this vulnerability [36, 37].

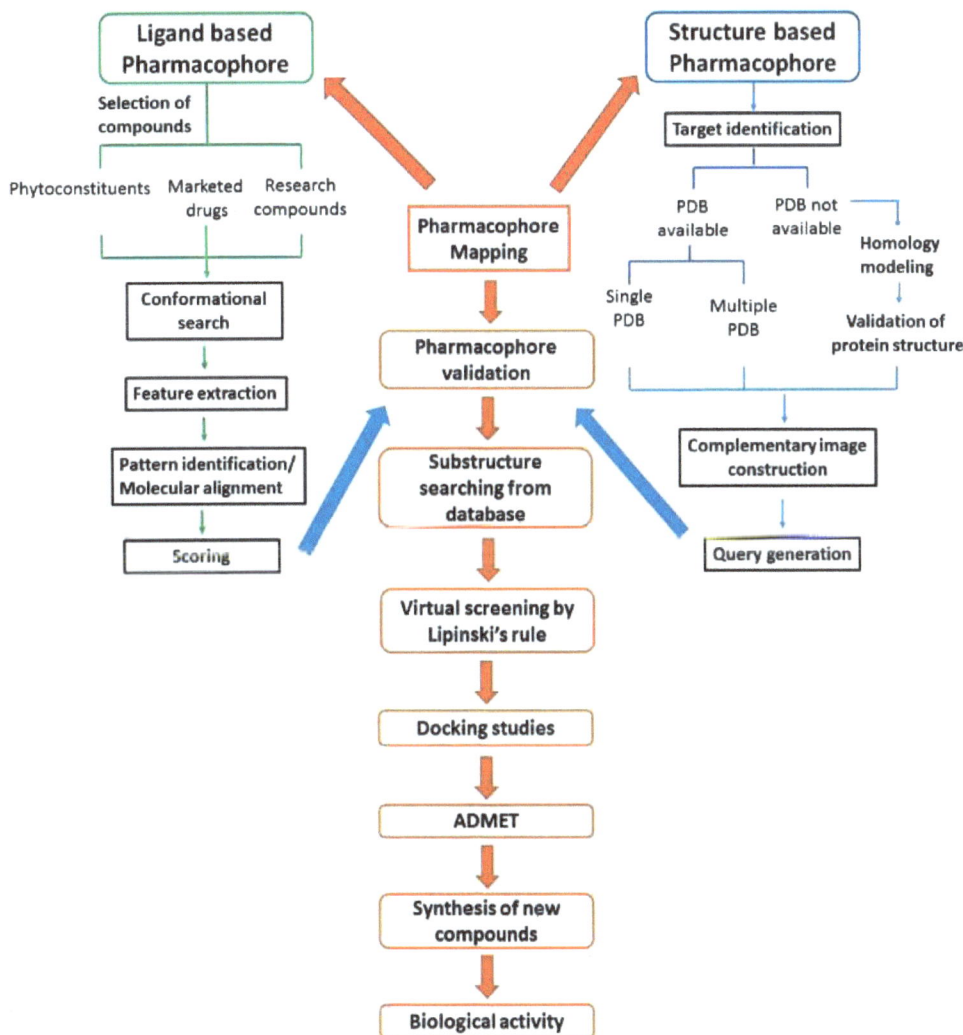

Fig. (2). Workflow of pharmacophore mapping.

Ligand-Based Pharmacophore Modeling

Modeling of pharmacophores based on ligands describes the broad procedures that computer programmes use to identify a pharmacophore pattern from a collection, about the number of molecules that engage a single receptor and the various ways that the pharmacophore notion is used [36].

Selection of the Right Set of Compounds and their Initial Structure

Since the structural type and size variety among the involved ligands strongly influence the final pharmacophore models, choosing the collection of ligands that will take part in the process of creating pharmacophore models is crucial. A few algorithms accept the assumption that all composites in a set are active, whereas further programs/methods considering molecule-related statistics that are inactive to be important because it gives insight into the structural characteristics that are able to lessen the activities and the ones that are necessary to increase interaction. Several programmes can manage ligands up to 100 in a set, based on the size of the dataset. If there are numerous molecules in the dataset, the classification and categorization are based on the activity value range. The considerable structural diversity also makes it easier to spot the characteristics required for marked binding and the creation of superior models. Model development depends on accurate compound structure through precise atomic bond ordering, valences, and aromaticity, as well as the relevant stereochemical flags [36].

Conformational Search

As ligands are adaptable, they can adopt a variety of conformations that can attach to the target's binding site in different ways. Therefore, it is crucial to concentrate on each molecule's flexibility in developing a pharmacophore. Thus, the most difficult issue in creating pharmacophore models is representing ligand flexibility. The first approach is the numerous configurations of each molecule are determined accordingly and saved in a database as part of the pre-enumerating technique [25, 28]. The second strategy is known as the "on-the-fly technique," which involves conformational analysis [25]. The first method has the benefit of requiring less processing power to carry out molecular alignment, but it may also require more mass storage space. The second method does not require any mass storage, although it may require more CPU time to carry out thorough optimization. The pre-enumerating strategy has been shown to be superior to the on-the-fly computation methodology [40]. Many sophisticated algorithms have recently been developed to sample the conformational regions of tiny compounds. Numerous commercial and academic pharmacophore modeling tools have used some of these techniques, including poling restrains [41], systemic torsional grids [42], directed tweak [43], genetic algorithm [44], and monte carlo [45]. However, an effective conformational designer must satisfy the following requirements: (i) productively generating entirely of the known constrained conformations when minor ligands mingle with targets or proteins; (ii) maintaining the catalogue of conformations of low energy as close as probable to prevent combinational explosion difficulties; and (iii) being much quicker in terms of time for the conformational estimations [28].

Feature Representation and Extraction

After the completion of the conformational search, the ligands are separated into a series of characteristics, each of which is capable of forming a certain kind of non-covalent contact with the receptor. Here are primarily three stages of design to specify the attributes. First, it may be based on an atom, in which a feature would be a 3D atomic location associated with the type of atom. The second two options include the topological characteristics of atoms like a phenyl ring or a C = O group. Third is function based, wherever the atoms are summoned with resolute attributes, there are details sorted to the receptors of non-bonded interactions. Pharmacophore properties include HB donor (D), HB acceptor (A), base (positive charge, in which pH is 7), acid (negative charge, in which pH is 7), aromatic moiety (rings) (R), and hydrophobic group (H). Even if they serve the same chemical purpose, various topological characteristics might belong to the same category of chemical features. Consequently, no specific functional category has been allocated to the functional characteristics. A -OH oxygen, for example, has the ability to function as an HB acceptor, a donor, and occasionally even as a negatively charged feature. The centers of the functional groups, which are just the exact atom locations, are frequently used to represent groups, such as positively/negatively charged species, HB donors, and acceptors [36]. The vector that puts a constraint on the bond directionality between the feature of the receptor's binding site and the corresponding ligand feature is another way that HB donors and acceptors are repeatedly embodied. The centroid of the group is referred to as the center of a hydrophobic site or an aromatic ring.

Following feature extraction, the chosen features are merged to show the entire molecule's structure. These representations are typically produced as one of three formats: (i) a collection of feature points with their interpoint distances in a 3D point set that represents a ligand structure in a 3D space; (ii) a labeled graph in which the vertices represent the relations and the nodes represent the features; or (iii) a set of interpoint distances. A n×n distance matrix, where n is the number of atoms, is typically used to hold the third sort of representation [36].

Pattern Identification/Molecular Alignment

The structure correlated with the features is recognized as a collection of comparative sites in the 3D space after the features for each ligand in the dataset have been extracted. The majority of techniques rely on spatially superimposing minimum root mean square (RMS) alignment error pharmacophore locations on diverse chemical conformations [36]. The main characteristics of the alignment methods may be divided into two groups: point-based and property-based approaches [40]. In point-based algorithms, points can be distinguished as atoms, pieces, or chemical characteristics [46]. Pairs of atoms, pieces of matter, or chemical feature points are often aligned in point-based algorithms utilizing least-squares fitting and group detection techniques [47, 48]. The graph-theoretical tactic of molecular structure defines a clique as the most thoroughly linked subgraph, which identifies all possible amalgamations of atoms and functional groups to identify common substructures for the alignment. This method's primary flaw is the requirement for predetermined anchor points, which can be challenging to generate when dealing with ligands that are distinct from one another. The electron density, shape, electrostatic potentials, volume, and charge distribution of molecules are only a few examples of the molecular attributes used in property-based or field-based algorithms. The interaction energy components between the ligand and a probe placed at each grid point are calculated, and a 3D grid of ligands is produced. Later, the calculated grid attributes are transformed into a collection of Gaussian representations. Then, local optimizations using the almost resemblance metric of the intermolecular overlap of the Gaussians are performed, subsequently generating a number of starting configurations that are either randomly or deliberately sampled. Recent methods for molecular alignment include stochastic proximity embedding [49], atomic property fields [50], fuzzy pattern recognition [51], and grid-based interaction energy [52].

Scoring the Common Pharmacophore

It is known that the partition tree's remaining nodes contribute as a common component to the pharmacophore. However, this should not be accepted at face value since false positives might occur. For instance, the tree's mirror copy of an active node likewise produces a surviving node. In order to eliminate these false positives, the score must be compared to a reference score. In the pharmacophores, each tree's surviving node is referred to as a reference pharmacophore, and it is compared to all other pharmacophore pairs in that node, referred to as non-reference pharmacophores, in the partition tree's surviving node. In order to select a reliable reference pharmacophore and prevent false positives, the RMSD is computed, and a cut-off is defined (the default is 1.2 Å). To come up with a reference score, calculations are made. The user-adjustable parameters wvector, wsite, and cut-off RMSD are employed in equation number 1 to calculate the site vector score. The conformational value is an independent characteristic of the ligand supplying the locus of the pharmacophore and is proper in equation number 2 [37].

Each ligand in the surviving node has its reference score determined, and the ligand with the highest reference score is designated as the reference ligand. As a result, each partition tree is assessed, and the reference score is progressively determined. There may be many reference ligands with equal reference scores at this stage, in which case the user must make a wise pharmacophore selection based on selectivity scores or the overall various actives are matched. The pharmacophore hypothesis currently refers to the pharmacophoric regions of the selected reference ligand [37].

The pharmacophoric tools will carry out each of the aforementioned steps with a few minor modifications, such as alternative algorithms and extra procedures to boost the specificity of the theory.

$$\text{Site\_Vector\_Score}_i = w_{site}\text{Site\_Score}_i + w_{vector}\text{Vector\_Score}_i \qquad \text{(Equation 1)}$$

Where,

$$\text{Site\_Score}_i = 1 - \text{RMSD}_i/\text{cutoff}_{RMSD}$$

$$\text{Vector\_Score}_i = \frac{1}{n_v}\sum_{j=1}^{n_v}\cos\theta_{ij}$$

$$\text{Reference\_Score}_i = \text{Site\_vector\_Score}_i + w_{prop}\text{Prop}_{ref} \qquad \text{(Equation 2)}$$

Pharmacophore Tools and their Algorithms

The ultimate makeup of all ligand-based pharmacophore models hinges on the methodology, but they all rely on the training ligands shared chemical functionally reliant on the shared chemical functions of the training ligands. The virtual screening scoring functions and ligand-based pharmacophore models may generally be separated into overlay-based and RMSD-based techniques [26]. The pharmacophore suitability in RMSD-based approaches is determined by the separation between the compound's feature group and the pharmacophore model featured centers. The overlay-based approaches assess how effectively a component is replicated using the features of radius and atoms [38]. Superposition and the bioactive conformation are proposed by programmes like MPHIL [53], GAMMA [54], SCAMPI [55], Forge [56], Pharao [57], *etc.* Ligand based pharmacophore tools, their algorithm, merits, and demerits are provided in Table 2.

Table 2. Tools for ligand-based pharmacophore generation.

Program	Algorithm	Scoring Algorithm	Provider	Merits	Demerits	Refs.
Overlay Based methods						
Discovery Studio (CATALYST)	HipHop	FitValue	Biovia (formerly known as Accelrys)	1] Determines 3D spatial groupings of chemical characteristics in various configurations. 2], A lot of flexibility is provided.	1] The locations of protein atom collisions cannot be accurately predicted by the HipHop method. False positive results could result from this. 2] Does not have the facility of exclusion volume.	[58, 59]
	HypoGen			1] Using a set of compounds whose activity has been evaluated against specific biological targets, it generates predictive pharmacophores. 2] Based on a probability function, the algorithm accepts all improvements and some steps that are harmful. 3], A straightforward regression equation is contained in each pharmacophore and can be used to calculate the activity levels of unidentified substances.	1] The locations of atom collisions in proteins cannot be roughly predicted by the algorithm. Due to this circumstance, the activity prediction is overestimated. 2] In the subtractive phase, molecules that are inactive are typically owing to lacking functionality but may also be caused by the steric mismatch.	[25, 59]

(Table 2) cont.....

Program	Algorithm	Scoring Algorithm	Provider	Merits	Demerits	Refs.
				Overlay Based methods		
				4], It is one of the best pharmacophore software used to develop the QSAR model.	3] Does not have exclusion volume.	
LigandScout	Espresso	Pharmacophore Fit	Inte:Ligand	1] Utilizes the molecules' annotated conformation and pharmacophore models from the screening dataset for high performance. 2] Pattern matching alignment Algorithm to superimpose database molecule with the query pharmacophore model. 3] Complete tolerant sub-sampling results in great geometrical precision. 4] feature-based scoring. 5] There are no collisions between atom Van der Waals spheres and exclusion volume spheres. 6], The ability to describe numerous query pharmacophore models simultaneously that can be connected *via* Boolean expressions.	1] It cannot generate pharmacophore queries using large number of molecules.	[60 - 62]

(Table 2) cont.....

Program	Algorithm	Scoring Algorithm	Provider	Merits	Demerits	Refs.
Overlay Based methods						
GASP (Sybyl)	Genetic algorithm	Fitness score	Tripos	1] Due to GASP's intrinsic simplicity and its connection to Sybyl, many procedures can be automated using SPL (Sybyl Programming Language). 2] Whenever a pharmacophore model is constructed, it takes into account both the steric interaction of the ligands and the pharmacophoric properties. 3], The pharmacophore model generated can be optimised.	1] Input phase labour demands are increased. 2] It frequently necessitates segmenting the data set into groups of two or three ligands. The selection of subsets can steer the generated pharmacophores in a certain direction.	[44, 63]
GALAHAD (Sybyl)	Genetic programming Algorithm LAMDA, or linear assignment for molecular dataset alignment	Fitness score	Tripos	1] Align ligands which shows comparable interaction motifs and geometries in a dynamic way. 2] It works in two independent processes, unlike other pharmacophore elucidation approaches. A genetic algorithm and a rigid-body alignment algorithm. 3] having prepared to operate with ionic, hydrogen-bonding, hydrophobic, and steric properties using an atom-based approach.	1], It cannot generate queries from molecules having very diversified chemical structures. 2], It generates a query from a database containing a few molecules.	[64]

(Table 2) cont.....

Program	Algorithm	Scoring Algorithm	Provider	Merits	Demerits	Refs.
			Overlay Based methods			
				4] Using hyperactive molecules to create partial match search queries. 5] For a variety of literature datasets, the method may provide pharmacophores and pharmacosteres with crystal structure alignments.		
PharmaGist	-	-	Tel Aviv University	1] Again, for the purpose of pharmacophore identification, it is the first publicly accessible web server. 2] Extremely effective, with a distinctive run on a normal PC taking seconds to a few minutes containing up to 32 compounds. 3] The ability to find pharmacophores that various groups of input molecules share. 4] The technique basically and deterministically manages the input ligands adaptability during the alignment phase.	1] Only applicable for ligand-based pharmacophores.	[65]

(Table 2) cont.....

Program	Algorithm	Scoring Algorithm	Provider	Merits	Demerits	Refs.
Overlay Based methods						
-	-	-		5] The deterministic and efficient algorithm behind the server allows fast and reliable detection of pharmacophores. 6] It is straightforward and convenient to navigate the web interface. The user simply needs to submit the input ligands because the algorithm is totally computerized.		
QUASI	-	-	DeNovo Pharmaceuticals	1] It is an easy and convenient method to generate a pharmacophore model.	1] Only applicable for ligand-based pharmacophores.	[66, 67]
RMSD based methods						
MOE	-	RMSD	Chemical Computing Group	1] Pre-computed conformer assembly is required while being on documentation is permitted. 2] Methodology for orientation, dimensional precision, and scoring clique identification with acceptability testing *via* a low-level state machine, scoring by characteristic RMSD. 3] Explanation of the exemption volume sphere for heavy atoms lacking atom radii.	1] Only applicable for ligand-based pharmacophores.	[62, 68]

(Table 2) cont.....

Program	Algorithm	Scoring Algorithm	Provider	Merits	Demerits	Refs.
Overlay Based methods						
PHASE	tree-based partitioning algorithm	Fitness	Schrödinger	1] Conformer generation and flexibility search: 2] Alignment strategy: Considering feature-independent proximity allowance, a decision tree. 3] Geometric and feature point alignment contributions are used to calculate the score. 4] Approximation of the exemption volume sphere using massive atoms and hydrogens of atom radii.	1] It can generate a model using only active molecules.	[34, 62, 69]
DISCOTech (Sybyl)	Bron-Kerbosh clique-detection algorithm	D_{mean}	Tripos	1] Position sites and ligand points give molecules their unique characteristics. 2] Pre-calculating the number of low energy conformers for each molecule allows for the handling of structural adaptability. 3] A geometry alignment algorithm called clique identification is applied. 4] It is one of the best pharmacophore software used to develop the QSAR model.	1] Only applicable for ligand-based pharmacophores.	[44, 63, 70]

(Table 2) cont.....

Program	Algorithm	Scoring Algorithm	Provider	Merits	Demerits	Refs.
			Overlay Based methods			
Pharmer	-	Rmsd	Camacho Lab	Rather than scaling with the amount of the drug catalog getting scanned, Pharmer is a new computational method for pharmacophore screening that grows in range and complications as the query does. 1] A specific pharmacophore can be found quickly using Pharmer by searching through a database of millions of structures. 2] SMARTS22 expressions with user-configurable options are used to identify pharmacophore features. 3] It is already feasible for an explanation to add targeted collection features. 4], Pharmer is far better than current technology.	1] It cannot generate queries from molecules having very diversified chemical structures.	[71]

Pharmacophore Validation

The optimal pharmacophore model needs to have a high correlation coefficient and low RMS value, hence, it ought to be verified utilizing a variety of techniques [72]. The most established pharmacophore models are validated using Fisher's cost analysis test set and randomization test prophecy [73]. The leave-one-out approach [74], is another specialized method for LBP validation. Metrics used to evaluate the statistical test include correlation value, pharmacophore feature composition, and predicted activity of the eliminated compound's quality [75]. New metrics have been developed, including the Rm^2 metric and the Kruskal-

Wallis [76]. The accuracy of the anticipated and observed response data was verified using the rm^2 metrics test (average rm^2 and delta rm^2) [77]. The suggested values for "Average rm^2" and "Delta rm^2" are 0.5 and 0.2, respectively [78].

Cost Analysis

The discrepancy between projected and actual outcomes, as well as model prejudice and intricacies predicated on chemistry properties and weighting, can be examined using the cost function. Better models are available at lower costs. To assess the quality of each hypothesis, two cost values—a fixed cost, where predicted outcomes and definite outcomes are equal, and the null cost, somewhere the anticipated results are produced by a pharmacophore to no features are utilized [79]. In order to choose the optimal model, the variance amongst the two cost values (null cost and total cost) had to be larger than 60. If it is higher than 60, then the prototypical has a remarkably accurate correlation. If the difference is between 40 and 60, the model has a prognosis correlation of 70 to 90%; however, if the difference is less than 40, the model may be challenging to forecast [80]. A hypothesis is seen to be more advantageous if its cost value is lower than the null cost and closer to the fixed cost. Additionally, there is a configuration cost that is determined by the intricacy of the space. A decent model should cost less than 17 to configure [79].

Fisher's Randomization Test

If there is a correlation between the anticipated and actual activities, Fisher's cross-validation or randomization test is performed to determine the same. There are three phases in this process; (i) Random assignments are made to compound activities, (ii) Both total cost values and correlation values are estimated using an initial and a new training set, respectively. The result values show how the expected activities and actual activities relate to one another and (iii) a unique training set is competed with the original set's variation. The model is said to be improved if the initial training set has a lesser parameter and a good correlation value [79]. This technique shows the relationship between chemical structure and biological activity [80].

Test Set Prediction

External validation of the pharmacophore model is achieved through test set prediction [81]. Using this technique, it can be determined if the created pharmacophore model is able to categorize chemicals outside of the training dataset and predict related activities [74]. This technique is intended to clarify if the created pharmacophore model can be a prediction of a compound's activity other than the training set and appropriately classify them on their activity scale

[73]. According to the geometric fit of these compounds to the pharmacophore model, the estimated activity values were projected for each molecule in the test set [74]. Simple regression analysis is used to validate the pharmacophore model's ability to predict outcomes for the chemicals in the test set, and the correlation coefficient value between experimental and prediction activities is determined [82]. The experimental and calculated IC_{50} values showed a strong correlation, demonstrating the pharmacophore model's strong predictive power [74].

Leave-one-out Method

Cross-validating model employs the leave-one-out strategy [74, 83]. One molecule from the training set is removed at a time to recompute pharmacophore hypotheses using the statistical method known as the leave-one-out method. The purpose of this test is to make sure that the correlation of pharmacophore hypotheses does not exclusively depend on one molecule [74, 83, 84]. The test is successful if the associated one-missing hypothesis can accurately forecast each omitted compound's behavior. The statistical test was evaluated using the correlation coefficient, the pharmacophore's feature composition, and the estimated level of activity for the eliminated compounds [74, 84].

3D-QSAR

Quantitative structure-activity relationship (QSAR) models may be useful in ligand-based approaches for the liability of targets with little or no structural knowledge. The goal of QSAR is to create significant correlations (models) between independent variables (such as molecular descriptors and structural characteristics of compounds) and a dependent variable, which is often the value that one intends to predict. Such modeling is conceptually based on the theory that substances with similar structural properties may have comparable biological activity. Statistical methods, molecular descriptors, and activity data are some examples of the variables that affect QSAR models. QSAR has been widely employed in different elements of drug discovery because of the benefits in throughput, cost savings (labor and reagents), turn-around time, and the ability to test compounds even before they are synthesized [85]. Several steps must be taken in order to develop a QSAR model subsuming: (1) gathering ligands known to be effective against the exact target; (2) extracting descriptors for the molecule; (3) selecting the most appropriate descriptors from a wider range of descriptors; (4) mapping the metabolic processes to the structural descriptions; and (5) internal as well as external validation of the QSAR model [86]. In recent times, 3D QSAR has been preferred over standard 2D QSAR because of its robustness and better predictive quality [87, 88]. The incorporation of 3D descriptors, such as molecule, electrostatic, and steric properties of coupling fields, is what makes 3D QSAR predictions of high quality. The inclusion of 3D descriptors makes the 3D QSAR extremely impressionable to the analyzed chemical's 3D alignment. Therefore, the primary prerequisite of the 3D-QSAR approach, which involves Comparative Molecular Field Alignment (CoMFA), is a relevant and well-considered orientation theory in 3D space. There are numerous alignment techniques, including alignment with a suitable docking-based alignment, template conformation, database alignment, pharmacophore-based alignment, field fit alignment, and atom fit alignment [88]. For linear QSAR modeling, statistical tools like PCA (principal component analysis), PLS (partial least squares analysis), and multivariable linear regression analysis (MLR) may be employed. However, there are additionally numerous non-linear models that can be created using machine learning techniques like neural networks and Bayesian networks. Internal cross-validation is utilized to confirm the QSAR model and the cross-validated R^2 or Q^2 is calculated. Additionally, it is important to do an objective evaluation, which is done by forecasting the behavior of outside test sets not applied to the model creation [89]. Whenever the compounds utilized for model development are based on a congenerous sequence of molecules, these 3D QSAR models perform well [90, 91].

If there are enough molecules with different levels of activity, a 3D QSAR model for each hypothesis could be created using training set structures which complement the pharmacophore on three or more sites and use fitness as the parameter to determine which alignment is optimal for each molecule. While the pharmacophore-based QSAR models might be better suited whenever the structures are extremely versatile and show considerable diversity [86].

Pharmacophore Based 3D QSAR

Pharmacophore-based QSAR models are only interested in the molecular locations that match the theory. These sites are handled as spheres with viewer radii, and attribute type-based categories are assigned to each one (A, D, H, N, P, R, X, Y, Z). The virtue of pharmacophore-based models is that the models produced have a wider application to molecules from various chemical families since they are less based on how consistently the total molecular superpositions are maintained. A pharmacophore-based QSAR, however, is unable to take into consideration the pharmacophore model directly. This necessitates an atom-based QSAR, which takes into account the full molecule structure [34]. The dataset was first split into a training set (containing 70% of the compounds) and a test set (containing 30% of the compounds) in order to create a 3D QSAR model. Molecules from the training set exhibited a variety of biological functions and structural characteristics. As a result, it is used to create the 3D QSAR model, whereas the test set, which contains less potent and moderately active substances, is used to verify the QSAR model [92, 93]. The training set molecules' pharmacophore characteristics shall be arranged together into the standard cubic matrix with 1.0 Å spacing during QSAR development [34, 92]. As compensation for various pharmacophoric properties in the training set, each cube will be given a 0 or 1 bit. As a result, multiple pharmacophoric sites may coexist in a single cube, eventually leading to one or more volume bits. Last but not least, a single molecule will be represented by a string of binary numbers based on occupying various cube sites. As an independent variable, these binary values will be used for the creation of the 3D model with PLS regression and dependent variables for biological activities of the training set. As a result, a number of regression models will be built, each of which will have a maximum number of components of N/5, where N is the number of ligands in the training set. The 3D QSAR model was then verified by forecasting the actions of the test molecules that used a variety of statistical indicators, including consistency, RMSE (root mean square error), standard deviation (SD), regression coefficient (R^2), variance ratio (F), and Pearson-R [34, 91, 94].

STRUCTURE BASED PHARMACOPHORE

The known 3D structure of a macromolecule or macromolecule-ligand complex is used for receptor-based (structure-based pharmacophore mapping to ascertain the pharmacophore characteristics and geometric limitations [28, 39]. As structural genomics advances, there are more protein targets with known X-ray structures, making this conceivable. An assessment of the complementary chemical properties of the active site and associated spatial correlations are part of the process for the receptor-based pharmacophore method, which is thereafter coming to the development of a pharmacophore model with a few chosen features. There are two subcategories of receptor-based pharmacophore modeling: based on macromolecule-ligand complexes and based on macromolecules (without ligand). The relationship between the ligand and the macromolecular target is determined by the macromolecule-ligand complex, which does this by accurately locating the target's ligand binding site. The flaw in this strategy is that it cannot be used when the compounds target-binding sites are unknown since it requires the 3D structure of the macromolecule-ligand complex. Macromolecular-based pharmacophores can be employed to address this issue [28]. The greatest solution to this issue is a pharmacophore based on a hot-spot-guided receptor. Using machine learning methods, this strategy prioritizes the atom cavity for targeted ligand binding [95]. The prevalence of receptor-based pharmacophore approaches is dwindling due to the inefficiency of existing docking and scoring algorithms and the availability of quick, simple methods for lead generation and optimization. Additionally, docking tools and structure-based pharmacophore approaches complement one another and are commonly used in a hybrid manner [39].

Structure Based Pharmacophore Model Generation

A structure-based technique for pharmacophore development analyses the active site's characteristics and their spatial interactions before using an active picture of that to build the pharmacophore model. It is crucial to identify which of the many properties that result from this are genuinely a part of the pharmacophore [96].

Active Site Identification

The three-dimensional receptor structure, often in the format of a Protein Data Bank (PBD), serves as the basis for receptor-based pharmacophore mapping. A spherical probe with a programmable ambit and position is used to pinpoint the pocket for the binding receptors, revealing both the binding site and the important residues that interact with the ligands. On the basis of the geometry of the surface, numerous programmes are accessible for the detection of clefts, fissures, and binding pockets as well as to provide suggestions about potential active site sites. By analyzing the protein's functionality following a single residue being mutated, one might infer the essential residues. If a specific residue's mutation impairs a protein's ability to function, that residue may be found at the active site. By correlating a protein's active site with that of another protein with a known active site, computational analysis, like multiple protein structure alignment methods, may also be used to determine a protein's active site [97].

Complementary Image Construction

By understanding which properties of the molecule shares with the active site, it becomes easier to design an effective interaction between them. To put it another way, the foundation for building an input pharmacophore model is the creation of a complement of the receptor binding site. The key components that are found in the binding site are functional elements like HB donor/acceptor and hydrophobic groups, which are then followed by the complementary features that are necessary for binding to occur [36, 97, 98].

Query Generation, Searching and Hit Analysis

The process of using the binding site map to generate pharmacophore models is challenging to discover and chemically confirm the dynamic regions. The interaction map frequently generates a large number of features because the receptor binding site has the capacity to engage a range of fragments in variability of the requisite conformations. All other characteristics are removed, and the feature that is closest to the geometric centre of the cluster is preserved as the cluster representative in order to solve this issue. There are a lot of features, and none of them can be used in a single model, even after the features have been clustered. This is due to the fact that models with all of these properties might be incapable to find any hits in the database [97, 99]. These are used by a number of pharmacophore algorithms, including CATALYST, to search chemical databases and evaluate the accuracy of models (pharmacophore "queries") to screen and disqualify exceedingly dynamic composites. Currently, compounds have screened that map to all enquiry criteria similarly, including a colossal substitutable that causes steric hindrance in addition to that, it prevents the chemical as opposed to slipping into the binding point. Because of this, it is sometimes advised to include some excluded volume characteristics, which lower the molecules score if certain atoms or groups are positioned in a way that increases the likelihood that they would collide with the atoms in the active site [97]. Tools for the generation of structure-based pharmacophores are given in Table **3**.

Table 3. Tools for ligand-based pharmacophore generation.

Software	Module	Developer	Mode	Merits	Demerits	Refs.
LigandScout	-	Inte:ligand GmbH	Ligand-target complex-based	1) Geometric, lexical, and rule-based automated PDB ligand interpretation. 2) Sophisticated 3D visuals and an undo option in a cutting-edge user interface. 3) Direct links between a 2D view and a hierarchical view and a 3D interface. 4) Comprehensive 2D depiction of protein-ligand interaction. 5) More sophisticated management of co-factors, ions, water molecules, and metal binding sites. 6) The ability to quickly create pharmacophores from docking postures and an advanced docking conclusion visualization.	1), A pharmacophore cannot be generated using multiple PDB structures of the same receptor/enzymes.	[60, 100 - 102]

(Table 3) cont.....

Software	Module	Developer	Mode	Merits	Demerits	Refs.
				7) Enabling virtual screening, pharmacophore export to Catalysttm, MOEtm, and Phasetm.		
LigBuilder	Pocket v.2	University of Peking's Institute of Physical Chemistry.	Ligand-target complex-based	1) Algorithm for automatically developing pharmacophores. 2) Comparable to DS Catalyst, SBP and LigandScout. 3) Additionally, Pocket v.2 efficiently condenses the primary components of the pharmacophore model to a manageable quantity. 4) When various ligands are bound, it may withstand slight deformation of the protein component to produce a reliable pharmacophore model. 5), Its capacity to produce analogous pharmacophore models for many proteins that adhere to the same	1) It is devoid of a screening module and is not capable of exporting pharmacophores in popular file types. 2) Since only the target structure is accessible, there are insufficiencies to create a pharmacophore.	[100, 103 - 105]
				ligand allows one to classify protein molecules in accordance with key binding characteristics. It will also be useful for anticipating probable detrimental impacts of known pharmaceutically significant substances.		

(Table 3) cont.....

Software	Module	Developer	Mode	Merits	Demerits	Refs.
Discovery Studio	Catalyst	Accelrys	Ligand-target complex/target structure-based	1) Providing access to a plethora of pharmacophore models to employ in activity assessment, which include the filtered pharmacophore database, HypoDB, from Inte: Ligand. 2) To quickly generate hit lists that are specially adapted to meet your particular receptor, generate pharmacophore models using protein structures. 3) Simple integration of ligand and protein structural properties will result in a more accurate model of the binding-relevant aspects. It can be particularly effective when the protein structure is not entirely understood.	1) A pharmacophore cannot be generated using multiple PDB structures ofthe same receptor/enzymes.	[100, 106 - 108]
Schrodinger	PHASE/e-pharmacophore	Schrödinger, LLC	Ligand-target complex-based	1) Structure-based pharmacophores that have had their energy optimised to produce new pharmacophores. 2) It has a variety of capabilities, including the Glide XP scoring mechanism. 3) Allows receptor-based excluded volumes.	-----	[100, 109 - 111]

(Table 3) cont.....

Software	Module	Developer	Mode	Merits	Demerits	Refs.
				4) There are single and fragment modes for the generation of e-pharmacophores. 5) It gives a greater level of diversity, speed, and performance.		
Shape4	-	Department of Pharmaceutical Sciences, Biomanufacturing Research Institute Technology Enterprise (BRITE), North Carolina Central University	Target structure-based	1) By accounting for and implementing the spatial restrictions of the target binding site into the pharmacophore model, database retrieval can be made more effective. 2) For structure-based pharmacophore creation and virtual screening, Shape4 is a quick, efficient, and user-friendly tool. 3) Uses computational geometry algorithms to find the binding location and provide a picture of the intended binding site that is not there.	1), A pharmacophore model is generated with only a single PDB structure.	[100, 112 - 114]
Snooker	-	Computational Drug Discovery Group, CMBI, Radboud University Nijmegen	Target structure-based	1) It is mainly designed for pharmacophore mapping of G-protein coupled receptor.	1) Only drugs targeting the tiny class of G-protein-coupled receptor subfamilies are appropriate for this method of pharmacophore discovery and chemical design.	[100, 115, 116]

Validation

Prior to screening for chemicals, a multistep technique called pharmacophore model validation is crucial. If the outcomes of experimental validation following virtual screening do not agree with the screening score derived from the pharmacophore model, the method can be repeated and revalidated [115, 116]. A pharmacophore model can be enhanced by utilizing validation information as input based on validation outcomes. More refining techniques must be used if the model's output does not match the *in vitro* results. To verify the effectiveness of the model, a test set of known chemicals with known activities is used for theoretical validation. It is vital to have a decoy set if early results from the various experimental outputs do not show inactive substances [115, 117]. A pharmacophore model's capacity to reliably distinguish between an array of active chemicals, comprising equally inactive and active portals, is demonstrated by the substantiations of the model using active complexes. Contrarily, testing a pharmacophore model with inactive substances examines its capacity to discriminate between inactive and active molecules [118]. There will be a hit list of n compounds that match the model as a consequence of observing through the verification of a dataset containing N entries. True-positive (TP) hits are active substances that model accurately identified. FP hits are inactive substances discovered by the models. Undiscovered by the model is an active substance known as a false-negative (FN) molecule. True-negative (TN) chemicals are inert substances that have been correctly categorized [119].

Test Set

The goal of the testing dataset is to assess the ability of a newly developed pharmacophore model to contain most of the effective medications but exclude inert alternatives. Using this test set, the pharmacophore model created with the training compounds will be theoretically verified and improved [120]. There are a maximum number of both active and inert substances in it. The compounds in this group benefit from having different structural compositions as well. If there are substances that have been proven to be inactive through experimentation, they should be added to the test set. The test set screening findings may also be used to finalize the model's refinement; based on the turns, spatial limit functionalities could be added or removed. More information about inactive chemicals that have been empirically verified has been accessible in recent years [115, 120]. List of databases are listed in Table **4**.

Table 4. List of databases.

S. No.	Databases	Data Provided	Refs.
1	PubChem	Results from high throughput screening of a significant number of chemicals are provided. 50 million no unique compounds with annotated biological data.	[119, 120]
2	ChEMBL	There are few yet unreleased (in-) activity statistics from academia and industry. Annotated biological data for over 1.3.	[119 - 121]
3	The DrugMatrix project from the NIH National Toxicology Program	132 drug- and side-effect-related targets are covered by *in vitro* screening findings for 870 therapeutic, industrial, and environmental chemicals (accessible *via* ChEMBL and ftp:/anonftp.niehs.nih.gov/drugmatrix).	[119, 120]
4	The Tox21 program by several US governmental organizations	Gives activity information for over 8100 distinctive, commonly used compounds that have been examined, including over 800 high throughput screening assays (Tox21 assay data are also available in PubChem)	[119, 120]

Decoy Set

Because specific knowledge regarding biological function is lacking, a decoy set is a collection of chemicals that are considered inert [38, 115]. A random screening in this kind of chemical database will only uncover a tiny subset of compounds that are physiologically active against a target. The validation of the model should also take this circumstance into account. Therefore, for the theoretical model authentication, it is necessary to have a substantial percentage of passive molecules and a limited number of active dataset molecules. An extra decoy set can be used to test and find one or more active molecules among many other (possibly) inert compounds when there is only a minuscule portion of recognized inactive compounds. As a standard guide, the test set should contain 40 decoys per active molecule, and all these decoys ought to be inherently different again from chemical ingredients yet physico-chemically comparable to them [38, 120]. Because there is extremely little chance that a random compound would be active, it is possible to create a decoy set with unknown activity toward the target [122]. A list of decoy databases is enlisted in Table **5**.

By counting the number of active or inactive (decoy) compounds in the test compounds, certain crucial factors can be found. The metrics include Sensitivity (Se), Specificity (Sp), Accuracy (Acc), Active Yield (Ya), and Enrichment Factor (EF), among others. These parameters of validation are used to validate ligand-based as well as structure-based pharmacophore models.

Table 5. List of decoy databases.

Database	Information	Refs.
Directory of Useful Decoys (DUD)	To create and test models, we used 106,200 decoys and 2950 active chemicals (version 2.0).	[38]
ChEMBL or Virtual Library	It includes 8.9 million drug-like molecules to produce the appropriate amount of dummies. A vast database might be filtered to create sets.	[121, 123]
DUD-E	A stronger impression of the DUD database, serves as the foundation for the DUD-E, an online decoy-generating service offered by University of San Francisco in California at the Shoichet lab. The DUD-E also offers a free, online decoy set generating module; this module requires a text file containing a set of active molecules as text files (SMILES codes) as fed, and it produces a decoy set that is being sent to the user.	[124]

Sensitivity (Se)

In a virtual screening experiment, a pharmacophore model's Sensitivity (Se) indicates its capacity to locate TP compounds. Its ratio including all active ingredients in the validation dataset is given as TP compounds [125]. Its value is between zero and one, where one means that all TP molecules are discovered by the pharmacophore model during a virtual screening procedure and zero means that none of them are discovered.

$$Se = TP/(TP + FN)$$

Specificity (Sp)

Specificity (Sp) is the ability of a pharmacophore model to eliminate inactive drugs from a computer-simulated screening assay. It is determined as the proportion of every compound involved in the dataset to the rejected TN compounds. Their values are in the zero-to--to-one range, much as the sensitivity. In a virtual screening process, one denotes the proper rejection of all inactive compounds, whereas zero denotes that almost all inactive chemicals reflect the model [125].

$$Sp = \frac{TN}{TN + FP}$$

Accuracy (Acc)

The accuracy of a dataset is measured by the proportion of TP and TN to active (P) and inactive (N) chemicals. It figures out how many chemicals are accurately categorized [116, 126].

$$Acc = (TP + TN)/(P + N)$$

Yield of Actives (Ya)

The "yield of actives" (Ya) of a virtual screening method is computed as either the ratio of TP compounds discovered to all hits (n) or the length of the hit list. As a result, it establishes the hit percentage list of TP hits [126, 127].

$$Ya = TP/n$$

Enrichment Factor (EF)

The enrichment factor (EF) is the proportion of TP hits to all active compounds in a database.

$$EF = \frac{TP/n}{n_{act}/N}$$

N_{act} = total number of molecules that are active in the database

The enhancement of active chemicals in the hit-list, as opposed to pure random selection, as opposed to pure random selection, is measured by the EF. This measure connects the authentication of the database's makeup, with the retrieval of active chemicals. The EF, however, has two significant flaws. First, how big and how many datasets there are utilized in the study have a significant impact. The maximum EF might increase with the size of the inactive dataset relative to the active dataset. Nevertheless, datasets with about equal numbers of active and inactive chemicals can never go beyond only tiny EFs. The grade of active chemicals inside a hit list is the second flaw. It is unknown whether the active compounds appear among the most popular hits because the EF provides no information regarding the ranking of active versus inactive hits. A comparative EF (considered EF split across the determined EF for the utilized dataset) can be considered to solve the first problem. The following measure provided gives a solution for the later problem [127].

Goodness of Hit-List (GH) Scoring/Güner–Henry (GH) Scoring Method

The Se, Sp, and Ya are combined in the goodness of hit list (GH). Compared to se, sp, or ya alone, it is better to assess the preferential power of a pharmacophore model. It takes into account both TP and TN ratios [128]. Its value is between 0 and 1. One means a pharmacophore model is excellent and is putting out a precise hit list containing just active ingredients [127].

$$GH = \left(\frac{3}{4}Ya + \frac{1}{4}Se\right) \times (Sp)$$

Receiver Operating Characteristic Curve (ROC)

To determine the effectiveness of a pharmacophore model, it utilizes ROC analysis [35, 129]. The hit tables containing rankings false positive and true positive rates are plotted on a curve. A ROC curve is a crucial criterion when comparing two or more models. It provides information on how accurately a model ranks the active chemicals. A plateau after the curve's sharp ascent indicates that there are more active chemicals than inactive ones. A user can select a threshold value for the hits if the curve becomes level after a rapid climb. If two curves must not intersect one another when comparing two models, the curve with the sharp beginning is predicted to perform much better [130]. The Area under Curve (AUC) is indeed a useful metric since it offers ratings based on the model's effectiveness, excluding visual examination. The AUC determined by the sensitivity and specificity of each active chemical is generated by the numerical integration of all rectangles. The AUC score runs from 0 to 1, with 0 indicating that almost all inactive molecules would rank first and 1 indicating that all active molecules will rank first. The AUC value of 0.5 showed a random selection of compounds, whereas high AUC values indicate higher model performance [131]. The ROC curve should have a steep slope and a high AUC value in order to provide a model of the highest quality; this will give all active hits without the inactive ones, with a specificity and sensitivity value of 1, along with a high EF; this is known as an ideal curve. A perfect ROC curve is depicted in Fig. (**3**) along with specificity and sensitivity values of 1. Lastly, the combinational model with all parameters is verified. The finished pharmacophore model is now prepared for virtual screening [132].

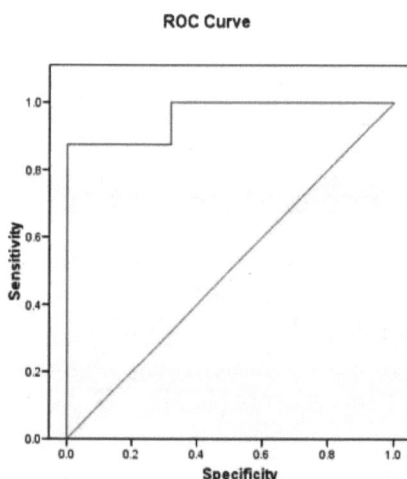

Fig. (3). Graphical representation of ROC curve with respect to specificity and sensitivity.

Virtual Screening

Database Searching

The most popular way to implement database search is a series of filtering steps. The process begins with a quick feature that will allow phases known as feature categories, feature quantity, and quick proximity assessments, and also allow for the already substantial elimination of compounds that are categorically nonmatching. Second, accurate 3D matching algorithms, which are typically slower but more constrained, are used.

Prefiltering

Prefiltering is essential, meanwhile the authentic three-dimensional configuration of the probing molecules and models of the pharmacophore is the phase that consumes one of the most time in the screening procedure [133, 134]. Prefiltering seeks to rapidly identify and exclude any configuration of molecules that cannot be matched in three dimensions to query the pharmacophore model. Now the molecules which are successfully navigating this clarify stages which prerequisite to proceed to the highly precise but computationally pricey stage of three-dimensional synchronization. Since minimal verification is required, comparison computations are quick, and the inference of biological resemblance from known structures, which is typically true descriptor-based similarity approaches is taken to show a suitable isolation [61, 133, 135]. Aspect count matching is a fairly straightforward yet efficient filtering technique that may efficiently dispose of a significant portion of the intricacy of the query and will determine the number of molecules in the database [133, 136]. Only molecules with component values for the query pharmacophore model as well as the database molecules are already computed, the very same (or higher) feature count as the demand must be transmitted to the time-consuming matching step. The idea of "pharmacophore keys" is another rather sophisticated way of exploring 3D pharmacophore databases [133, 137]. Pharmacophore keys essentially represent the spatial arrangement of characteristics in 3D pharmacophores through the use of straightforward binary fingerprints. Binning interfeature distances and computing middle encryption for every conceivable two-, three-, or four-point feature subset of an input pharmacophore yield a set of particular bit indices in a given size bitset. Each of these bits then indicates the existence or absenteeism of a specific n-point pharmacophore. The program then reduces to a straightforward connection assessment to find molecules that don't match the query. With only minor variations, such as the incorporation of sample feature restrictions, various feature descriptions, varied binning requirements, *etc.*, nearly all currently available software programmes employ comparable methodologies that adhere to this idea [27, 133, 138]. The majority of algorithms even have checkpoints which might eliminate compounds that theoretically might match the pharmacophore model in the search, however, this compromise in screening is tolerated in return for greater effectiveness. In conclusion, prefiltering aims to speed up the search by removing significant portions of the total search space. However, this acceleration technique can essentially guarantee that the sheer value of the screening output is upheld and that the objectives' similar actives improvement and the discovery of new scaffolds may still be achieved [133, 139].

3D Pharmacophore Model Matching

The conformation-specific pharmacophores of all database molecules that have a good possibility of matching the requested pharmacophore need to be looked at more carefully to determine how well they reflect the spatial arrangement of the query characteristics. It's going to be evaluated at this phase as the final hit list or if a database combination is denied. Generally speaking, considerable consideration must be given to this decision-making process because it directly affects the caliber of the screening findings acquired. Finding an appropriate collection of characteristics that satisfies all possible combinations of the query's n-point distances can be used to simplify the challenge of such a querying pharmacophore model as well as the derived pharmacophore from a database molecule conformation, there must be a symmetrical sequence. Early proposals of greedy algorithms that tackle this problem include three-dimensional group identification techniques with the gradual accumulation of greater and bigger appropriate feature arrangements [140]. Pure feature pair distance comparisons (two-point pharmacophores) cannot distinguish between a pharmacophore and its different versions, hence an explicit base layer in three-dimensional reality is required to properly identify a matching to the query inside the defined feature tolerances [47]. Additionally, overlaying is required to assess and/or score further limitations placed by planar aromatic rings, hydrogen bond acceptors and donors, and exclusion/inclusion volume spheres. In the 3D pharmacophore corresponding stage, software programs for pharmacophore modeling that encompasses cutting-edge screening functionality, such as phase, LigandScout, catalyst, and MOE, all perform a type of geometric alignment, which is typically accomplished by minimalizing the RMSD among the accompanying feature sets [58]. Despite a few similarities across the broad approaches to hit detection, there are a number of specifics where they diverge. These details include how conformational flexibility is handled, how query feature limitations are interpreted, and how search parameters may be altered.

Strategies for Selection of Compounds for Biological Testing From Hit Lists

Despite a few similarities across the broad approaches to hit detection, there are a number of specifics where they diverge. These details include how conformational flexibility is handled, how query feature limitations are interpreted, and how search parameters may be altered.

Lipinski Filter

For choosing compounds from the hit list based on their anticipated oral bioavailability, the Lipinski filter is a well-known criterium [31]. It was shown that 90% of orally active medications that advanced clinical trials of phase II and so had a higher chance of receiving FDA approval had a number of characteristics. In accordance with the identified qualities, these constraints—which he dubbed the "rule of 5" (RO5)—are molecular weight $\leq$ 500, clogP value $\leq$ 5, HBDs $\leq$ 5, and HBAs (sum of O and N atoms) $\leq$ 10 [31, 141]. This restriction, however, only applies to medications that are used directly and are assimilated passively. Consequently, medicine applicants that fail to meet Lipinski's RO5 might have subpar penetration or captivation capabilities. Less lipophilic area and a range of 200–350 of molecular weight will allow for the potential of additional lead compound modification when choosing hits from potential lead compounds during target identification [142].

Molecular Similarity Evaluation

Evaluation of molecular similarity is a crucial method for choosing compounds that are lead or drug-like in nature. A result could have a high possibility of becoming an active drug molecule if its chemical composition is similar to known active drug molecules. The dimension of the hit list is another issue. Using a 2D similarity-based clustering approach can help to focus the experimental investigation on structurally varied drugs if the modeler returns a large number of similar hits. But not all clustering techniques are completely trustworthy. Molecules in a single scaffold are frequently organized into several clusters of them. Additionally, the clustering algorithm continuously requests that the number of clusters is computed. Several scaffolds will indeed be overlooked if a user simply calculates a small number of clusters from a lengthy hit list. In contrast, creating several clusters for short hit lists would produce cluster centers that are quite similar, frequently using the same chemical scaffold [143]. Hit clustering is difficult, but new methods are continually being developed for this crucial stage [144]. A specific property that seems to be advantageous or detrimental to the biological activity can be refined using two-dimensional fingerprinting technologies. These fingerprints may be identified, as well as the hit lists are filtered based on previously established criteria [145].

Consensus Hits

Consensus techniques are increasingly being used to improve the probability of choosing bioactive components for biological testing or even to whittle down the number of chemicals that really are targeted. This way, chemicals are examined against two or more target models. But just those that pass the evaluations for some or all models are given the green light to undergo an experimental inquiry. Either the same programmes or alternative ones might be used to produce the applicable models. It is also demonstrated that pharmacophore models may generate complementary hit-lists with little connection, and that, respectively, hit-lists can contain distinct functional substances [145]. In addition, pharmacophore models are regularly integrated with various techniques, such as shape-based screening [146] or docking [147].

APPLICATION OF PHARMACOPHORE MAPPING

Pharmacophore mapping is an important rational drug design tool to find innovative active molecules which are progressive in contrast to a receptor. It can predict the activity as well as the selectivity of molecules to build more specific ligands for the target. Pharmacophore mapping is also applicable in finding the ADMET properties of the drug. Applications of pharmacophore mapping are provided in Fig. (**4**).

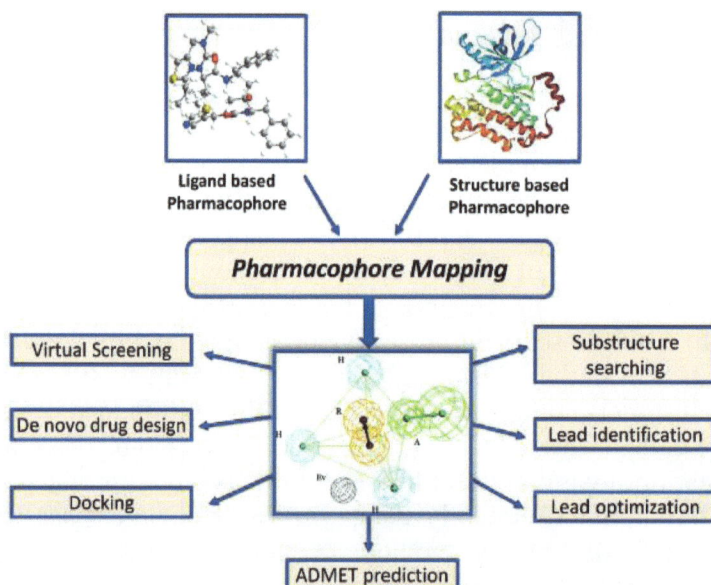

Fig. (4). Applications of Pharmacophore mapping.

Designing new compounds successfully using ligand- and structure-based pharmacophore mappings are enlisted in Tables **6** and **7**, respectively.

Table 6. Successful examples of molecules derived from ligand-based pharmacophore mapping.

Target	Derivative/Structure	Software Used	Database From	Active Molecules	Biological Activity	Refs.
Rho-Kinase-II	NSC2488 / NSC2888 / NSC4231	Discovery Studio (v2.0) (Hypogen)	National Cancer Institute (NCI) database	3 hits were found NSC2488 NSC2888 NSC4231. Among them, NSC2488 is the most active compound.	Rho-kinase inhibitory activity: NSC2488 (IC_{50}=8.02 nM) NSC2888 (IC_{50}= 8.4nM) NSC4231 (IC_{50}=9.01 nM) Vasodilat-ory activity; NSC2488 (IC_{50}=8.4 nM) NSC2888 (IC_{50}= 8.5nM) NSC4231 (IC_{50}=9.0nM)	[148]
Histone Methyltransferase Disruptor Of Telomeric silencing 1-Like (DOT1L) Inhibitor	Massonianoside B (MA)	PHASE	-	Massonianoside B (MA).	IC_{50}=399 nM	[149]
Organic Cation Transporter (OCT1)	Dextrorphan (Morphine analogue with lack of ethe linkage between C4 and C5)	LigandScout 4.4	ChEMBL database (version 23)	Dextrorphan is the most active.	With an inhibitory potency varying from 6 to 25 M, morphinan opioids without the ether bridge are active. Dextrorphan (IC_{50}=6.4 μM)	[150, 151]
Glyoxalase-I inhibitor	Dicarboxylic (or Diacidic) moieties	Biovia2017 ®'s versions of Discovery Studio (DS) and Pipeline Pilot	Maybridge Screening Collection database (2017) and the Aldrich MyriaScreenII database	32 hits are selected for biological evaluation. ST018515 is the most active hit.	IC_{50}= 340nM	[152]

(Table 6) cont.....

Target	Derivative/Structure	Software Used	Database From	Active Molecules	Biological Activity	Refs.
Soluble epoxide hydrolase (sEH) Inhibitor	HTS07656 (% inhibition= 69.3±4.3) CD11292 (% inhibition= 55.9±4.6) BTB04690 (% inhibition= 62.3±2.4)	Phase	DUD-E database (for decoy set) Maybridge database	3 hits are active. HTS07656 is the most active compound.	HTS07656 and CD11292 IC_{50}= < 5μM BTB04690 IC_{50}= < 25 μM	[153]
Takeda G-protein-coupled receptor 5 (TGR5) agonists	V12 V14	Catalyst (HipHop) DS 3.0	ZINC database	20 hits were selected for biological evaluation. Among them, 2 hits selected having more than 40 to 80% receptor activation are V12 and V14.	EC_{50}= 19.5 μM (V12) EC_{50}= 7.7 μM (V14)	[154]
CXCR2 Inhibitor	CX25 CX4152	Catalyst (Hypogen)	Asinex (Moscow, Russia), Vitas-M Laboratory (Apeldoorn, Netherlands), and Enamine (Kiev, Ukraine)	CX25 is the most active hit compound on CXCR2 and CXCR4 receptor. CX25 is chosen for further lead optimization. CX25 is selective on CXCR2 and CXCR4. Whereas CX4152 obtained from shape-based pharmacophore mapping is more selective over CXCR2 than CXCR4, which is the most active compound.	CX25 (non-selective) IC_{50}= 360nM CX4152 (selective CXCR2 Inhibitor) IC_{50}= 7.6±6.2 μM	[155]

(Table 6) cont.....

Target	Derivative/Structure	Software Used	Database From	Active Molecules	Biological Activity	Refs.
Signal transducer and activator of transcription 3 (STAT3)	Compound-1	Discovery Studio 3.0 (Catalyst-Hypogen)	National Cancer Institute database	The four highest-scoring reported compounds 1–4. Comp. 1 is the most active comp.	$IC_{50}= 10\mu M$ (Comp. 1)	[156]
Human Epidermal Growth Factor Receptor-2 (HER2)	Compound 120 Compound 126	Catalyst/Hypogen	National Cancer Institute (NCI) and Drug Bank database	4 hits have activity on HER2 receptor Compound 120,123,125,126 among of them, 120 and 126 are the most active compounds.	All four hits having $IC_{50} <$ 5μM. But the most active compound 120 and 126 are having IC_{50} 1.43 μM and 1.76 μM, respectively.	[157]
RA2B antagonist	Z1139491704	GALAHAD (SYBYL-X 2.0)	ZINC database	18 compounds were selected for biological evaluation, among them, Z1139491704 is the most active.	$pEC_{50}=$ 7.77±0.17	[158]
CXCR2	Scaffold F Compound 1a Derivatives of 1a : 4H-1,2,4-triazol scaffold Compound 1e	Discovery Studio 2.5 (HipHop)	Maybridge database	Scaffold F containing compound 1a is selected as a hit. For the database screening, chemical compound 1a with the unique framework was chosen and put through an *in vitro* biological test, which revealed moderate CXCR2 antagonist activity. With additional SAR study and customization, compound 1e demonstrated improved CXCR2 antagonist activity.	$IC_{50}= 74\mu M$. (1a) $IC_{50}= 14.8\mu M$. (1e comp.) 72% inhibition of cell migration at 50 μg ml−1 (1e comp.)	[159]

(Table 6) cont.....

Target	Derivative/Structure	Software Used	Database From	Active Molecules	Biological Activity	Refs.
Ribonucleotide Reductase (RR) Inhibitor	Thiosemicarbazone derivative; Thiosemicarbazone (Red) Compound Ig; R = CN R'= -N(CH₂CH₂Cl)₂ Bis(2-chloroethyl)amino moiety.	Discovery Studio 3.5	NA	Ig has the higgest cyotostatic activity in the assay on HeLA cells, CEM, and L1210 cells.	$IC_{50}=$ 0.3- 2.5 µg/ml. (0.71-5.95µM)	[160]

A Successful Example of Pharmacophore-based Drug Design: An Example of How Anthranilamide Derivatives Were Successfully Shown to be Promising Factor Xa Inhibitors [163]

Prothrombin is converted by factor Xa into thrombin, which is involved in a number of thromboembolic problems. Factor Xa inhibitors potentially have an anticoagulant effect by preventing prothrombin to thrombin conversion while not affecting typical haemostasis (decreased the risk of bleeding). For the development of factor Xa inhibitors, Junhao Xing *et al* have employed the combined docking, structure-based pharmacophores, and 3D fragment-based drug design (FBDD). By using FBDD, by using chemical space analysis, 305 pieces were converted into 11858 compounds. After that, molecules were filtered using ADMET profile and Lipinski's rule using discovery studio 3.0. 814. Thereafter, models of pharmacophores based on structure were created and 12 pharmacophore models were generated. Among them, hypothesis 1 was employed to identify 380 compounds which have excellent mapping of features for pharmacophoric (aromatic ring, hydrogen bond donor and hydrogen bond acceptor and hydrophobic). 380 compounds were used in docking investigations. Compounds 3780 and 319 were identified as hit molecules based on the docking data. A series of molecules were designed using these hit compounds and *in vitro* inhibition of factor Xa study (selectivity in comparison with the thrombin and prothrombin time (PT) test) was performed. As a result, one promising compound (9b) was identified as a factor Xa inhibitor with a high selectivity versus thrombin (IC_{50} = 40 M) and an IC_{50} value of 23 nM (Refer to Table **7** for the structure of compound 9b).

Table 7. Successful examples of molecules derived from structure-based pharmacophore mapping.

Target	Derivative/Structure	Software Used	Active Molecules	Biological Activity	Refs.
GlyT2 inhibitor (Glysine transporter 2)	Hit-1/Lead-1 ZINC 6620309 Hit-2/Lead-2 ZINC 6865169	Schrodinger Suite 2017-3	4	1) IC$_{50}$=480 nM 2) IC$_{50}$=520 nM	[161, 162]
Factor Xa Inhibitor	Anthranilamide derivatives Compound 3780 Analogue 9b R = Cl	Discovery Studio 3.0	2 hit molecules. 1) 319 and it's 12a-g analogues 2) 3780 and it's 9a-e analogues. 9b is the most potent and active.	IC$_{50}$= 23±8 nM	[163]
Aiibβ3 receptor	4c- open form of A$_{IIb}$β$_3$ antagonist 4d- A$_{IIb}$β$_3$ antagonist of open form 12b- A$_{IIb}$β$_3$ antagonist of closed form	LigandScout	In contrast to commercially available antithrombotic Tirofiban, 2 out of 4 hits for accessible ligands and two out of four hits for closed-form ligands showed greater propensity and anti-aggregation action.	4c) IC$_{50}$-6.2±0.9 nM 4d) IC50= 25.0±5.0 nM 12b) IC50= 11 ± 1 nM	[164]

(Table 7) cont.....

Target	Derivative/Structure	Software Used	Active Molecules	Biological Activity	Refs.
Selective histone methyltransferase SET7 inhibitor	DC-S100 DC-S238 DC-S239	Accelrys Discovery Studio 3.0	Hit compound DC-S100. Active analogue DC-S238 and DC-S239	-DC-S100 (IC50= 30.04 μM) -DC-S238 (IC50= 4.88 μM) Inhibition ratio= 94.71% at 100 μM -DC-S239 (IC50 = 4.59 μM) Inhibition ratio= 90.51% at 100 μM.	[165]
Iinterlukin-15 inhibitor	Lead 80	DS 3.0: Discovery Studio (San Diego, CA Accelrys Software Inc.,)	Compound -1 is an HIT compound. Lead-80,82, and 76 are active lead compounds. 76 is the most efficacious compound.	HIT-1 IC50(prolif) =17.3 μM IC50(p-Stat)= 9.11 μM Lead 80 IC50(prolif) =1.6 μM IC50(p-Stat)= 40 nM	[166]
	Lead 82 Lead 76 Hit-1			Lead 82 IC50(prolif) =4.8 μM IC50(p-Stat)= 55 nM **Lead 76 IC50(prolif) =0.8 μM IC50(p-Stat)= 57 nM (best compound)**	

(Table 7) cont.....

Target	Derivative/Structure	Software Used	Active Molecules	Biological Activity	Refs.
DNA Gyrase Inhibitors	4'-methyl-N2-phenyl-[4,5'-bithiazole]-2,2'-diamine H$_3$CH$_2$COC COOH Compound 18	Ligand Scout software	1 active lead	IC50 = 1.1 μM	[167]

Applications of Artificial Intelligence in Pharmacophore Mapping

Researchers from all around the world are always creating new techniques and algorithms to get the right molecules quickly and affordably. A considerable influence on the success rate of drug development has been shown with the advancement of deep learning (DL), machine learning (ML), artificial intelligence (AI), and computational chemistry. These techniques, used singly or collectively and can create new tactics that utilise a variety of effective algorithms to improve predictions [168]. ML has a history of more than 20 years in medicinal chemistry, backed by cheminformatics, notably for molecular property prediction and virtual compound screening [169]. As a result of this achievement, a significant research field has developed that focuses on (a) developing methods for describing chemical structures more precisely and for capturing the factors that affect their properties, such as pharmacophores and three-dimensional structure, as well as autonomously learned representations and (b) developing mathematical equations that are ever-more sophisticated to describe the nexus between these chemical qualities and the physiological activity of interest for prediction purposes [170 - 172]. First attempts at more complicated machine learning models were viable thanks to an increase in structural knowledge, data production through combinatorial libraries, and high-throughput screening. The anticipation and promise, nevertheless, were quickly followed by disillusionment. In the 1990s, the developing area of QSAR had to face several difficult lessons concerning model validation, control trials, and other hazards [173]. These confident predictions allowed computational drug hunters must modify expectations while increasing the clarity of their instruments capabilities. As a result, machine learning was more successfully used in drug discovery and design in academia and business in the 2000s, gradually regained the public's trust and resulted in a steady increase in their utilisation. By 2015, computational developments made it possible to train larger and deeper neural nets, such as the widespread use of GPUs in contemporary computing frameworks and the rise in RAM availability [174].

Modern methods involve using a predetermined set of chemical transformations for optimization, including matching molecular pairings or using general guidelines to modify functional groups and molecular frameworks [175, 176].

To create innovative, powerful molecules without using information from reference compounds, many software and procedures have been established in *de novo* drug creation. Contrary to other structure-based screening approaches, these *de novo* methods are unfortunately not used as frequently in drug creation. Molecules that really are challenging to synthesize were produced using this approach. Encoder and decoder networks are indeed the two neural networks that make up the variational autoencoder [177]. The active task of the encoder network is to convert the SMILE notation of the chemical structure of molecules into a real-value continuous vector. The majority of the dominant molecule's reverse translation predominates, and minor conformational changes are less likely to occur. In a different research, the performance of the adversarial autoencoder and the variant autoencoder were examined [178]. Generating a model for adversarial autoencoders manufacturing synthetic organic molecules using an *in silico* model in conjunction with the prediction of new structures reveals more potent drugs that block type 2 dopamine receptors. Similar to this, Kadurin *et al.* recommended drugs with powerful and safe anticancer activities using a generative adversarial network (GAN) [179, 180]. Recursive neural networks (RNNs) are broadly employed in the *de novo* drug design process. It was first used in natural language processing, and it has been used to enter sequential data. The RNN is used to construct the chemical structures since SMILES notation strings represent the chemical structure format as a letter sequence. To train the neural network for the SMILES string, RNNs are trained on a sizable data set of chemical compounds acquired from the collection, such as ChEMBL, or a sizable set of commercially accessible chemicals [181]. This method has also been applied to the creation of brand-new peptides, whether they are in sequence or structural form. By using reinforcement learning, the synthesised chemicals are skewed toward forecasting their significance. Transfer learning, on the other hand, is used as a distinct tactic for creating powerful new compounds with anticipated biological activity. Additionally, several architectural model types are applied in ML techniques that can create powerful innovative structures. By using this method and training pharmacological molecules with comparable characteristics, novel chemical features may be studied [182].

Limitations of Pharmacophore Modeling

Pharmacophore inquiries need a trustworthy grading metric unlike docking and virtual screening methods. Due to this limitation, pharmacophore mapping is not able to give an effective matches. This method also has conformational limitations. A more number of validation parameters are needed to validate the pharmacophore so that it can be used to search correct structures from databases. Pharmacophore database contains an inadequate number of low-slung Oomph conformations within each molecule which may exclude active molecules. When screening the identical database, it can be challenging to build a query whereby two comparable pharmacophores on the same target return distinct compounds as hits.

CONCLUSION

The modern computational techniques in drug design have helped to reduce the cost in drug discovery. Various approaches are available in both ligand as well as structure-based drug design which can be able to find lead and hits. Pharmacophore mapping is a unique tool which is used in both ligand and structure-based drug design. The present review emphasized on the generation of both pharmacophores, their validation by various statistical parameters. Database searching after generation of pharmacophore queries has also been included. Commercially available software tools which are used currently have also been discussed. The detailed application of pharmacophore mapping including success stories in modern drug discovery have been also described. The present review will help researchers to generate a pharmacophore to design and discover potent and bioactive molecules targeting to a specific receptors or enzymes in the treatment of various diseases.

ACKNOWLEDGEMENTS

The authors thank L.J. Institute of Pharmacy, L.J. University for providing necessary support.

REFERENCES

[1]   Newman DJ, Cragg GM. Natural products as sources of new drugs over the last 25 years. J Nat Prod 2007; 70(3): 461-77.
[http://dx.doi.org/10.1021/np068054v] [PMID: 17309302]

[2]   Kalva S, Agrawal N, Skelton AA, Saleena LM. Identification of novel selective MMP-9 inhibitors as potential anti-metastatic lead using structure-based hierarchical virtual screening and molecular dynamics simulation. Mol Biosyst 2016; 12(8): 2519-31.
[http://dx.doi.org/10.1039/C6MB00066E] [PMID: 27250644]

[3]   Purohit D, Makhija M, Pandey P, *et al.* Role of computer-aided drug design in the discovery and development of new medicinal agents a review. Int J Pharm Sci 2018; 1405-15.

[4]   Ehrlich P. Über die constitution des diphtheriegiftes. Dtsch Med Wochenschr 1898; 24(38): 597-600.
[http://dx.doi.org/10.1055/s-0029-1204471]

[5]   Güner OF, Bowen JP. Setting the record straight: The origin of the pharmacophore concept. J Chem Inf Model 2014; 54(5): 1269-83.
[http://dx.doi.org/10.1021/ci5000533] [PMID: 24745881]

[6]   Wermuth CG. Pharmacophores: Historical perspective and viewpoint from a medicinal chemist. Pharmacophores and pharmacophore searches. Wiley Online Library 2006; 32: pp. 1-13.

[7]   Wermuth CG, Robin Ganellin C, Lindberg P, Mitscher LA. Glossary of terms used in medicinal chemistry (IUPAC Recommendations 1997). Annu Rep Med Chem 1998; 33: 385-95.
[http://dx.doi.org/10.1016/S0065-7743(08)61101-X]

[8]   Evans BE, Rittle KE, Bock MG, *et al.* Methods for drug discovery: Development of potent, selective, orally effective cholecystokinin antagonists. J Med Chem 1988; 31(12): 2235-46.
[http://dx.doi.org/10.1021/jm00120a002] [PMID: 2848124]

[9]   Thompson LA, Ellman JA. Synthesis and applications of small molecule libraries. Chem Rev 1996; 96(1): 555-600.
[http://dx.doi.org/10.1021/cr9402081] [PMID: 11848765]

[10]   Wermuth CG. Search for new lead compounds: The example of the chemical and pharmacological dissection of aminopyridazines. J Heterocycl Chem 1998; 35(5): 1091-100.
[http://dx.doi.org/10.1002/jhet.5570350508]

[11]   Ariens EJ. Molecular pharmacology, a basis for drug design. Fortschritte der Arzneimittelforschung/Progress Drug Res des Rech Pharm. Springer 1966; pp. 429-529.
[http://dx.doi.org/10.1007/978-3-0348-7059-7_8]

[12]   Langley JN. On the reaction of cells and of nerve-endings to certain poisons, chiefly as regards the reaction of striated muscle to nicotine and to curari. J Physiol 1905; 33(4-5): 374-413.
[http://dx.doi.org/10.1113/jphysiol.1905.sp001128] [PMID: 16992819]

[13]   Ehrlich P, Morgenroth J. Über Hämolysine. Dritte Mitheilung. Berl Klin Wschr 1900; 37: 453-8.

[14]   Fischer E. Influence of configuration on the action of enzymes. Ber Dtsch Chem Ges 1894; 27(3): 2985-93.
[http://dx.doi.org/10.1002/cber.18940270364]

[15]   Perutz MF, Mazzarella L. Structure of hæmoglobin: A preliminary x-ray analysis of haemoglobin H. Nature 1963; 199(4894): 639.
[http://dx.doi.org/10.1038/199639a0] [PMID: 14074547]

[16]   Marshall GR. Binding-site modeling of unknown receptors. 3D QSAR drug Des Theory, methods Appl 1993; 80.

[17]   Woods DD, Fildes P. The anti-sulphanilamide activity *(in vitro)* of p-aminobenzoic acid and related compounds. Chem Ind 1940; 59: 133-4.

[18]   Easson LH, Stedman E. Studies on the relationship between chemical constitution and physiological action: Molecular dissymmetry and physiological activity. Biochem J 1933; 27(4): 1257-66.
[http://dx.doi.org/10.1042/bj0271257] [PMID: 16745220]

[19]   Beckett AH. Stereochemical factors in biological activity. Fortschritte der Arzneimittelforschung/Progress in Drug Research/Progrès des recherches pharmaceutiques. Springer 1959; pp. 455-530.
[http://dx.doi.org/10.1007/978-3-0348-7035-1_6]

[20]   Barry CD, Ellis RA, Graesser SM, Marshall GR. Pertinent Concepts in Computer Graphics. Chicago: University of Illinois Press 1969.

[21]   Kier LB, Aldrich HS. A theoretical study of receptor site models for trimethylammonium group interaction. J Theor Biol 1974; 46(2): 529-41.
[http://dx.doi.org/10.1016/0022-5193(74)90013-7] [PMID: 4419765]

[22]   Peroutka SJ, U'Prichard DC, Greenberg DA, Snyder SH. Neuroleptic drug interactions with norepinephrine alpha receptor binding sites in rat brain. Neuropharmacology 1977; 16(9): 549-56.
[http://dx.doi.org/10.1016/0028-3908(77)90023-5] [PMID: 21357]

[23] Pullman B, Coubeils JL, Courrière P, Gervois JP. Quantum mechanical study of the conformational properties of phenethylamines of biochemical and medicinal interest. J Med Chem 1972; 15(1): 17-23.
      [http://dx.doi.org/10.1021/jm00271a006] [PMID: 5007090]

[24] Gund P, Wipke WT, Langridge R. Computer searching of a molecular structure file for pharmacophoric patterns. Comput Chem Res Educ Technol 1974; 3: 5-21.

[25] Poptodorov K, Luu T, Hoffmann RD. Pharmacophore model generation software tools. Pharma Pharmaco Sea 2006; 32: 15-47.
      [http://dx.doi.org/10.1002/3527609164.ch2]

[26] Sanders MPA, Barbosa AJM, Zarzycka B, *et al.* Comparative analysis of pharmacophore screening tools. J Chem Inf Model 2012; 52(6): 1607-20.
      [http://dx.doi.org/10.1021/ci2005274] [PMID: 22646988]

[27] Leach AR, Gillet VJ, Lewis RA, Taylor R. Three-dimensional pharmacophore methods in drug discovery. J Med Chem 2010; 53(2): 539-58.
      [http://dx.doi.org/10.1021/jm900817u] [PMID: 19831387]

[28] Yang SY. Pharmacophore modeling and applications in drug discovery: Challenges and recent advances. Drug Discov Today 2010; 15(11-12): 444-50.
      [http://dx.doi.org/10.1016/j.drudis.2010.03.013] [PMID: 20362693]

[29] Mohan CG. Structural bioinformatics: Applications in preclinical drug discovery process. Springer 2019; 27.
      [http://dx.doi.org/10.1007/978-3-030-05282-9]

[30] Cauët E, Rooman M, Wintjens R, Liévin J, Biot C. Histidine− aromatic interactions in proteins and protein− ligand complexes: Quantum chemical study of X-ray and model structures. J Chem Theory Comput 2005; 1(3): 472-83.
      [http://dx.doi.org/10.1021/ct049875k] [PMID: 26641514]

[31] Greene J, Kahn S, Savoj H, Sprague P, Teig S. Chemical function queries for 3D database search. J Chem Inf Comput Sci 1994; 34(6): 1297-308.
      [http://dx.doi.org/10.1021/ci00022a012]

[32] Wang T, Zhou J. 3DFS: A new 3D flexible searching system for use in drug design. J Chem Inf Comput Sci 1998; 38(1): 71-7.
      [http://dx.doi.org/10.1021/ci970070y] [PMID: 9461644]

[33] Zuccotto F. Pharmacophore features distributions in different classes of compounds. J Chem Inf Comput Sci 2003; 43(5): 1542-52.
      [http://dx.doi.org/10.1021/ci034068k] [PMID: 14502488]

[34] Dixon SL, Smondyrev AM, Knoll EH, Rao SN, Shaw DE, Friesner RA. PHASE: A new engine for pharmacophore perception, 3D QSAR model development, and 3D database screening: 1. Methodology and preliminary results. J Comput Aided Mol Des 2006; 20(10-11): 647-71.
      [http://dx.doi.org/10.1007/s10822-006-9087-6] [PMID: 17124629]

[35] Gund P, Güner OF, Beusen DD, Marshall GR. Pharmacophore perception, development and use in drug design. In: Osman FG, Ed. Molecules 2000; 5(7): 987-9.

[36] Choudhury C, Narahari Sastry G. Pharmacophore Modelling and Screening: Concepts, Recent Developments and Applications in Rational Drug Design. In: Mohan C (Eds) Structural Bioinformatics: Applications in Preclinical Drug Discovery Process. Challenges and Advances in Computational Chemistry and Physics, 2019, vol 27. Springer, Cham.
      [http://dx.doi.org/10.1007/978-3-030-05282-9_2]

[37] Swaminathan P. Advances in pharmacophore modeling and its role in drug designing. Computer-aided drug design. Springer 2020; pp. 223-43.
      [http://dx.doi.org/10.1007/978-981-15-6815-2_10]

[38]   Vuorinen A, Schuster D. Methods for generating and applying pharmacophore models as virtual screening filters and for bioactivity profiling. Methods 2015; 71: 113-34.
[http://dx.doi.org/10.1016/j.ymeth.2014.10.013] [PMID: 25461773]

[39]   Dong X, Ebalunode JO, Yang S-Y, Zheng W. Receptor-based pharmacophore and pharmacophore key descriptors for virtual screening and QSAR modeling. Curr Computeraided Drug Des 2011; 7(3): 181-9.
[http://dx.doi.org/10.2174/157340911796504332] [PMID: 21726192]

[40]   Wolber G, Seidel T, Bendix F, Langer T. Molecule-pharmacophore superpositioning and pattern matching in computational drug design. Drug Discov Today 2008; 13(1-2): 23-9.
[http://dx.doi.org/10.1016/j.drudis.2007.09.007] [PMID: 18190860]

[41]   Smellie A, Teig SL, Towbin P. Poling: Promoting conformational variation. J Comput Chem 1995; 16(2): 171-87.
[http://dx.doi.org/10.1002/jcc.540160205]

[42]   Gippert GP, Wright PE, Case DA. Distributed torsion angle grid search in high dimensions: A systematic approach to NMR structure determination. J Biomol NMR 1998; 11(3): 241-63.
[http://dx.doi.org/10.1023/A:1008209806860] [PMID: 9691275]

[43]   Hurst T. Flexible 3D searching: The directed tweak technique. J Chem Inf Comput Sci 1994; 34(1): 190-6.
[http://dx.doi.org/10.1021/ci00017a025]

[44]   Jones G, Willett P, Glen RC. A genetic algorithm for flexible molecular overlay and pharmacophore elucidation. J Comput Aided Mol Des 1995; 9(6): 532-49.
[http://dx.doi.org/10.1007/BF00124324] [PMID: 8789195]

[45]   Li KF, Pahlevan K, Kirschvink JL, Yung YL. Proc Nat Acad Sci USA: 2009; 106: p. 9576.

[46]   Dror O, Shulman-Peleg A, Nussinov R, Wolfson H. Predicting molecular interactions *in silico*: I. A guide to pharmacophore identification and its applications to drug design. Curr Med Chem 2004; 11(1): 71-90.
[http://dx.doi.org/10.2174/0929867043456287] [PMID: 14754427]

[47]   Brint AT, Willett P. Algorithms for the identification of three-dimensional maximal common substructures. J Chem Inf Comput Sci 1987; 27(4): 152-8.
[http://dx.doi.org/10.1021/ci00056a002]

[48]   Bron C, Kerbosch J. Algorithm 457: Finding all cliques in an undirected graph, Community. ACM 1973; 16(9): 575-7.

[49]   Bandyopadhyay D, Agrafiotis DK. A self-organizing algorithm for molecular alignment and pharmacophore development. J Comput Chem 2008; 29(6): 965-82.
[http://dx.doi.org/10.1002/jcc.20854] [PMID: 17999384]

[50]   Totrov M. Atomic property fields: Generalized 3D pharmacophoric potential for automated ligand superposition, pharmacophore elucidation and 3D QSAR. Chem Biol Drug Des 2008; 71(1): 15-27.
[http://dx.doi.org/10.1111/j.1747-0285.2007.00605.x] [PMID: 18069986]

[51]   Nettles JH, Jenkins JL, Williams C, *et al.* Flexible 3D pharmacophores as descriptors of dynamic biological space. J Mol Graph Model 2007; 26(3): 622-33.
[http://dx.doi.org/10.1016/j.jmgm.2007.02.005] [PMID: 17395510]

[52]   Baroni M, Cruciani G, Sciabola S, Perruccio F, Mason JS. A common reference framework for analyzing/comparing proteins and ligands. Fingerprints for Ligands and Proteins (FLAP): Theory and application. J Chem Inf Model 2007; 47(2): 279-94.
[http://dx.doi.org/10.1021/ci600253e] [PMID: 17381166]

[53]   Holliday JD, Willett P. Using a genetic algorithm to identify common structural features in sets of ligands. J Mol Graph Model 1997; 15(4): 221-32.

[http://dx.doi.org/10.1016/S1093-3263(97)00080-6] [PMID: 9524931]

[54] Handschuh S, Wagener M, Gasteiger J. Superposition of three-dimensional chemical structures allowing for conformational flexibility by a hybrid method. J Chem Inf Comput Sci 1998; 38(2): 220-32.
[http://dx.doi.org/10.1021/ci970438r] [PMID: 9538519]

[55] Chen X, Rusinko A III, Tropsha A, Young SS. Automated pharmacophore identification for large chemical data sets. J Chem Inf Comput Sci 1999; 39(5): 887-96.
[http://dx.doi.org/10.1021/ci990327n] [PMID: 10529987]

[56] Clark DE, Willett P, Kenny PW. Pharmacophoric pattern matching in files of three-dimensional chemical structures: Use of bounded distance matrices for the representation and searching of conformationally flexible molecules. J Mol Graph 1992; 10(4): 194-204.
[http://dx.doi.org/10.1016/0263-7855(92)80068-O] [PMID: 1476991]

[57] Jakes SE, Willett P. Pharmacophoric pattern matching in files of 3-D chemical structures: Selection of interatomic distance screens. J Mol Graph 1986; 4(1): 12-20.
[http://dx.doi.org/10.1016/0263-7855(86)80088-1]

[58] Barnum D, Greene J, Smellie A, Sprague P. Identification of common functional configurations among molecules. J Chem Inf Comput Sci 1996; 36(3): 563-71.
[http://dx.doi.org/10.1021/ci950273r] [PMID: 8690757]

[59] Sutter J, Li J, Maynard AJ, Goupil A, Luu T, Nadassy K. New features that improve the pharmacophore tools from Accelrys. Curr Comput Aided Drug Des 2011; 7(3): 173-80.
[http://dx.doi.org/10.2174/157340911796504305] [PMID: 21726193]

[60] Wolber G, Langer T. LigandScout: 3-D pharmacophores derived from protein-bound ligands and their use as virtual screening filters. J Chem Inf Model 2005; 45(1): 160-9.
[http://dx.doi.org/10.1021/ci049885e] [PMID: 15667141]

[61] Wolber G, Dornhofer AA, Langer T. Efficient overlay of small organic molecules using 3D pharmacophores. J Comput Aided Mol Des 2007; 20(12): 773-88.
[http://dx.doi.org/10.1007/s10822-006-9078-7] [PMID: 17051340]

[62] Seidel T, Ibis G, Bendix F, Wolber G. Strategies for 3D pharmacophore-based virtual screening. Drug Discov Today Technol 2010; 7(4): e221-8.
[http://dx.doi.org/10.1016/j.ddtec.2010.11.004] [PMID: 24103798]

[63] Güner OF. Pharmacophore perception, development, and use in drug design. Internat'l University Line 2000; 2.

[64] Richmond NJ, Abrams CA, Wolohan PRN, Abrahamian E, Willett P, Clark RD. GALAHAD: 1. Pharmacophore identification by hypermolecular alignment of ligands in 3D. J Comput Aided Mol Des 2006; 20(9): 567-87.
[http://dx.doi.org/10.1007/s10822-006-9082-y] [PMID: 17051338]

[65] Schneidman-Duhovny D, Dror O, Inbar Y, Nussinov R, Wolfson HJ. PharmaGist: A webserver for ligand-based pharmacophore detection. Nuc Acid Res 2008; 36: W223W228.

[66] Available from: www.denovobiopharma.com Available from: https://www.denovobiopharma.com/en/index_English.html (accessed on Sep. 27, 2022)

[67] Todorov NP, Alberts IL, de Esch IJP, Dean PM. QUASI: A novel method for simultaneous superposition of multiple flexible ligands and virtual screening using partial similarity. J Chem Inf Model 2007; 47(3): 1007-20.
[http://dx.doi.org/10.1021/ci6003338] [PMID: 17497844]

[68] Chemical Computing Group (CCG), Computer-Aided Molecular Design. Available from: http://www.chemcomp.com (accessed on Sep. 27, 2022).

[69] Tsiaka T, Kritsi E, Tsiantas K, Christodoulou P, Sinanoglou VJ, Zoumpoulakis P. Design and

development of novel nutraceuticals: Current trends and methodologies. Nutraceuticals 2022; 2(2): 71-90.
[http://dx.doi.org/10.3390/nutraceuticals2020006]

[70]　Tripos Inc., Certara. Available from: http://www.tripos.com (accessed on Sep. 27, 2022)

[71]　Koes DR, Camacho CJ. Pharmer: Efficient and exact pharmacophore search. J Chem Inf Model 2011; 51(6): 1307-14.
[http://dx.doi.org/10.1021/ci200097m] [PMID: 21604800]

[72]　Jiang Y, Gao H. Pharmacophore-based drug design for the identification of novel butyryl-cholinesterase inhibitors against Alzheimer's disease. Phytomedicine 2019; 54: 278-90.
[http://dx.doi.org/10.1016/j.phymed.2018.09.199] [PMID: 30668379]

[73]　Pal S, Kumar V, Kundu B, *et al.* Ligand-based pharmacophore modeling, virtual screening and molecular docking studies for discovery of potential topoisomerase I inhibitors. Comput Struct Biotechnol J 2019; 17: 291-310.
[http://dx.doi.org/10.1016/j.csbj.2019.02.006] [PMID: 30867893]

[74]　Niu M, Qin J, Tian C, *et al.* Tubulin inhibitors: Pharmacophore modeling, virtual screening and molecular docking. Acta Pharmacol Sin 2014; 35(7): 967-79.
[http://dx.doi.org/10.1038/aps.2014.34] [PMID: 24909516]

[75]　Niu M, Wang K, Zhang C, *et al.* The discovery of potential tubulin inhibitors: A combination of pharmacophore modeling, virtual screening, and molecular docking studies. J Taiwan Inst Chem Eng 2014; 45(5): 2111-21.
[http://dx.doi.org/10.1016/j.jtice.2014.07.016]

[76]　Noha SM, Fischer K, Koeberle A, Garscha U, Werz O, Schuster D. Discovery of novel, non-acidic mPGES-1 inhibitors by virtual screening with a multistep protocol. Bioorg Med Chem 2015; 23(15): 4839-45.
[http://dx.doi.org/10.1016/j.bmc.2015.05.045] [PMID: 26088337]

[77]　Roy K, Chakraborty P, Mitra I, Ojha PK, Kar S, Das RN. Some case studies on application of r_m^2 metrics for judging quality of quantitative structure-activity relationship predictions: Emphasis on scaling of response data. J Comput Chem 2013; 34(12): 1071-82.
[http://dx.doi.org/10.1002/jcc.23231] [PMID: 23299630]

[78]　Mitra I, Saha A, Roy K. Exploring quantitative structure–activity relationship studies of antioxidant phenolic compounds obtained from traditional Chinese medicinal plants. Mol Simul 2010; 36(13): 1067-79.
[http://dx.doi.org/10.1080/08927022.2010.503326]

[79]　Hung CL, Chen CC. Computational approaches for drug discovery. Drug Dev Res 2014; 75(6): 412-8.
[http://dx.doi.org/10.1002/ddr.21222] [PMID: 25195585]

[80]　Kandakatla N, Ramakrishnan G. Ligand based pharmacophore modeling and virtual screening studies to design novel HDAC2 inhibitors. Adv Bioinformatics 2014; 2014

[81]　Bagga V, Silakari O, Ghorela VS, Bahia MS, Rambabu G, Sarma J. A three-dimensional pharmacophore modelling of ITK inhibitors and virtual screening for novel inhibitors. SAR QSAR Environ Res 2011; 22(1-2): 171-90.
[http://dx.doi.org/10.1080/1062936X.2010.510480] [PMID: 21391146]

[82]　Gupta CL, Babu Khan M, Ampasala DR, Akhtar S, Dwivedi UN, Bajpai P. Pharmacophore-based virtual screening approach for identification of potent natural modulatory compounds of human Toll-like receptor 7. J Biomol Struct Dyn 2019; 37(18): 4721-36.
[http://dx.doi.org/10.1080/07391102.2018.1559098] [PMID: 30661449]

[83]　Modi P, Patel S, Chhabria MT. Identification of some novel pyrazolo[1,5- *a*]pyrimidine derivatives as InhA inhibitors through pharmacophore-based virtual screening and molecular docking. J Biomol Struct Dyn 2019; 37(7): 1736-49.

[http://dx.doi.org/10.1080/07391102.2018.1465852] [PMID: 29663870]

[84]   Zampieri D, Mamolo MG, Laurini E, *et al*. Synthesis, biological evaluation, and three-dimensional *in silico* pharmacophore model for σ(1) receptor ligands based on a series of substituted Benzo[d]oxazol-2(3H)-one derivatives. J Med Chem 2009; 52(17): 5380-93.
[http://dx.doi.org/10.1021/jm900366z] [PMID: 19673530]

[85]   Fan F, Toledo Warshaviak D, Hamadeh HK, Dunn RT II. The integration of pharmacophore-based 3D QSAR modeling and virtual screening in safety profiling: A case study to identify antagonistic activities against adenosine receptor, A2A, using 1,897 known drugs. PLoS One 2019; 14(1): e0204378.
[http://dx.doi.org/10.1371/journal.pone.0204378] [PMID: 30605479]

[86]   Zhou N, Xu Y, Liu X, *et al*. Combinatorial pharmacophore-based 3D-QSAR analysis and virtual screening of FGFR1 inhibitors. Int J Mol Sci 2015; 16(12): 13407-26.
[http://dx.doi.org/10.3390/ijms160613407] [PMID: 26110383]

[87]   Verma J, Khedkar VM, Coutinho EC. 3D-QSAR in drug design : A review. Curr Top Med Chem 2010; 10(1): 95-115.
[http://dx.doi.org/10.2174/156802610790232260] [PMID: 19929826]

[88]   Uddin R, Saeed M, Ul-Haq Z. Molecular docking-and genetic algorithm-based approaches to produce robust 3D-QSAR models. Med Chem Res 2014; 23(5): 2198-206.
[http://dx.doi.org/10.1007/s00044-013-0812-0]

[89]   Acharya C, Coop A, Polli JE, Mackerell AD Jr. Recent advances in ligand-based drug design: Relevance and utility of the conformationally sampled pharmacophore approach. Curr Comput Aided Drug Des 2011; 7(1): 10-22.
[http://dx.doi.org/10.2174/157340911793743547] [PMID: 20807187]

[90]   Moonsamy S, Dash RC, Soliman ME. Integrated computational tools for identification of CCR5 antagonists as potential HIV-1 entry inhibitors: Homology modeling, virtual screening, molecular dynamics simulations and 3D QSAR analysis. Molecules 2014; 19(4): 5243-65.
[http://dx.doi.org/10.3390/molecules19045243] [PMID: 24762964]

[91]   Shih KC, Lin CY, Zhou J, *et al*. Development of novel 3D-QSAR combination approach for screening and optimizing B-Raf inhibitors *in silico*. J Chem Inf Model 2011; 51(2): 398-407.
[http://dx.doi.org/10.1021/ci100351s] [PMID: 21182293]

[92]   Matt C, Hess T, Benlian A. Digital transformation strategies. Bus Inf Syst Eng 2015; 57(5): 339-43.
[http://dx.doi.org/10.1007/s12599-015-0401-5]

[93]   Golbraikh A, Shen M, Xiao Z, Xiao YD, Lee KH, Tropsha A. Rational selection of training and test sets for the development of validated QSAR models. J Comput Aided Mol Des 2003; 17(2-4): 241-53.
[http://dx.doi.org/10.1023/A:1025386326946] [PMID: 13677490]

[94]   Pan Y, Wang Y, Bryant SH. Pharmacophore and 3D-QSAR characterization of 6-arylquinazolin-4-amines as Cdc2-like kinase 4 (Clk4) and dual specificity tyrosine-phosphorylation-regulated kinase 1A (Dyrk1A) inhibitors. J Chem Inf Model 2013; 53(4): 938-47.
[http://dx.doi.org/10.1021/ci300625c] [PMID: 23496085]

[95]   Barillari C, Marcou G, Rognan D. Hot-spots-guided receptor-based pharmacophores (HS-Pharm): A knowledge-based approach to identify ligand-anchoring atoms in protein cavities and prioritize structure-based pharmacophores. J Chem Inf Model 2008; 48(7): 1396-410.
[http://dx.doi.org/10.1021/ci800064z] [PMID: 18570371]

[96]   Khedkar SA, Malde AK, Coutinho EC, Srivastava S. Pharmacophore modeling in drug discovery and development: An overview. Med Chem 2007; 3(2): 187-97.
[http://dx.doi.org/10.2174/157340607780059521] [PMID: 17348856]

[97]   Le Guilloux V, Schmidtke P, Tuffery P. Fpocket: An open source platform for ligand pocket detection. BMC Bioinformatics 2009; 10(1): 168.

[http://dx.doi.org/10.1186/1471-2105-10-168] [PMID: 19486540]

[98]   Schmidtke P, Bidon-Chanal A, Luque FJ, Barril X. MDpocket: Open-source cavity detection and characterization on molecular dynamics trajectories. Bioinformatics 2011; 27(23): 3276-85.
       [http://dx.doi.org/10.1093/bioinformatics/btr550] [PMID: 21967761]

[99]   Carlson HA, Masukawa KM, Rubins K, *et al.* Developing a dynamic pharmacophore model for HIV-1 integrase. J Med Chem 2000; 43(11): 2100-14.
       [http://dx.doi.org/10.1021/jm990322h] [PMID: 10841789]

[100] Gaurav A, Gautam V. Structure-based three-dimensional pharmacophores as an alternative to traditional methodologies. J Rec Lig Chan Res 2014; 7: 27-38.
       [http://dx.doi.org/10.2147/JRLCR.S46845]

[101] Seidel T, Bryant SD, Ibis G, Poli G, Langer T. 3D Pharmacophore modeling techniques in computer-aided molecular design using ligandScout. Tutorials in Chemoinformatics. Wiley Online Library 2017; pp. 279-309.
       [http://dx.doi.org/10.1002/9781119161110.ch20]

[102] Alamri MA, Tahir ul Qamar M, Afzal O, Alabbas AB, Riadi Y, Alqahtani SM. Discovery of anti-MERS-CoV small covalent inhibitors through pharmacophore modeling, covalent docking and molecular dynamics simulation. J Mol Liq 2021; 330: 115699.
       [http://dx.doi.org/10.1016/j.molliq.2021.115699] [PMID: 33867606]

[103] Wang R, Gao Y, Lai L. LigBuilder: A multi-purpose program for structure-based drug design. J Mol Model 2000; 6(7): 498-516.
       [http://dx.doi.org/10.1007/s0089400060498]

[104] Yuan Y, Pei J, Lai L. Lig Builder 2: A practical *de novo* drug design approach. J Chem Inf Model 2011; 51(5): 1083-91.
       [http://dx.doi.org/10.1021/ci100350u] [PMID: 21513346]

[105] Chen J, Lai L. Pocket v.2: Further developments on receptor-based pharmacophore modeling. J Chem Inf Model 2006; 46(6): 2684-91.
       [http://dx.doi.org/10.1021/ci600246s] [PMID: 17125208]

[106] B. D. S. Systèmes, "nd. Available from: https://www. 3ds. com/products-services/biovia/ products/molecular-modeling-simulation/biovia-discovery-studio (Accessed on May, vol. 5, 2021).

[107] Steindl T, Langer T. Influenza virus neuraminidase inhibitors: Generation and comparison of structure-based and common feature pharmacophore hypotheses and their application in virtual screening. J Chem Inf Comput Sci 2004; 44(5): 1849-56.
       [http://dx.doi.org/10.1021/ci049844i] [PMID: 15446845]

[108] Barreca ML, De Luca L, Iraci N, *et al.* Structure-based pharmacophore identification of new chemical scaffolds as non-nucleoside reverse transcriptase inhibitors. J Chem Inf Model 2007; 47(2): 557-62.
       [http://dx.doi.org/10.1021/ci600320q] [PMID: 17274611]

[109] e-Pharmacophores Schrödinger. Available from: https://www.schrodinger.com/science-articles/--pharmacophores (Accessed on 20th Oct. 2022)

[110] Salam NK, Nuti R, Sherman W. Novel method for generating structure-based pharmacophores using energetic analysis. J Chem Inf Model 2009; 49(10): 2356-68.
       [http://dx.doi.org/10.1021/ci900212v] [PMID: 19761201]

[111] Dixon SL, Smondyrev AM, Rao SN. PHASE: A novel approach to pharmacophore modeling and 3D database searching. Chem Biol Drug Des 2006; 67(5): 370-2.
       [http://dx.doi.org/10.1111/j.1747-0285.2006.00384.x] [PMID: 16784462]

[112] Ebalunode JO, Ouyang Z, Liang J, Zheng W. Novel approach to structure-based pharmacophore search using computational geometry and shape matching techniques. J Chem Inf Model 2008; 48(4): 889-901.
       [http://dx.doi.org/10.1021/ci700368p] [PMID: 18396858]

[113] Ebalunode JO, Dong X, Ouyang Z, Liang J, Eckenhoff RG, Zheng W. Structure-based shape pharmacophore modeling for the discovery of novel anesthetic compounds. Bioorg Med Chem 2009; 17(14): 5133-8.
[http://dx.doi.org/10.1016/j.bmc.2009.05.060] [PMID: 19520579]

[114] Edelsbrunner H, Facello M, Liang J. On the definition and the construction of pockets in macromolecules. Discrete Appl Math 1998; 88(1-3): 83-102.
[http://dx.doi.org/10.1016/S0166-218X(98)00067-5]

[115] Sanders MPA, Verhoeven S, de Graaf C, *et al.* Snooker: A structure-based pharmacophore generation tool applied to class A GPCRs. J Chem Inf Model 2011; 51(9): 2277-92.
[http://dx.doi.org/10.1021/ci200088d] [PMID: 21866955]

[116] Roland WSU, Sanders MPA, van Buren L, *et al.* Snooker structure-based pharmacophore model explains differences in agonist and blocker binding to bitter receptor hTAS2R39. PLoS One 2015; 10(3): e0118200.
[http://dx.doi.org/10.1371/journal.pone.0118200] [PMID: 25729848]

[117] Tyagi R, Singh A, Chaudhary KK, Yadav MK. Pharmacophore modeling and its applications. Bioinformatics. Elsevier 2022; pp. 269-89.
[http://dx.doi.org/10.1016/B978-0-323-89775-4.00009-2]

[118] Akram M, Waratchareeyakul W, Haupenthal J, Hartmann RW, Schuster D. Pharmacophore modeling and *in silico/in vitro* screening for human cytochrome P450 11B1 and cytochrome P450 11B2 inhibitors. Front Chem 2017; 5: 104.
[http://dx.doi.org/10.3389/fchem.2017.00104] [PMID: 29312923]

[119] John S, Thangapandian S, Arooj M, Hong JC, Kim KD, Lee KW. Development, evaluation and application of 3D QSAR pharmacophore model in the discovery of potential human renin inhibitors. BMC Bioinformatics 2011; 12(Suppl 14): 1-14.
[http://dx.doi.org/10.1186/1471-2105-12-S14-S4] [PMID: 22372967]

[120] Vuorinen A, Nashev LG, Odermatt A, Rollinger JM, Schuster D. Pharmacophore model refinement for 11β-hydroxysteroid dehydrogenase inhibitors: Search for modulators of intracellular glucocorticoid concentrations. Mol Inform 2014; 33(1): 15-25.
[http://dx.doi.org/10.1002/minf.201300063] [PMID: 27485195]

[121] Akram M, Kaserer T, Schuster D. Pharmacophore modeling and pharmacophore-based virtual screening. Silico Drug Discov Des 2015; 123-53.

[122] Gaulton A, Bellis LJ, Bento AP, *et al.* ChEMBL: A large-scale bioactivity database for drug discovery. Nucleic Acids Res 2012; 40(D1): D1100-7.
[http://dx.doi.org/10.1093/nar/gkr777] [PMID: 21948594]

[123] Huang N, Shoichet BK, Irwin JJ. Benchmarking sets for molecular docking. J Med Chem 2006; 49(23): 6789-801.
[http://dx.doi.org/10.1021/jm0608356] [PMID: 17154509]

[124] Fox S, Farr-Jones S, Sopchak L, *et al.* High-throughput screening: Update on practices and success. SLAS Discov 2006; 11(7): 864-9.
[http://dx.doi.org/10.1177/1087057106292473] [PMID: 16973922]

[125] Virshup AM, Contreras-García J, Wipf P, Yang W, Beratan DN. Stochastic voyages into uncharted chemical space produce a representative library of all possible drug-like compounds. J Am Chem Soc 2013; 135(19): 7296-303.
[http://dx.doi.org/10.1021/ja401184g] [PMID: 23548177]

[126] Triballeau N, Acher F, Brabet I, Pin JP, Bertrand HO. Virtual screening workflow development guided by the receiver operating characteristic curve approach. Application to high-throughput docking on metabotropic glutamate receptor subtype 4. J Med Chem 2005; 48(7): 2534-47.
[http://dx.doi.org/10.1021/jm049092j] [PMID: 15801843]

[127] Gao H, Williams C, Labute P, Bajorath J. Binary quantitative structure-activity relationship (QSAR) analysis of estrogen receptor ligands. J Chem Inf Comput Sci 1999; 39(1): 164-8.
[http://dx.doi.org/10.1021/ci980140g] [PMID: 10094611]

[128] Jacobsson M, Lidén P, Stjernschantz E, Boström H, Norinder U. Improving structure-based virtual screening by multivariate analysis of scoring data. J Med Chem 2003; 46(26): 5781-9.
[http://dx.doi.org/10.1021/jm030896t] [PMID: 14667231]

[129] Güner O, Clement O, Kurogi Y. Pharmacophore modeling and three dimensional database searching for drug design using catalyst: Recent advances. Curr Med Chem 2004; 11(22): 2991-3005.
[http://dx.doi.org/10.2174/0929867043364036] [PMID: 15544485]

[130] Bradley AP. The use of the area under the ROC curve in the evaluation of machine learning algorithms. Pattern Recognit 1997; 30(7): 1145-59.
[http://dx.doi.org/10.1016/S0031-3203(96)00142-2]

[131] Metz CE. Basic principles of ROC analysis. Semin Nucl Med 1978; 8(4): 283-98.
[http://dx.doi.org/10.1016/S0001-2998(78)80014-2] [PMID: 112681]

[132] Schuster D, Waltenberger B, Kirchmair J, *et al.* Predicting cyclooxygenase inhibition by three-dimensional pharmacophoric profiling. Part I: Model generation, validation and applicability in ethnopharmacology. Mol Inform 2010; 29(1-2): 75-86.
[http://dx.doi.org/10.1002/minf.200900071] [PMID: 27463850]

[133] Moussa N, Hassan A, Gharaghani S. Pharmacophore model, docking, QSAR, and molecular dynamics simulation studies of substituted cyclic imides and herbal medicines as COX-2 inhibitors. Heliyon 2021; 7(4): e06605.
[http://dx.doi.org/10.1016/j.heliyon.2021.e06605] [PMID: 33889764]

[134] Gomes MN, Muratov EN, Pereira M, *et al.* Chalcone derivatives: Promising starting points for drug design. Molecules 2017; 22(8): 1210.
[http://dx.doi.org/10.3390/molecules22081210] [PMID: 28757583]

[135] Seidel T, Schuetz DA, Garon A, Langer T. The pharmacophore concept and its applications in computer-aided drug design. Prog Chem Org Nat Prod 2019; 110: 99-141.
[http://dx.doi.org/10.1007/978-3-030-14632-0_4] [PMID: 31621012]

[136] Sheridan RP, Kearsley SK. Why do we need so many chemical similarity search methods? Drug Discov Today 2002; 7(17): 903-11.
[http://dx.doi.org/10.1016/S1359-6446(02)02411-X] [PMID: 12546933]

[137] Johnson MA, Maggiora GM. Concepts and applications of molecular similarity. Wiley 1990.

[138] Whitehead TL. Molecular modeling: Basic principles and applications, (Hans-Dieter Höltje, Wolfgang Sippl, Didier Rognan, and Gerd Folkers). ACS Publications 2006.

[139] Zhu F, Agrafiotis DK. Recursive distance partitioning algorithm for common pharmacophore identification. J Chem Inf Model 2007; 47(4): 1619-25.
[http://dx.doi.org/10.1021/ci7000583] [PMID: 17547387]

[140] Evers A, Hessler G, Matter H, Klabunde T. Virtual screening of biogenic amine-binding G-protein coupled receptors: Comparative evaluation of protein-and ligand-based virtual screening protocols. J Med Chem 2005; 48(17): 5448-65.
[http://dx.doi.org/10.1021/jm050090o] [PMID: 16107144]

[141] Lemmen C, Lengauer T. Computational methods for the structural alignment of molecules. J Comput Aided Mol Des 2000; 14(3): 215-32.
[http://dx.doi.org/10.1023/A:1008194019144] [PMID: 10756477]

[142] Lipinski CA, Lombardo F, Dominy BW, Feeney PJ. Experimental and computational approaches to estimate solubility and permeability in drug discovery and development settings. Adv Drug Deliv Rev 2001; 46(1-3): 3-26.

[http://dx.doi.org/10.1016/S0169-409X(00)00129-0] [PMID: 11259830]

[143] Lin LT, Hsu WC, Lin CC. Antiviral natural products and herbal medicines. J Tradit Complement Med 2014; 4(1): 24-35.
[http://dx.doi.org/10.4103/2225-4110.124335] [PMID: 24872930]

[144] Oprea TI, Davis AM, Teague SJ, Leeson PD. Is there a difference between leads and drugs? A historical perspective. J Chem Inf Comput Sci 2001; 41(5): 1308-15.
[http://dx.doi.org/10.1021/ci010366a] [PMID: 11604031]

[145] Reddy AS, Pati SP, Kumar PP, Pradeep HN, Sastry GN. Virtual screening in drug discovery : A computational perspective. Curr Protein Pept Sci 2007; 8(4): 329-51.
[http://dx.doi.org/10.2174/138920307781369427] [PMID: 17696867]

[146] Clark RD. Prospective ligand- and target-based 3D QSAR: State of the art 2008. Curr Top Med Chem 2009; 9(9): 791-810.
[http://dx.doi.org/10.2174/156802609789207118] [PMID: 19754395]

[147] Ruiz-Agudo E, Putnis CV, Rodriguez-Navarro C. Interaction between epsomite crystals and organic additives. Cryst Growth Des 2008; 8(8): 2665-73.
[http://dx.doi.org/10.1021/cg070442n]

[148] Kesar S, Paliwal S, Mishra P, *et al.* Identification of novel rho-kinase-II inhibitors with vasodilatory activity. ACS Med Chem Lett 2020; 11(9): 1694-703.
[http://dx.doi.org/10.1021/acsmedchemlett.0c00126] [PMID: 32944136]

[149] Chen J, Park HJ. Computer-aided discovery of massonianoside B as a novel selective DOT1L inhibitor. ACS Chem Biol 2019; 14(5): 873-81.
[http://dx.doi.org/10.1021/acschembio.8b00933] [PMID: 30951287]

[150] Yang K, Nong K, Gu Q, Dong J, Wang J. Discovery of N-hydroxy-3-alkoxybenzamides as direct acid sphingomyelinase inhibitors using a ligand-based pharmacophore model. Eur J Med Chem 2018; 151: 389-400.
[http://dx.doi.org/10.1016/j.ejmech.2018.03.065] [PMID: 29649738]

[151] Meyer MJ, Neumann VE, Friesacher HR, Zdrazil B, Brockmöller J, Tzvetkov MV. Opioids as substrates and inhibitors of the genetically highly variable organic cation transporter OCT1. J Med Chem 2019; 62(21): 9890-905.
[http://dx.doi.org/10.1021/acs.jmedchem.9b01301] [PMID: 31597043]

[152] Al-Shar'i NA, Al-Rousan EK, Fakhouri LI, Al-Balas QA, Hassan MA. Discovery of a nanomolar glyoxalase-I inhibitor using integrated ligand-based pharmacophore modeling and molecular docking. Med Chem Res 2020; 29(3): 356-76.
[http://dx.doi.org/10.1007/s00044-019-02486-3]

[153] Bhagwati S, Siddiqi MI. Identification of potential soluble epoxide hydrolase (sEH) inhibitors by ligand-based pharmacophore model and biological evaluation. J Biomol Struct Dyn 2020; 38(16): 4956-66.
[http://dx.doi.org/10.1080/07391102.2019.1691659] [PMID: 31701805]

[154] Zhao S, Li X, Peng W, *et al.* Ligand-based pharmacophore modeling, virtual screening and biological evaluation to identify novel TGR5 agonists. RSC Advances 2021; 11(16): 9403-9.
[http://dx.doi.org/10.1039/D0RA10168K] [PMID: 35423434]

[155] Ha H, Debnath B, Odde S, *et al.* Discovery of novel CXCR2 inhibitors using ligand-based pharmacophore models. J Chem Inf Model 2015; 55(8): 1720-38.
[http://dx.doi.org/10.1021/acs.jcim.5b00181] [PMID: 26153616]

[156] Leung KH, Liu LJ, Lin S, *et al.* Discovery of a small-molecule inhibitor of STAT3 by ligand-based pharmacophore screening. Methods 2015; 71: 38-43.
[http://dx.doi.org/10.1016/j.ymeth.2014.07.010] [PMID: 25160651]

[157] Zalloum H, Tayyem R, Irmaileh BA, *et al.* Discovery of new human epidermal growth factor receptor-

2 (HER2) inhibitors for potential use as anticancer agents *via* ligand-based pharmacophore modeling. J Mol Graph Model 2015; 61: 61-84.
[http://dx.doi.org/10.1016/j.jmgm.2015.06.008] [PMID: 26188796]

[158] Paz OS, de Jesus Pinheiro M, do Espirito Santo RF, Villarreal CF, Castilho MS. Nanomolar anti-sickling compounds identified by ligand-based pharmacophore approach. Eur J Med Chem 2017; 136: 487-96.
[http://dx.doi.org/10.1016/j.ejmech.2017.05.035] [PMID: 28528302]

[159] Che J, Wang Z, Sheng H, *et al.* Ligand-based pharmacophore model for the discovery of novel CXCR2 antagonists as anti-cancer metastatic agents. R Soc Open Sci 2018; 5(7): 180176.
[http://dx.doi.org/10.1098/rsos.180176] [PMID: 30109074]

[160] Karaküçük-İyidoğan A, Aydınöz B, Taşkın-Tok T, Oruç-Emre EE, Balzarini J. Synthesis, biological evaluation and ligand based pharmacophore modeling of new aromatic thiosemicarbazones as potential anticancer agents. Pharm Chem J 2019; 53(2): 139-49.
[http://dx.doi.org/10.1007/s11094-019-01968-3] [PMID: 32214540]

[161] Fratev F, Miranda-Arango M, Lopez AB, Padilla E, Sirimulla S. Discovery of GlyT2 inhibitors using structure-based pharmacophore screening and selectivity studies by FEP+ calculations. ACS Med Chem Lett 2019; 10(6): 904-10.
[http://dx.doi.org/10.1021/acsmedchemlett.9b00003] [PMID: 31223446]

[162] Fratev F, Miranda-Arango M, Padilla E, Sirimulla S. Discovery of new classes of glycine transporter 2 (GlyT2) inhibitors and study of GlyT2 selectivity by combination of novel structural based virtual screening approach and free energy perturbation (FEP+) calculations. bioRxiv 2019; 510487.
[http://dx.doi.org/10.1101/510487]

[163] Xing J, Yang L, Li H, *et al.* Identification of anthranilamide derivatives as potential factor Xa inhibitors: Drug design, synthesis and biological evaluation. Eur J Med Chem 2015; 95: 388-99.
[http://dx.doi.org/10.1016/j.ejmech.2015.03.052] [PMID: 25839438]

[164] Polishchuk PG, Samoylenko GV, Khristova TM, *et al.* Design, virtual screening, and synthesis of antagonists of αIIbβ3 as antiplatelet agents. J Med Chem 2015; 58(19): 7681-94.
[http://dx.doi.org/10.1021/acs.jmedchem.5b00865] [PMID: 26367138]

[165] Meng F, Cheng S, Ding H, *et al.* Discovery and optimization of novel, selective histone methyltransferase SET7 inhibitors by pharmacophore-and docking-based virtual screening. J Med Chem 2015; 58(20): 8166-81.
[http://dx.doi.org/10.1021/acs.jmedchem.5b01154] [PMID: 26390175]

[166] Quéméner A, Maillasson M, Arzel L, *et al.* Discovery of a small-molecule inhibitor of interleukin 15: Pharmacophore-based virtual screening and hit optimization. J Med Chem 2017; 60(14): 6249-72.
[http://dx.doi.org/10.1021/acs.jmedchem.7b00485] [PMID: 28657314]

[167] Brvar M, Perdih A, Renko M, Anderluh G, Turk D, Solmajer T. Structure-based discovery of substituted 4,5′-bithiazoles as novel DNA gyrase inhibitors. J Med Chem 2012; 55(14): 6413-26.
[http://dx.doi.org/10.1021/jm300395d] [PMID: 22731783]

[168] Mak KK, Pichika MR. Artificial intelligence in drug development: Present status and future prospects. Drug Discov Today 2019; 24(3): 773-80.
[http://dx.doi.org/10.1016/j.drudis.2018.11.014] [PMID: 30472429]

[169] Varnek A, Baskin I. Machine learning methods for property prediction in chemoinformatics: Quo Vadis? J Chem Inf Model 2012; 52(6): 1413-37.
[http://dx.doi.org/10.1021/ci200409x] [PMID: 22582859]

[170] Todeschini R, Consonni V, Mannhold R. Methods and principles in medicinal chemistry.Handb Mol descriptors. 2000// K. Yang *et al.*, Are learned molecular representations ready for prime time? 2019.

[171] Vamathevan J, Clark D, Czodrowski P, *et al.* Applications of machine learning in drug discovery and development. Nat Rev Drug Discov 2019; 18(6): 463-77.

[http://dx.doi.org/10.1038/s41573-019-0024-5] [PMID: 30976107]

[172] Chen H, Engkvist O, Wang Y, Olivecrona M, Blaschke T. The rise of deep learning in drug discovery. Drug Discov Today 2018; 23(6): 1241-50.
[http://dx.doi.org/10.1016/j.drudis.2018.01.039] [PMID: 29366762]

[173] Dearden JC, Cronin MTD, Kaiser KLE. How not to develop a quantitative structure–activity or structure–property relationship (QSAR/QSPR). SAR QSAR Environ Res 2009; 20(3-4): 241-66.
[http://dx.doi.org/10.1080/10629360902949567] [PMID: 19544191]

[174] Ma J, Sheridan RP, Liaw A, Dahl GE, Svetnik V. Deep neural nets as a method for quantitative structure-activity relationships. J Chem Inf Model 2015; 55(2): 263-74.
[http://dx.doi.org/10.1021/ci500747n] [PMID: 25635324]

[175] Jiménez-Luna J, Grisoni F, Weskamp N, Schneider G. Artificial intelligence in drug discovery: Recent advances and future perspectives. Expert Opin Drug Discov 2021; 16(9): 949-59.
[http://dx.doi.org/10.1080/17460441.2021.1909567] [PMID: 33779453]

[176] Brown N, Ertl P, Lewis R, Luksch T, Reker D, Schneider N. Artificial intelligence in chemistry and drug design. J Comput Aided Mol Des 2020; 34(7): 709-15.
[http://dx.doi.org/10.1007/s10822-020-00317-x] [PMID: 32468207]

[177] Lin E, Lin CH, Lane HY. Relevant applications of generative adversarial networks in drug design and discovery: Molecular *de novo* design, dimensionality reduction, and *de novo* peptide and protein design. Molecules 2020; 25(14): 3250.
[http://dx.doi.org/10.3390/molecules25143250] [PMID: 32708785]

[178] Grisoni F, Moret M, Lingwood R, Schneider G. Bidirectional molecule generation with recurrent neural networks. J Chem Inf Model 2020; 60(3): 1175-83.
[http://dx.doi.org/10.1021/acs.jcim.9b00943] [PMID: 31904964]

[179] Pogány P, Arad N, Genway S, Pickett SD. *De novo* molecule design by translating from reduced graphs to SMILES. J Chem Inf Model 2019; 59(3): 1136-46.
[http://dx.doi.org/10.1021/acs.jcim.8b00626] [PMID: 30525594]

[180] Kell DB, Samanta S, Swainston N. Deep learning and generative methods in cheminformatics and chemical biology: Navigating small molecule space intelligently. Biochem J 2020; 477(23): 4559-80.
[http://dx.doi.org/10.1042/BCJ20200781] [PMID: 33290527]

[181] Prykhodko O, Johansson SV, Kotsias PC, *et al.* A *de novo* molecular generation method using latent vector based generative adversarial network. J Cheminform 2019; 11(1): 74.
[http://dx.doi.org/10.1186/s13321-019-0397-9] [PMID: 33430938]

[182] Schneider P, Walters WP, Plowright AT, *et al.* Rethinking drug design in the artificial intelligence era. Nat Rev Drug Discov 2020; 19(5): 353-64.
[http://dx.doi.org/10.1038/s41573-019-0050-3] [PMID: 31801986]

CHAPTER 4

Up-to-Date Developments in Homology Modeling

Muhammed Tilahun Muhammed[1,*] and **Esin Aki-Yalcin**[2]

[1] *Department of Pharmaceutical Chemistry, Faculty of Pharmacy, Suleyman Demirel University, Isparta, Turkey*

[2] *Department of Pharmaceutical Chemistry, Faculty of Pharmacy, Cyprus Health and Social Sciences University, Guzelyurt, Northern Cyprus*

Abstract: Homology modeling is used to predict protein 3D structure from its amino acid sequence. It is the most accurate computational approach to estimate 3D structures. It has straightforward steps that save time and labor. There are several homology modeling tools under use. There is no sole tool that is superior in every aspect. Hence, the user should select the most appropriate one carefully. It is also a common practice to use two or more tools at a time and choose the best model among the resulting models.

Homology modeling has various applications in the drug design and development process. Such applications need high-quality 3D structures. It is widely used in combination with other computational methods including molecular docking and molecular dynamics simulation. Like the other computational methods, it has been influenced by the involvement of artificial intelligence. In this regard, homology modeling tools, like AlphaFold, have been introduced. This type of method is expected to contribute to filling the gap between protein sequence release and 3D structure determination.

This chapter sheds light on the history, relatively popular tools and steps of homology modeling. A detailed explanation of MODELLER is also given as a case study protocol. Furthermore, homology modeling's application in drug discovery is explained by exemplifying its role in the fight against the novel Coronavirus. Considering the new advances in the area, better tools and thus high-quality models are expected. These, in turn, pave the way for more applications of it.

Keywords: Computer Aided Drug Design, 3D Structure, Drug Discovery, Homology Modeling, Modeller, Molecular Modeling.

* **Corresponding author Muhammed Tilahun Muhammed:** Department of Pharmaceutical Chemistry, Faculty of Pharmacy, Suleyman Demirel University, Isparta, Turkey; E-mail: muh.tila@gmail.com

Igor José dos Santos Nascimento (Ed.)

INTRODUCTION

Proteins' 3D (3-dimensional) structures play a crucial role in defining their functions [1]. Hence, investigations about protein structures have an important contribution to understanding the mechanism of diseases [2]. Thus, knowledge about protein 3D structure has a vital role in rational drug design and discovery [3]. As a result, a number of Nobel Prizes have been awarded to researchers in this area. Scientists have been awarded the prize for elucidating the structures of myoglobin, lysozyme, integral membrane protein, HIV (human immunod-eficiency virus) protease, ion channels, RNA (ribonucleic acid) polymerase, and GPCR (G protein-coupled receptor). In addition to this, the prize was awarded to researchers who pioneered in using X-ray crystallography, NMR (nuclear magnetic resonance), and Cryo-electron microscopy (Cryo-EM) for protein structure determination [4].

The quality of protein 3D structures solved has been improved as the available techniques improved [5]. Together with this, the experimental methods are not applicable to solving the structure of each protein. In this regard, NMR is used to solve the 3D structure of relatively small molecules, which are dissolvable [6]. Similarly, X-ray crystallography is used to solve protein 3D structures in a crystal state [7]. Cryo-EM is preferred to large macromolecule complexes with low resolution [8]. In addition to this, the experimental methods take a long time, labor and resource [9]. As a result, the experimental protein 3D structure determination could not keep pace with the protein sequence release. Consequently, the gap between the protein sequences available and the experimentally solved protein structures has been widening. Hence, computational protein 3D structure prediction methods can play a substantial role in filling this gap [10].

Homology (comparative) modeling is protein 3D structure prediction from its amino acid sequence. Homology modeling is used when the query sequence and templates selected share a common ancestor. In comparative modeling, there is just sequence similarity without shared ancestral history [11]. Homology modeling yields 3D structures with better reliability than the other computational structure prediction approaches [12, 13]. In addition to this, it has straightforward steps that take relatively less time. Hence, homology modeling is used to generate high quality structures that have the potential to convert the applications of the other computational methods in case they require 3D structures [14].

In this chapter, the brief history and the general procedures of homology modeling are presented. Homology modeling tools that are widely used these days are also given. Together with this, a case study protocol with MODELLER is included.

Furthermore, applications of homology modeling in the drug discovery process is summarized with a special focus on the latest ones. So, this chapter is expected to provide updated information on homology modeling.

BRIEF HISTORY OF HOMOLOGY MODELING

The idea of protein structure prediction has a long history since 1894 when Emin Fischer suggested that a protein's 3D structure determines its function [11]. Thereafter, Christian Anfinsen suggested that among the possible conformations, the native conformation has the lowest energy. In the 1970s, he proposed that the structure of a protein is determined by its amino acid sequence in a particular physiological condition [15]. This is the basis for the concept of homology modeling. The α-lactalbumin 3D structure, which was built based on the structure of lysozyme in 1969, is considered the first homology model [16]. After this time, various homology modeling programs and servers were developed. In this regard, MODELLER was revealed in 1993 [17]. In the same year, the concept of a server for automated homology modeling was introduced through SWISS-MODEL [18]. The milestones in the history of homology modeling are summarized in Fig. (**1**).

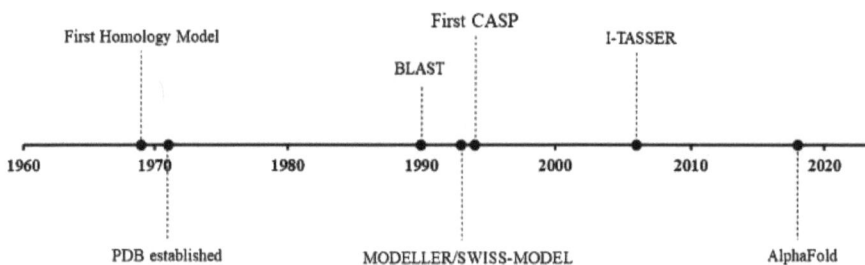

Fig. (1). Milestones in the history of homology modeling.

HOMOLOGY MODELING PROCEDURE

Homology modeling has straightforward major steps (Fig. **2**). General information about each step is presented in this section.

Identification and Selection of templates

In the first step of the process, the target (query) sequence is used to identify template structures in the worldwide PDB (https://www.wwpdb.org/) or other structural databases [19]. First, the protein basic local alignment search tool (BLASTp) search is performed by using the target sequence as a query and PDB as a database in NCBI (national center for biotechnology information) (https://blast.ncbi.nlm.nih.gov/Blast.cgi) [20, 21]. BLASTp search gives the 3D structures inside the PDB with high identity and coverage of the query. In case homology is low, methods like position-specific iterated BLAST (PSI-BLAST), profile-profile alignments, and Hidden Markov Models (HMMs) are used to identify eligible templates with reduced shifts and gaps [22, 23].

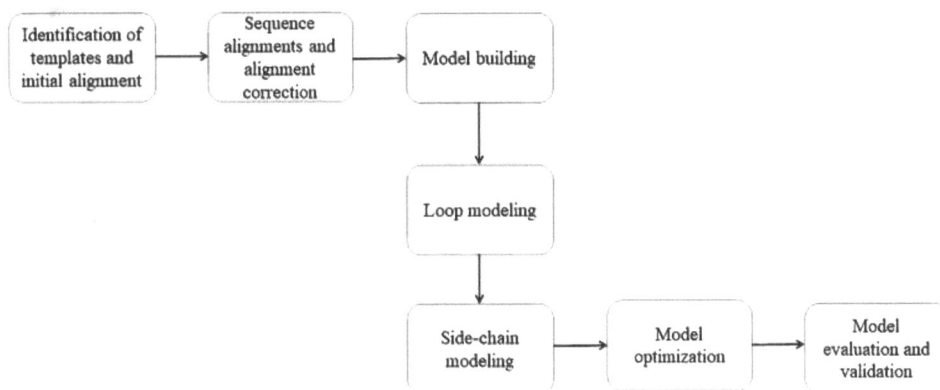

Fig. (2). The homology modeling procedure.

Among the template candidates, 3D structures with good properties are chosen. Sequence identity and coverage level is important to generate a reliable model. In general, identity similarity of above 25% is acceptable for homology modeling [24]. Homology modeling with templates from similar phylogenetic trees is expected to be more successful. Similarly, it is recommended to choose template structures with high resolution [10]. The basic information about the template structure is given in the PDB entry. To get the details about the template structure in the PDB, reading the article that published the template structure is beneficial. After the eligible template/s are selected, unnecessary elements, solvents, ligands, and ions are removed from the structure. The structure might also need refinement and adding of missing hydrogens [25].

Sequence Alignments and Alignment Correction

After the best eligible templates are selected, sequence alignments and alignment corrections are performed. In this regard, target–template alignment and template–template alignment, in case multiple templates are used, take place. Early detection of alignment gaps or shifts is crucial in homology modeling [26]. Otherwise, sequence alignment errors will be reflected in the backbone. In general, it is recommended to utilize profile HMMs if sequence percentage identities are below 35% [27]. Careful inspections and corrections during alignments would improve the quality of the 3D structures to be built.

Model Building

The 3D structures of a target sequence are built by using model building approaches called rigid-body assembly, segment matching, spatial restraint, and artificial evolution methods.

In the rigid-body assembly method, the aligned regions are treated as rigid bodies that are assembled to generate the model. Backbone coordinates are picked from the template's aligned regions. This method is applied in COMPOSER [28], 3D-JIGSAW [29], and SWISS-MODEL tools [18].

In the segment matching method, a subset of atomic coordinates obtained from the template structures is used as guiding positions. The model is built by comparing and matching the most suitable fragment from known protein structures in a database. This method is applied in SegMod/ENCAD program [30].

In the spatial restraint method, the model is generated by meeting restraints that came from the template's aligned regions together with stereochemical restraints. The model with minor violations of these constraints is considered the best model. This method is applied in MODELLER [31].

In the artificial evolution method, rigid-body assembly and stepwise template evolutionary mutations are used together to build the model. This is applied in NEST [32].

Loop Modeling

Insertions and deletions are often occurred in the less conserved loops rather than α-helix and β-sheet [33]. In addition to this, there might be a difference between query and template loop conformations. Loops play an important role in functional specificity and contribute to active sites [34]. Hence, loop modeling needs extensive care so that the generated models would be utilized in further applications [35]. The way of performing loop modeling is decided based on the loop size and residue composition [36].

Loop modeling is done by using the *ab initio* method, database search method or a combination of them. The *ab initio* method depends on an exhaustive search for minimal energy conformation through a scoring function optimization. The Mod-Loop program available in MODELLER is a widely used example of this method [37]. The database search method browses the known databases to find homologous conformations for a query loop sequence. ProMod3, available in the SwissModel server, is a commonly used example for this method [38].

Furthermore, a combination of *ab initio* and database search methods can be used to achieve higher accuracy [39].

Side-Chain Modeling

Side-chain modeling is done by the addition of side chains onto the major backbone. Side chains exist in rotamers. Side-chain modeling needs rotamer library selection, scoring function, and screening method [40]. Among the tools that are used in side-chain addition, OPUS-Rota2 [41], SCWRL [42], and FASPR [43] are some to mention.

Model Optimization

Model optimization is performed to increase the quality of the resulting model. This is done by energy minimization through molecular mechanics force fields [44]. The energy minimization reduces atomic clashes and excludes major and minor errors. Further model optimization can be performed by using molecular dynamics and Monte Carlo simulations [45].

Model Evaluation and Validation

In this step, the model is evaluated and validated through various methods. The model accuracy is correlated with the value and function of the model. The model is evaluated based on stereochemistry, physical parameters, knowledge-based parameters, statistical mechanics, and many more criteria [46]. To evaluate these parameters, tools like DOPE [47], MolProbity [48], PROSAII [49], PROCHECK [50], QMEAN [51], Ramachandran plot [52], Verify 3D [53], WHATCHECK [54], WHAT IF [55], and global distance test (GDT) [56] are used. Each tool evaluates the model from a different perspective. Some quality evaluation tools assess query-template alignments. Some of them evaluate stereochemical irregularities in model predictions. Some others integrate several features into an evaluation score. Hence, it is recommended to use several tools together for good evaluation and validation [57]. SAVES (Structure Analysis and VErification Server) is a crucial server that consists of several programs used in model validation [58]. PROCHECK and PDBSUM are programs that can be used for validation [50].

OVERVIEW OF HOMOLOGY MODELING TOOLS

Various tools are used to generate the 3D structure of proteins. Some of them are programs, whereas some others are servers. Similarly, there are freely available tools as well as commercial ones. In this part, general information about popular tools is presented with special attention to their advantages and disadvantages. In addition to the tool used in homology modeling, there are some limitations related to the method. The quality of the model relies on sequence similarity between the query and target. The modeling tools may generate models with errors in case the sequence alignment is incorrect. Therefore, sequence alignment is important in building backbone, loop, and side chain modeling that is good enough [59].

MODELLER

MODELLER is a standalone homology modeling program that builds a 3D model of the target by satisfaction of spatial restraints. It defines many constraints that describe atomic distances and dihedral angles based on template-target alignments [31]. It is freely available and results in reliable outcomes. It provides powerful features and also combines multiple templates to build a model [11]. MODELLER is based on command lines and thus needs knowledge of Python scripting. Therefore, a graphical user interface for MODELLER named EasyModeller has been developed. EasyModeller provides model building, assessment, visualization, and optimization of the models in a straightforward manner [60]. Herein, it is possible to inspect the sequence alignments. When there is an alignment problem that will affect the end model, it is possible to correct it in time. Especially this is crucial in case the sequence identity is below %50. It provides the opportunity to select the best model among the generated models *via* its internal assessment parameter, the DOPE score [61]. However, the user is expected to install the program, select the templates, and thus needs to have some specialty. MODELLER also has convergence problems as it might give models with enlarged structures that bear side chains under the standard [62]. Furthermore, a previous study has reported that MODELLER predicts flexible regions of proteins poorly [63].

I-TASSER

I-TASSER (Iterative Threading ASSEmbly Refinement) is a hierarchical approach server used to predict protein structure [64]. First, homologs of the target sequence are identified through BLAST (Basic Local Alignment Search Tool). Then, templates are identified from protein structure data banks by a multiple threading approach LOMETS. Finally, models are built using a modified replica-exchange Monte Carlo simulation [65]. It is also possible to assign additional templates or exclude templates and specify a secondary structure for a region. I-TASSER became the best server in some CASP (Critical Assessment of protein Structure Prediction) experiments. It is also an easy-to-use server, but it needs an internet connection to get the service [66]. However, the use of a fully automated approach has its own disadvantages. In this case, it is not possible to do inspection and necessary manual manipulation on sequence alignments. As sequence alignment is important in homology modeling, this might result in poor-quality models, especially if the homology is low [64]. It might not also provide good quality models for transmembrane proteins. A study has also reported that I-TASSER might give side chain orientations that could lead to steric clashes [63].

SWISS-MODEL

SWISS-MODEL is the first fully automated homology modeling server accessible through the Expasy web server or from DeepView [67]. In this server, the template is searched in the Swiss Model Template Library (SMTL) using BLAST and HHblits [68]. Then, template-target alignments are performed with the selected templates and multiple templates are selected when necessary. If there are insertions and deletions in the alignments, they are modeled by using suitable structures from databases. If there are no suitable structures in the structural databases, Monte Carlo simulations are used for modeling such regions. Finally, the models generated are optimized by energy minimization [69]. SWISS-MODEL provides user-friendly web interface. It also gives users the opportunity to determine the templates to be used. On the other hand, users are expected to have internet access to utilize the server. In addition to this, it is expected to install DeepView to access it *via* this program.

Prime

Prime is part of the Schrodinger package that consists of new methods and algorithms to provide protein structure prediction with high accuracy. It simulates the flexibility of targets and identifies alternative binding modes of ligands. In this package, homology modeling and fold recognition are merged. Prime builds model regions that are not derived from the templates by an *ab initio* method, whereas side chain conformations are obtained from a rotamer library [70]. It has a loop refinement module and OPLS default force field. Research revealed that prime could provide the best results in loop modeling relative to similar tools [71]. It includes an easy-to-use interface. It also allows users to specify and adjust parameters to increase the quality of the structure predictions [70]. Some tools use an integrated energy minimization and/or molecular dynamics simulation to increase the accuracy of models by overcoming kinetic barriers. In this regard, prime can incorporate glide to achieve a better accuracy [72]. On the other hand, prime is a commercially available package.

Phyre2

Phyre2 (Protein Homology/analogY Recognition Engine 2) is a server that uses various advanced detection tools to generate 3D structures. First, the target sequence is aligned with template structures by PSI-BLAST and secondary structure prediction algorithms. After multiple sequence alignments are created, secondary structures are predicted with PSIPRED. Then, the alignment and secondary structures are used as query-hidden Markov models (HMMs). The best alignments are utilized to generate a model from a known HMM database. In the final step, loop and side-chain modeling is undertaken. Phyre2 is a freely available server even for commercial users [73, 74]. Phyre2 is widely used as it completes a submitted modeling job in hours [74]. On the other hand, users are expected to have the experience of using its intensive model. Phyre2 has limitations in generating homology models of proteins with low sequence identity and estimating the structural effects of point mutations, like many other tools [74].

HHPRED

HHPRED is the first server to generate 3D structures using pairwise comparison of profile HMMs from single or multiple query sequences [75]. It is different from other servers in its speed of homolog searching. In this server, it is possible to do the searching against broad databases, PDB being the default one [76]. It returns search outputs of a single sequence or multiple alignment inputs within minutes. Alignment that excludes non-homolog sequences is incorporated to reduce false positives. The secondary structure similarity included in this method might lead to a non-homologous but analogous structure [11].

RosettaCM

RosettaCM (Comparative Modeling with Rosetta) generates 3D structures with high accuracy even if the sequence identity is low [77]. In this method, target-sequence alignments are done with HHSearch [78], SPARKS-X [79] and RaptorX [80]. RosettaCM algorithm inserts fragments in unaligned regions, replaces fragments with segments from a different template, and employs Cartesian-space energy minimization, and optimization. It also refines the models by using the FastRelax protocol [81] and the best model is chosen accordingly. As the RosettaCM protocol is based on scripts, it is not easy [77].

Alpha Fold

AlphaFold, which is developed by Google's AI (artificial intelligence) project Alpha and DeepMind, is based mainly on *ab initio* modeling principles [82]. This project ranked first in the CASP experiments of 2018 and 2020. DeepMind and the European Bioinformatics Institute (EMBL-EBI) collaborated to establish the platform AlphaFold DB. This platform is a freely accessible repository of structures that represent the majority of the curated UniProt database entries. Most entries of this platform are whole-protein structures and are developed with the same computational approach. Furthermore, there is a high potential to refine the structures due to AI. Its prediction accuracy is found to be competitive with experimentally determined structures [83]. Though the availability of such structures provides a high opportunity in drug design, the areas to benefit from this are yet to be established [84].

CASE STUDY

Homology modeling with MODELLER is demonstrated as a practical protocol example. This program is chosen since it is freely available, gives models with high accuracy, and is popular. Homology modeling of RhlR, a quorum sensing regulator, was performed to demonstrate the workflow of MODELLER. EasyModeller was used for this purpose.

The sequence of RhlR, with the accession number P54292, from UniProt was utilized [85]. Protein BLAST was performed *via* NCBI (National Center for Biotechnology Information) [86]. PDB was selected as the database and PSI-BLAST was performed. Among the structures that were found to be homologs to RhlR, the one with a PDB code of 4Y17 was chosen [87]. The structure had %98 query coverage and %40.76 identity. The template structure, 4Y17, was retrieved from UCSB PDB.

Firstly, the query sequence and template were loaded into EasyModeller 4.0 [60]. The query sequence was checked for gaps (Fig. **3**). Then, the templates were aligned by implementing the salign command of MODELLER. This tab is available because MODELLER allows the use of multiple templates. Thereafter, the query sequence was aligned to the template. At last, the models were built. The generated models were saved for further optimization, validation, and verification.

As it is possible to generate several models, the one with better characteristics is chosen. The models can be compared with each other by using the DOPE values, energies, TM (template modeling) score, and RMSD (root mean square deviation). In MODELLER, DOPE is the primary parameter used in selecting the best model. After the best model is selected, it is possible to perform loop modeling over this model. Then, the best model was optimized (Fig. **4**) [3].

In the final step, validation and verification of the modeling process outputs should be performed to evaluate the process. This step was done by using SAVES [88]. The server outcome gave an overall quality factor of 93.8326 and 92.44% of the residues had an averaged 3D-1D score of ≥ 0.2. According to the Ramachandran plot outcome, 92.9% of the residues were in the most favored regions, 6.6% of the residues were in the additional allowed regions, 0.5% residues were in the generously allowed regions, and no residue in the disallowed regions (Fig. **5**). Hence, the overall quality assessment demonstrated that a reliable model was generated through MODELLER.

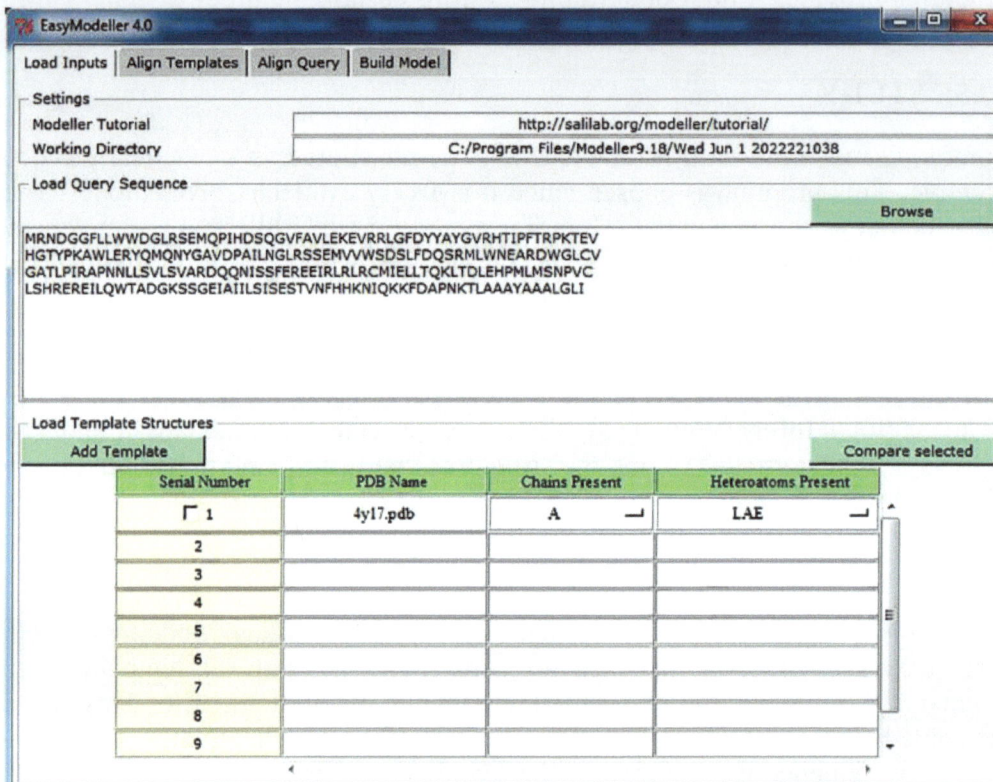

Fig. (3). The query sequence and template loaded into Easy Modeller.

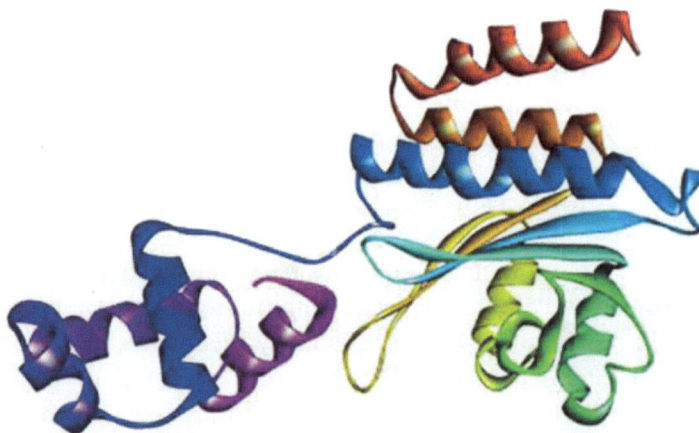

Fig. (4). 3D structure of the model generated *via* MODELLER.

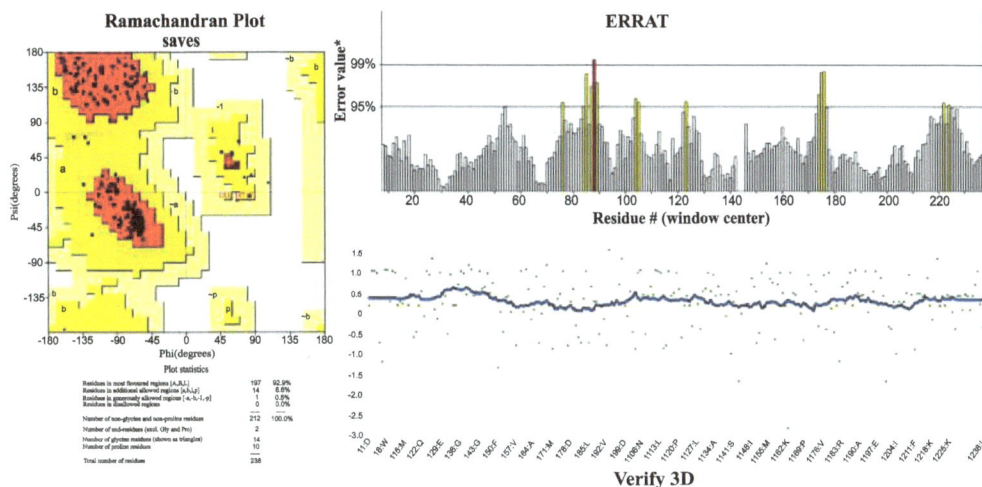

Fig. (5). Ramachandran plot (100% in the allowed regions), ERRAT (93.8326), and Verify 3D (92.44%) values of the model.

APPLICATIONS OF HOMOLOGY MODELING IN DRUG DISCOVERY

In the absence of experimentally determined 3D structures, homology modeling plays a crucial role in characterizing macromolecule properties, insight into mode of binding, designing of new ligands, and structure-based drug design [12]. Homology modeling is applied in drug discovery to reduce the labor, time, and cost required in the experimental part of this process. High quality 3D structures are needed for such applications. Hence, each step of the homology modeling should be undertaken with high care [10].

Homology modeling is applied in drug discovery in combination with other computational methods like molecular docking and MD simulation [25]. Reliable structures of G-protein-coupled receptors (GPCRs), which are crucial drug targets, were predicted [89]. In other studies, target structures that are used in the design and discovery of novel chemotherapeutic agents were generated [90, 91]. In the last few years, the world has been combating with the COVID-19 (Corona Virus Disease-19) pandemic [92]. In the endeavor to discover novel drugs that can be used against novel Coronavirus, homology modeling played its role together with the other computational methods [93]. Especially in the first few months of the pandemic the structures of the target proteins were not revealed. Hence, homology modeling played a crucial role. Some of the applications of homology modeling in this effort are summarized in Table **1**. Latest publications that focus on the discovery of novel drugs through homology modeling by targeting angiotensin converting enzyme 2 (ACE2), main protease (Mpro), non-structural proteins (Nsps), papain-like protease (PLpro), RNA-dependent RNA polymerase

(RdRp), spike (S) protein, and transmembrane protease serine 2 (TMPRSS2) were presented.

Table 1. Applications of homology modeling in the fight against COVID-19.

Target	Major Finding	Homology Modeling Tool	Other Computational Methods
Mpro, S protein	Hits that target Mpro and S proteins were determined [94].	MODELLER	Molecular docking
RdRp	The target of remedesivir in the Ebola virus and novel Coronavirus was investigated [95].	Prime	-
TMPRSS2	The mechanism of action for some inhibitors of the target was revealed [96].	SWISS-MODEL	Molecular docking, MD simulation
Nsp13, Nsp14	Lead compounds that could be dual inhibitors of the two targets were identified [97].	MODELLER	Molecular docking
PLpro	Three promising PLpro inhibitors were identified [98].	MODELLER	Molecular docking, MD simulation
TMPRSS2	Eight potential inhibitors of the target were identified [99].	SWISS-MODEL	Molecular docking
S protein, ACE2	The interactions between S protein and ACE2 were elucidated [100].	MODELLER	MD simulation, Binding energy calculation
RdRp	Mechanism of the binding of remdesivir to RdRp was revealed [101].	Prime	Molecular docking, pharmacophore screening, MD simulation
S protein, ACE2	Binding affinities of S protein-ACE2 across species were compared [102].	MODELLER	Molecular docking, MD simulation
RdRp, Nsp15	Potential lead compounds that might inhibit RdRp and Nsp15 were identified [103].	MODELLER	Molecular docking, MD simulation

CONCLUSION

The increase in the number of 3D structures with high quality is anticipated to increase the contribution of homology modeling in the structure determination of macromolecules with similar sequences [104]. As discussed earlier, various methods are used in model building. There is no single method that is superior in every aspect [89]. Hence, the user is recommended to select the suitable method based on the need. New approaches have also been introduced in the last decade [105]. In addition to this, the endeavor to integrate homology modeling with subsequent applications has been progressing [106].

In recent years, ML (machine learning) and AI have contributed to the improvement of homology modeling [107]. ML gives the programs the ability to learn from their own experience. The ability of ML to learn the function that converts the input to the output is acquired through training [108]. Similarly, AI has contributed to the prediction efficiency of the modeling [109]. In short, with the contribution of the state of the art approaches, more efficient modeling will be performed.

REFERENCES

[1]   Sun PD, Foster CE, Boyington JC. Overview of protein structural and functional folds. Curr Protoc Protein Sci 2004; 35: 1711.
[http://dx.doi.org/10.1002/0471140864.ps1701s35]

[2]   Bergendahl LT, Gerasimavicius L, Miles J, *et al.* The role of protein complexes in human genetic disease. Protein Sci 2019; 28: 1400.
[http://dx.doi.org/10.1002/pro.3667]

[3]   Muhammed MT, Son ÇD, İzgü F. Three dimensional structure prediction of panomycocin, a novel exo-β-1,3-glucanase isolated from Wickerhamomyces anomalus NCYC 434 and the computational site-directed mutagenesis studies to enhance its thermal stability for therapeutic applications. Comput Biol Chem 2019; 80: 270-7.

[4]   NobelPrize.org. Available from: https://www.nobelprize.org/prizes/lists/all-nobel-prizes/ (cited 2022 Jun 20)

[5]   Dill KA, MacCallum JL. The protein-folding problem, 50 years on. Science 2012; 338(6110): 1042-6.
[http://dx.doi.org/10.1126/science.1219021] [PMID: 23180855]

[6]   Puthenveetil R, Vinogradova O. Solution NMR: A powerful tool for structural and functional studies of membrane proteins in reconstituted environments. J Biol Chem 2019; 294: 15914.

[7]   Maveyraud L, Mourey L. Protein X-ray crystallography and drug discovery. Mol 2020; 25(5): 1030. Available from: https://www.mdpi.com/1420-3049/25/5/1030/htm

[8]   Zhou ZII. Chapter 1 - Atomic resolution cryo electron microscopy of macromolecular complexes. In: Advances in Protein Chemistry and Structural Biology. Elsevier 2011; 82: 1-35.
[http://dx.doi.org/10.1016/B978-0-12-386507-6.00001-4]

[9]   Kalman M, Ben-Tal N. Quality assessment of protein model-structures using evolutionary conservation. Bioinformatics 2010; 26: 1299-307.

[10]   Muhammed MT, Aki-Yalcin E. Homology modeling in drug discovery: Overview, current applications, and future perspectives. Chem Biol Drug Des 2019; 93: 12-20. https://onlinelibrary.wiley.com/doi/full/10.1111/cbdd.13388 (cited 2023 May 17)

[11]   Jisna VA, Jayaraj PB. Protein structure prediction: Conventional and deep learning perspectives. Protein J 2021; 40: 522-44.
[http://dx.doi.org/10.1007/s10930-021-10003-y]

[12]   Cavasotto CN, Phatak SS. Homology modeling in drug discovery: Current trends and applications. Drug Discov Today 2009; 14(13-14): 676-83.
[http://dx.doi.org/10.1016/j.drudis.2009.04.006] [PMID: 19422931]

[13]   Werner T, Morris MB, Dastmalchi S, Church WB. Structural modelling and dynamics of proteins for insights into drug interactions. Adv Drug Deliv Rev 2012; 64: 323-43.

[14]   Macalino SJY, Gosu V, Hong S, Choi S. Role of computer-aided drug design in modern drug discovery. Arch Pharm Res 2015; 38(9): 1686-701.
[http://dx.doi.org/10.1007/s12272-015-0640-5] [PMID: 26208641]

[15]   Anfinsen CB. Principles that govern the folding of protein chains. Science 1973; 181: 223-30.
[http://dx.doi.org/10.1126/science.181.4096.223]

[16]   Browne WJ, North ACT, Phillips DC, Brew K, Vanaman TC, Hill RL. A possible three-dimensional structure of bovine α-lactalbumin based on that of hen's egg-white lysozyme. J Mol Biol 1969; 42: 65-86.

[17]   Šali A, Blundell TL. Comparative protein modelling by satisfaction of spatial restraints. J Mol Biol 1993; 234(3): 779-815.

[18]   Schwede T, Kopp J, Guex N, Peitsch MC. SWISS-MODEL: An automated protein homology-modeling server. Nucleic Acids Res 2003; 31: 3381.
[http://dx.doi.org/10.1093/nar/gkg520]

[19]   Burley SK, Berman HM, Bhikadiya C, *et al.* Protein Data Bank: The single global archive for 3D macromolecular structure data. Nucleic Acids Res 2019; 47: D520-8.
https://academic.oup.com/nar/article/47/D1/D520/5144142

[20]   SayersRicha A, Tanya B, Jeff B, *et al.* Database resources of the national center for biotechnology information. Nucleic Acids Res 2015; 43: D6-D17. Available from: https://academic.oup.com/nar/article-lookup/doi/10.1093/nar/gku1130

[21]   Boratyn GM, Thierry-Mieg J, Thierry-Mieg D, Busby B, Madden TL. Magic-BLAST, an accurate RNA-seq aligner for long and short reads. BMC Bioinformatics 2019; 20(1): 405.

[22]   Altschul SF, Madden TL, Schäffer AA, *et al.* Gapped BLAST and PSI-BLAST: A new generation of protein database search programs. Nucleic Acids Res 1997; 25(17): 3389-402.
[http://dx.doi.org/10.1093/nar/25.17.3389] [PMID: 9254694]

[23]   Söding J. Protein homology detection by HMM–HMM comparison. Bioinformatics 2005; 21(7): 951-60.
[http://dx.doi.org/10.1093/bioinformatics/bti125] [PMID: 15531603]

[24]   Rost B, Sander C. Bridging the protein sequence-structure gap by structure predictions. Ann Rev 2003; 25: 113–36. https://www.annualreviews.org/doi/abs/10.1146/annurev.bb.25.060196.000553

[25]   Haddad Y, Adam V, Heger Z. Ten quick tips for homology modeling of high-resolution protein 3D structures. PLoS Comput Biol 2020; 16(4): e1007449.
[http://dx.doi.org/10.1371/journal.pcbi.1007449] [PMID: 32240155]

[26]   Haddad Y, Heger Z, Adam V. Guidelines for homology modeling of dopamine, norepinephrine, and serotonin transporters. ACS Chem Neurosci 2016; 7: 1607-13.

[27]   Venclovas Č. Methods for sequence–structure alignment. Methods Mol Biol 2011; 857: 55-82.
[http://dx.doi.org/10.1007/978-1-61779-588-6_3]

[28]   Sutcliffe MJ, Haneef I, Carney D, Blundell TL. Knowledge based modelling of homologous proteins, part I: three-dimensional frameworks derived from the simultaneous superposition of multiple structures. Protein Eng Des Sel 1987; 1(5): 377-84.
[http://dx.doi.org/10.1093/protein/1.5.377] [PMID: 3508286]

[29]   Bates PA, Kelley LA, MacCallum RM, Sternberg MJE. Enhancement of protein modeling by human intervention in applying the automatic programs 3D-JIGSAW and 3D-PSSM. Proteins Struct Funct Genet 2001; 45: 39-46. https://onlinelibrary.wiley.com/doi/full/10.1002/prot.1168

[30]   Levitt M. Accurate modeling of protein conformation by automatic segment matching. J Mol Biol 1992; 226: 507-33.

[31]   Webb B, Sali A. Comparative protein structure modeling using modeller. Curr Protoc Bioinforma 2014; 47: 5.6.1-5.6.32.
[http://dx.doi.org/10.1002/0471250953.bi0506s47]

[32]   Wallner B, Elofsson A. Identification of correct regions in protein models using structural, alignment,

and consensus information. Protein Sci 2006; 15(4): 900-13.
[http://dx.doi.org/10.1110/ps.051799606] [PMID: 16522791]

[33]   Pascarella S, Argos P. Analysis of insertions/deletions in protein structures. J Mol Biol 1992; 224: 461-71.

[34]   Vyas VK, Ukawala RD, Chintha C, Ghate M. Homology modeling a fast tool for drug discovery: Current perspectives. Indian J Pharm Sci 2012; 74(1): 1-17.
[http://dx.doi.org/10.4103/0250-474X.102537] [PMID: 23204616]

[35]   Lee GR, Shin WH, Park HB, Shin SM, Seok CO. Conformational sampling of flexible ligand-binding protein loops. Bull Korean Chem Soc 2012; 33(3): 770-4.
[http://dx.doi.org/10.5012/bkcs.2012.33.3.770]

[36]   Krieger E, Nabuurs SBVG, Manuscript A, *et al.* Homology modeling. In: Bourne E., Philip, Weissig H, Eds. Struct Bioinforma. 2nd ed. Wiley-Liss 2012; pp. 507–21. http://www.springerlink.com/index/10.1007/978-1-61779-588-6

[37]   Fiser A, Sali A. ModLoop: automated modeling of loops in protein structures. Bioinformatics 2003; 19: 2500-1.
[http://dx.doi.org/10.1093/bioinformatics/btg362]

[38]   Studer G, Tauriello G, Bienert S, *et al.* Modeling of protein tertiary and quaternary structures based on evolutionary information. Methods Mol Biol 2019; 1851: 301-16.
[http://dx.doi.org/10.1007/978-1-4939-8736-8_17]

[39]   Liang S, Zhang C, Sarmiento J, Standley DM. Protein loop modeling with optimized backbone potential functions. J Chem Theory Comput 2012; 8(5): 1820-7.
[http://dx.doi.org/10.1021/ct300131p] [PMID: 26593673]

[40]   Liang S, Grishin NV. Side-chain modeling with an optimized scoring function. Protein Sci 2002; 11: 322-31.
[http://dx.doi.org/10.1110/ps.24902]

[41]   Xu G, Ma T, Du J, Wang Q, Ma J. PUS-Rota2: An improved fast and accurate side-chain modeling method. J Chem Theory Comput [2019; 15: 5154-60.
https://pubs.acs.org/doi/abs/10.1021/acs.jctc.9b00309

[42]   Krivov GG, Shapovalov MV, Dunbrack RL Jr. Improved prediction of protein side-chain conformations with SCWRL4. Proteins 2009; 77(4): 778-95.
[PMID: 19603484]

[43]   Huang X, Pearce R, Zhang Y. FASPR: An open-source tool for fast and accurate protein side-chain packing. Bioinformatics 2020; 36(12): 3758-6.
[http://dx.doi.org/10.1093/bioinformatics/btaa234]

[44]   Han R, Leo-Macias A, Zerbino D, Bastolla U, Contreras-Moreira B, Ortiz AR. An efficient conformational sampling method for homology modeling. Proteins 2008; 71(1): 175-88.
[http://dx.doi.org/10.1002/prot.21672] [PMID: 17985353]

[45]   Hong SH, Joung IS, Flores-Canales JC, *et al.* Protein structure modeling and refinement by global optimization in CASP12. Proteins Struct Funct Bioinforma 2018; 86: 122-35.
[http://dx.doi.org/10.1002/prot.25426]

[46]   Kryshtafovych A, Monastyrskyy B, Fidelis K. CASP prediction center infrastructure and evaluation measures in CASP10 and CASP ROLL. Proteins 2014; 82 (2)(0 2): 7-13.
[http://dx.doi.org/10.1002/prot.24399] [PMID: 24038551]

[47]   Shen MV, Sali A. Statistical potential for assessment and prediction of protein structures. Protein Sci 2006; 15(11): 2507.
[http://dx.doi.org/10.1110/ps.062416606]

[48]   Chen VB, Arendall WB, Headd JJ, *et al.* MolProbity: All-atom structure validation for

macromolecular crystallography. Acta Crystallogr D Biol Crystallogr 2010; 66(Pt 1): 12-21. http://scripts.iucr.org/cgi-bin/paper?dz5180 (cited 2022 Jun 11)

[49]   Wiederstein M, Sippl MJ. ProSA-web: Interactive web service for the recognition of errors in three-dimensional   structures   of   proteins.   Nucleic   Acids   Res   2007;   35:   W407-410. https://academic.oup.com/nar/article/35/suppl_2/W407/2920938

[50]   Laskowski RA, MacArthur MW, Moss DS, Thornton JM. PROCHECK: A program to check the stereochemical   quality   of   protein   structures.   J   Appl   Cryst   1993;   26(2):   283-91. https://onlinelibrary.wiley.com/doi/full/10.1107/S0021889892009944

[51]   Benkert P, Tosatto SCE, Schomburg D. QMEAN: A comprehensive scoring function for model quality assessment. Proteins Struct Funct Bioinforma 2008; 71(1): 261-77.
[http://dx.doi.org/10.1002/prot.21715]

[52]   Carugo O, Djinovic Carugo K. Half a century of Ramachandran plots. Acta Crystallogr Sect D Biol Crystallogr 2013; 69: 1333-41.
[http://dx.doi.org/10.1107/S090744491301158X]

[53]   Bowie JU, Lüthy R, Eisenberg D. A method to identify protein sequences that fold into a known three-dimensional structure. Science 1991; 253(5016): 164-70.
[http://dx.doi.org/10.1126/science.1853201] [PMID: 1853201]

[54]   Hooft RWW, Vriend G, Sander C, Abola EE. Errors in protein structures. Nature 1996; 381(6580): 272.
[http://dx.doi.org/10.1038/381272a0] [PMID: 8692262]

[55]   Vriend G. WHAT IF: A molecular modeling and drug design program. J Mol Graph 1990; 8(1): 52-6, 29.
[http://dx.doi.org/10.1016/0263-7855(90)80070-V] [PMID: 2268628]

[56]   Li W, Dustin Schaeffer R, Otwinowski Z, Grishin NV. Estimation of uncertainties in the global distance   test   (GDT_TS)   for   CASP   models.   PLoS   One   2016;   11(5):   e0154786. https://journals.plos.org/plosone/article?id=10.1371/journal.pone.0154786

[57]   Eramian D, Shen M, Devos D, Melo F, Sali A, Marti-Renom MA. A composite score for predicting errors in protein structure models. Protein Sci 2006; 15(7): 1653.
[http://dx.doi.org/10.1110/ps.062095806]

[58]   Colovos C, Yeates TO. Verification of protein structures: Patterns of nonbonded atomic interactions. Protein Sci 1993; 2(9): 1511-9. https://onlinelibrary.wiley.com/doi/full/10.1002/pro.5560020916

[59]   Nayeem A, Sitkof D. A comparative study of available software for high-accuracy homology modeling: From sequence alignments to structural models. Protein Sci 2006; 15(4): 808-24. https://onlinelibrary.wiley.com/doi/full/10.1110/ps.051892906

[60]   Kuntal BK, Aparoy P, Reddanna P. EasyModeller: A graphical interface to MODELLER. BMC Res Notes 2010; 3: 226.
[http://dx.doi.org/10.1186/1756-0500-3-226] [PMID: 20712861]

[61]   Webb B, Sali A. Protein structure modeling with modeller. Methods Mol Biol 2021; 2199: 239-55.
[http://dx.doi.org/10.1007/978-1-0716-0892-0_14]

[62]   Pitman MR, Menz RI. Methods for protein homology modelling. Appl Mycol Biotechnol 2006; 6: 37-59.
[http://dx.doi.org/10.1016/S1874-5334(06)80005-5]

[63]   Nikolaev DM, Shtyrov AA, Panov MS, *et al.* A comparative study of modern homology modeling algorithms for rhodopsin structure prediction. ACS Omega 2018; 3(7): 7555-66.
[http://dx.doi.org/10.1021/acsomega.8b00721] [PMID: 30087916]

[64]   Zhang Y. I-TASSER server for protein 3D structure prediction. BMC Bioinformatics 2008; 9: 40.
[http://dx.doi.org/10.1186/1471-2105-9-40] [PMID: 18215316]

[65]   Hameduh T, Haddad Y, Adam V, Heger Z. Homology modeling in the time of collective and artificial intelligence. Comput Struct Biotechnol J 2020; 18: 3494-506.
[http://dx.doi.org/10.1016/j.csbj.2020.11.007] [PMID: 33304450]

[66]   Yang J, Yan R, Roy A, Xu D, Poisson J, Zhang Y. The I-TASSER suite: Protein structure and function prediction. Nat Methods 2014; 12(1): 7-8.

[67]   Arnold K, Bordoli L, Kopp J, Schwede T. The SWISS-MODEL workspace: A web-based environment for protein structure homology modelling. Bioinformatics 2006; 22(2): 195-201.
[http://dx.doi.org/10.1093/bioinformatics/bti770] [PMID: 16301204]

[68]   Remmert M, Biegert A, Hauser A, Söding J. HHblits: Lightning-fast iterative protein sequence searching by HMM-HMM alignment. Nat Methods 2011; 9(2): 173-5.
[http://dx.doi.org/10.1038/nmeth.1818] [PMID: 22198341]

[69]   Waterhouse A, Bertoni M, Bienert S, *et al.* SWISS-MODEL: Homology modelling of protein structures and complexes. Nucleic Acids Res 2018; 46(W1): W296-303.

[70]   Jacobson MP, Pincus DL, Rapp CS, *et al.* A hierarchical approach to all-atom protein loop prediction. Proteins: Struc Func Bioinform 2004; 55(2): 351-67.
[http://dx.doi.org/10.1002/prot.10613]

[71]   Rossi KA, Weigelt CA, Nayeem A, Krystek SR Jr. Loopholes and missing links in protein modeling. Protein Sci 2007; 16(9): 1999-2012.
[http://dx.doi.org/10.1110/ps.072887807] [PMID: 17660258]

[72]   Dolan MA, Noah JW, Hurt D. Comparison of common homology modeling algorithms: Application of user-defined alignments. Methods Mol Biol 2012; 857: 399-414.
[http://dx.doi.org/10.1007/978-1-61779-588-6_18] [PMID: 22323232]

[73]   Kelley LA, Sternberg MJE. Protein structure prediction on the Web: A case study using the Phyre server. Nat Protoc 2009; 4(3): 363-71.
[http://dx.doi.org/10.1038/nprot.2009.2] [PMID: 19247286]

[74]   The Phyre2 web portal for protein modeling, prediction and analysis. Nat Protoc 2015; 10(6): 845-58.
[http://dx.doi.org/10.1038/nprot.2015.053]

[75]   Söding J, Biegert A, Lupas AN. The HHpred interactive server for protein homology detection and structure prediction. Nucleic Acids Res 2005; 33(Web Server): W244-8.
[http://dx.doi.org/10.1093/nar/gki408]

[76]   Zimmermann L, Stephens A, Nam SZ, *et al.* A completely reimplemented MPI bioinformatics toolkit with a new HHpred server at its core. J Mol Biol 2018; 430(15): 2237-43.

[77]   Song Y, DiMaio F, Wang RYR, *et al.* High-resolution comparative modeling with RosettaCM. Structure 2013; 21(10): 1735-42.
[http://dx.doi.org/10.1016/j.str.2013.08.005] [PMID: 24035711]

[78]   Steinegger M, Meier M, Mirdita M, Vöhringer H, Haunsberger SJ, Söding J. HH-suite3 for fast remote homology detection and deep protein annotation. BMC Bioinformatics 2019; 20(1): 1-15.
https://bmcbioinformatics.biomedcentral.com/articles/10.1186/s12859-019-3019-7

[79]   Chen J, Long R, Wang XL, Liu B, Chou KC. dRHP-PseRA: Detecting remote homology proteins using profile-based pseudo protein sequence and rank aggregation. Sci Reports 2016; 6: 32333. Available from: https://www.nature.com/articles/srep32333 (cited 2022 May 29).

[80]   Källberg M, Wang H, Wang S, *et al.* Template-based protein structure modeling using the RaptorX web server. Nat Protoc 2012; 7(8): 1511-22.
[http://dx.doi.org/10.1038/nprot.2012.085] [PMID: 22814390]

[81]   Conway P, Tyka MD, DiMaio F, Konerding DE, Baker D. Relaxation of backbone bond geometry improves protein energy landscape modeling. Protein Sci 2014; 23(1): 47-55.
[http://dx.doi.org/10.1002/pro.2389] [PMID: 24265211]

[82]   Jumper J, Evans R, Pritzel A, *et al.* Highly accurate protein structure prediction with AlphaFold. Nature 2021; 596(7873): 583-9.
[http://dx.doi.org/10.1038/s41586-021-03819-2]

[83]   Varadi M, Anyango S, Deshpande M, *et al.* AlphaFold protein structure database: Massively expanding the structural coverage of protein-sequence space with high-accuracy models. Nucleic Acids Res 2022; 50: D439-44.
[http://dx.doi.org/10.1093/nar/gkab1061]

[84]   Rossi Sebastiano M, Ermondi G, Hadano S, Caron G. AI-based protein structure databases have the potential to accelerate rare diseases research: AlphaFoldDB and the case of IAHSP/Alsin. Drug Discov Today 2022; 27(6): 1652-60.
[http://dx.doi.org/10.1016/j.drudis.2021.12.018]

[85]   Apweiler R, Martin MJ, O'Donovan C, *et al.* Ongoing and future developments at the Universal Protein Resource. Nucleic Acids Res 2011; 39(Database issue): D214-9.
[PMID: 21051339]

[86]   Agarwala R, Barrett T, Beck J, *et al.* Database resources of the national center for biotechnology information. Nucleic Acids Res 2016; 44(D1): D7-D19.
[http://dx.doi.org/10.1093/nar/gkv1290] [PMID: 26615191]

[87]   Nguyen Y, Nguyen NX, Rogers JL, *et al.* Structural and mechanistic roles of novel chemical ligands on the SdiA quorum-sensing transcription regulator. mBio 2015; 6(2): e02429-14.
https://journals.asm.org/doi/full/10.1128/mBio.02429-14

[88]   Colovos C, Yeates TO. Verification of protein structures: Patterns of nonbonded atomic interactions. Protein Sci 1993; 2(9): 1511-9.

[89]   Schmidt T, Bergner A, Schwede T. Modelling three-dimensional protein structures for applications in drug design. Drug Discov Today 2014; 19(7): 890-7.
[http://dx.doi.org/10.1016/j.drudis.2013.10.027]

[90]   Nalini Chadha N, Bahia MS, Kaur M, Bahadur R, Silakari O. Computational design of new protein kinase D 1 (PKD1) inhibitors: Homology-based active site prediction, energy-optimized pharmacophore, docking and database screening. Mol Divers 2018; 22(1): 47-56.

[91]   Norouz Dizaji A, Yazdani Kohneshahri M, Gafil S, *et al.* Fluorescence labelled XT5 modified nano-capsules enable highly sensitive myeloma cells detection. Nanotechnology 2022; 33(26): 265101.
[http://dx.doi.org/10.1088/1361-6528/ac60dc]

[92]   Serafim MSM, Gertrudes JC, Costa DMA, Oliveira PR, Maltarollo VG, Honorio KM. Knowing and combating the enemy: A brief review on SARS-CoV-2 and computational approaches applied to the discovery of drug candidates. Biosci Rep 2021; 41(3): BSR20202616.
[http://dx.doi.org/10.1042/BSR20202616] [PMID: 33624754]

[93]   Battisti V, Wieder O, Garon A, Seidel T, Urban E, Langer T. A computational approach to identify potential novel inhibitors against the coronavirus SARS-CoV-2. Mol Inform 2020; 39(10): e2000090.
[http://dx.doi.org/10.1002/minf.202000090] [PMID: 32721082]

[94]   Hall DC, Ji HF. A search for medications to treat COVID-19 *via in silico* molecular docking models of the SARS-CoV-2 spike glycoprotein and 3CL protease. Travel Med Infect Dis 2020; 35: 101646.
[http://dx.doi.org/10.1016/j.tmaid.2020.101646]

[95]   Lo MK, Albariño CG, Perry JK, *et al.* Remdesivir targets a structurally analogous region of the Ebola virus and SARS-CoV-2 polymerases. Proc Natl Acad Sci USA 2020; 117(43): 26946-54.
[http://dx.doi.org/10.1073/pnas.2012294117] [PMID: 33028676]

[96]   Kishk SM, Kishk RM, Yassen ASA, *et al.* Molecular insights into human transmembrane protease serine-2 (TMPS2) inhibitors against SARS-CoV2: Homology modelling, molecular dynamics, and docking studies. Molecules 2020; 25(21): 5007.
[http://dx.doi.org/10.3390/molecules25215007] [PMID: 33137894]

[97] Gurung AB. *In silico* structure modelling of SARS-CoV-2 Nsp13 helicase and Nsp14 and repurposing of FDA approved antiviral drugs as dual inhibitors. Gene Reports 2020; 21: 100860.
[http://dx.doi.org/10.1016/j.genrep.2020.100860]

[98] Arwansyah A, Arif AR, Ramli I, *et al.* Molecular modelling on SARS-CoV-2 papain-like protease: An integrated study with homology modelling, molecular docking, and molecular dynamics simulations. SAR QSAR Environ Res 2021; 32(9): 699-718.
[http://dx.doi.org/10.1080/1062936X.2021.1960601]

[99] M P, Reddy GJ, Hema K, Dodoala S, Koganti B. Unravelling high-affinity binding compounds towards transmembrane protease serine 2 enzyme in treating SARS-CoV-2 infection using molecular modelling and docking studies. Eur J Pharmacol 2021; 890: 173688.
[http://dx.doi.org/10.1016/j.ejphar.2020.173688]

[100] Sakkiah S, Guo W, Pan B, *et al.* Elucidating interactions between SARS-CoV-2 trimeric spike protein and ACE2 using homology modeling and molecular dynamics simulations. Front Chem 2021; 8: 622632.
[http://dx.doi.org/10.3389/fchem.2020.622632] [PMID: 33469527]

[101] Arba M, Wahyudi ST, Brunt DJ, Paradis N, Wu C. Mechanistic insight on the remdesivir binding to RNA-Dependent RNA polymerase (RdRp) of SARS-cov-2. Comput Biol Med 2021; 129: 104156.
[http://dx.doi.org/10.1016/j.compbiomed.2020.104156]

[102] Piplani S, Singh PK, Winkler DA, Petrovsky N. *In silico* comparison of SARS-CoV-2 spike protein-ACE2 binding affinities across species and implications for virus origin. Sci Rep 2021; 11(1): 13063.
[http://dx.doi.org/10.1038/s41598-021-92388-5] [PMID: 34168168]

[103] Barage S, Karthic A, Bavi R, *et al.* Identification and characterization of novel RdRp and Nsp15 inhibitors for SARS-COV2 using computational approach. J Biomol Struct Dyn 2022; 40(6): 2557-74.
[http://dx.doi.org/10.1080/07391102.2020.1841026]

[104] Xiang Z. Advances in homology protein structure modeling zhexin. Curr Protein Pept Sci 2006; 7(3): 217-27.

[105] Gupta CL, Akhtar S, Bajpai P. *In silico* protein modeling: Possibilities and limitations. EXCLI J 2014; 13: 513-5.
[PMID: 26417278]

[106] Kajiwara Y, Yasuda S, Takamuku Y, Murata T, Kinoshita M. Identification of thermostabilizing mutations for a membrane protein whose three-dimensional structure is unknown. J Comput Chem 2017; 38: 211-23. https://onlinelibrary.wiley.com/doi/full/10.1002/jcc.24673

[107] Pearce R, Zhang Y. Toward the solution of the protein structure prediction problem. J Biol Chem 2021; 297(1): 100870.

[108] Heo L, Feig M. High-accuracy protein structures by combining machine-learning with physics-based refinement. Proteins Struct Funct Bioinforma 2020; 88(5): 637-42.
[http://dx.doi.org/10.1002/prot.25847]

[109] Si D, Nakamura A, Tang R, Guan H, Hou J, Firozi A, *et al.* Artificial intelligence advances for *de novo* molecular structure modeling in cryo-electron microscopy. Wiley Interdiscip Rev Comput Mol Sci 2022; 12(2): e1542.
[http://dx.doi.org/10.1002/wcms.1542]

CHAPTER 5

Anticancer Activity of Medicinal Plants Extract and Molecular Docking Studies

Serap ÇETINKAYA[1] and **Burak TÜZÜN**[2,*]

[1] *Department of Molecular Biology and Genetics, Science Faculty, Sivas Cumhuriyet University, Sivas, Turkey*

[2] *Plant and Animal Production Department, Technical Sciences Vocational School of Sivas, Sivas Cumhuriyet University, Sivas, Turkey*

Abstract: Molecular docking involves the interaction of a molecule with another place, usually in the protein structure, and simulating the placement of the molecule in the protein structure with certain score algorithms, taking into account many quantities, such as the electro-negativity of atoms, their positions to each other, and the conformation of the molecule to be inserted into the protein structure. Finally, the activity of the molecule with the highest percentage by mass against various cancer proteins was investigated according to the GC-MS results made on some medicinal and aromatic plants in order to set an example of molecular docking calculations.

Keywords: Activity, Aromatic plants, Cancer proteins, Molecular docking, Medicinal.

INTRODUCTION

Molecular docking involves the interaction of a molecule with another place, usually in the protein structure, and consists of simulating the placement of the molecule in the protein structure with certain score algorithms, taking into account many quantities such as the electro-negativity of the atoms, the positions of the atoms to each other, and the conformation of the molecule to be inserted into the protein structure [1, 2].

The docking process plays an important role in explaining the receptor-ligand, enzyme-ligand relationship. Finding suitable antagonist and agonist compounds has an important place in enzyme inhibition studies [3].

* **Corresponding author Burak TÜZÜN:** Plant and Animal Production Department, Technical Sciences Vocational School of Sivas, Sivas Cumhuriyet University, Sivas, Turkey; E-mail: theburaktuzun@yahoo.com

Igor José dos Santos Nascimento (Ed.)

Molecular docking studies are the theoretical method that plays an important role in determining whether millions of synthesized compounds are effective drug substances [4]. It is impossible to study each of the millions of chemical substances individually *in vitro*, and molecular docking studies have a very important role in selecting the most effective substances.

By converting the effect of a chemical substance on a protein structure into a numerical value, it saves money by preventing *in vitro* and *in vivo* studies of molecules that are impossible to be effective [5, 6]. It also lays the groundwork for the modification of the molecule with the correct estimation of the binding modes of the relevant molecule, and creates a strategic infrastructure by guiding the synthesis of molecules that are likely to be more effective.

The docking process plays an important role in explaining the receptor-ligand, enzyme-ligand relationship [7, 8]. It enables the comparison of the activities of molecules against proteins in studies to inhibit the enzyme in finding suitable antagonist compounds.

In addition to all these important and useful features of molecular docking studies, it also needs to be supported by molecular dynamics. Because the molecule clamped into the protein structure may have achieved good coupling and high scores in the first place, but both the enzyme and the relevant molecule in the solvent are in interaction [9]. This dynamic and synergetic state means that the chelating molecule cannot stay in the docked place for a long time, and its effect will be limited as it is related to the residence time in the attached area. Due to this situation, molecular docking calculations in computational chemistry are supported by molecular dynamics and the problem is solved.

Computer Aided Drug Design (CADD)

In silico methods are increasingly used for the development of new drugs. Computer-aided drug design (CADD) [10, 11] is a discipline that uses computational methods to simulate drug-receptor, drug-enzyme interactions. Calculations made by examining the 3-D properties of chemical molecules accelerate the optimization process of precursor compounds [12]. Thus, the success rate in drug research and development (R&D) studies increases, R&D costs decrease and R&D period shortens [13].

Computer-aided drug design programs require knowledge of ligands and receptors; bioinformatics develops depending on tools, applications and databases. If a target (receptor) exists, its 3D structure (by x-ray or NMR) together with its ligand must be known; If there is no experimental data, the 3D structure of the target molecule is tried to be created by homology modeling based on the sequence data.

There are two basic approaches to drug design: ligand-based and receptor-based molecular design methods.

Ligand-based Approach

The second and more branched approach to drug discovery is the ligand-based route. The general assumption of ligand-based methods [14, 15] is that the active site of a target protein may have similar atoms, functional groups, or moieties to have 11 similar functional properties. Nitrogen atoms in a Histidine residue at a particular position in the protein sequence must make a Hydrogen bond interaction with a polar Hydrogen atom on the ligand for the protein to lose its biological function (also called "inhibition") [16]. Of course, the change in the properties of a protein cannot be brought about by a single interaction on a single atom. However, if this approach is embodied for an entire molecule that has several interactions with more than one amino acid in the binding gap, the desired switch of function can be established. One of the first uses of ligand-based methods is seen as structure-activity relationship (SAR) studies, a method that has been used for decades [17].

However, the problem with the activity of small molecules in the body is that it cannot be predicted with sufficient accuracy. The reasons behind this disadvantage are [18]:

i) A full quantum-mechanical description of a ligand (*i.e.*, accurate calculation of the partial charges on each of its atoms) cannot be made,

ii) Actual activity depends on numerous factors such as: the character of the target, its environment, and the interactions established between a target and the ligand.

Perhaps the most promising avenues in a ligand-based approach are 3D pharmacophore modeling or 3D quantitative structure-activity relationship (QSAR) methods. Pharmacophore modeling encompasses the discovery of the spatial arrangement of pharmacophore groups in a molecule because that molecule is considered biologically active or relevant. The term "pharmacophore" was first defined by Schueler in the 1960s [19, 20] as functionalities in a molecule that determine its biological activity [21].

These chemical groups responsible for the activity of a drug molecule can be searched for and compared with desired activities through chemical libraries created by pre-established SAR studies. Similarly, in a 3D-QSAR analysis, the 3D positioning of the electron acceptor, electron donor, ring, and negatively charged or positively charged groups is obtained from a large number of input molecules with available biological activity information [22]. The most advantageous aspect of ligand-based drug discovery methods is that there is no need for structural information about the target. On the other hand, this advantage cannot sometimes be considered too easy or disadvantageous. This tradeoff makes ligand-based techniques for predicting structure-activity relationships often used for validation purposes rather than methods of discovering new molecules for a particular target that can be used clinically in the future. However, ligand-based methods can be considered reliable until the target structure is elucidated [23].

Structure (Receptor)-based Approach

As the name suggests, structure-based methods consist of applications derived from the structure of the target, namely protein-ligand docking [24, 25]. Molecular docking is a term used to computationally predict the binding modes and interaction energies between a target and its ligand [26]. To this end, the active site amino acid side chains of the target protein are usually kept immobile while the ligand in it rotates, stretches, and twists to fit properly while making non-covalent electronic interactions with the amino acid side chains of the target. In some algorithms, the binding gap of the target may be allowed to move to a certain extent. In a typical docking study, the biological target's active region [1, 27] is found divided into individual plots (or grid) in the 3D coordinate system, with atoms in this active region having the correct partial charges and a ligand probe atom or molecule (which also predicted partial charges) from this grid to calculate their interaction energies. This data is then converted into total interaction energy by different algorithms for each placement method, and this process [28] is called "scoring". The interaction energy value returned from such scoring functions [29] is called the "placement score". Although initially, all placement exercises perform similar procedures, their supposed difficulty or subtlety stems from their differences in scoring functions. In a drug discovery search, the docking and scoring functions are selected based on researchers' interests, the types of atoms in the protein and/or ligands, and the number of molecules to be screened [30].

Covalent Interactions in Biological Systems

Recent studies show that there is intense interest in the design of drugs that form covalent bonds with the target protein, and the first 30% of marketed drugs target enzymes known to act *via* covalent inhibition [31, 32]. Such inhibitors derive their activity from both non-covalent interactions and the formation of a covalent bond between the inhibitor and the target protein [33 - 38]. Covalent drugs typically have a much stronger binding affinity with targets due to the covalent bond formed, having a stronger effect between ligand (electrophilic) and target (nucleophilic), thus maintaining the pharmaceutically preferred small molecule size. Covalent interaction with the target protein has the advantage of prolonging the duration of the biological effect. However, these types of inhibitors tend to be associated with toxicity due to the difficulty of dissociation if off-target binding occurs. Therefore, highly determined selectivity profiles of covalent drugs are required. About 33% of the covalent drugs on the market are anti-infective (most notably β-lactam class antibiotics), 20% treat cancer, 15% treat gastrointestinal disorders, and ~ 15% treat the central nervous system and cardiovascular indications reported to be used [39].

Molecular Docking: Non-Covalent and Covalent Docking

Molecular docking is well known to be a computational procedure performed in structure-based rational drug design to determine the correct conformations of small molecule ligands and also to estimate the strength of protein-ligand interaction, usually a receptor and a ligand [40 - 42]. The most common nesting programs and software include Autodock [43], Autodoc k Vina [44], GOLD [45], Maestro [46] and FlexX [47]. Still, these and many other similar methods mainly focus on docking between two molecules through non-covalent interactions (van der Waals interaction, electrostatic interaction, and hydrogen bonding) or using other empirical or knowledge-based scoring functions to characterize these non-covalent interactions [47]. However, not all drugs bind non-covalently to the active site; there are other categories of drugs, namely covalent drugs [39].

Clamping of ligands that bind to a receptor *via* non-covalent interactions is now relatively conventional. Much of the research in insertion methods development has focused on the efficient prediction of the binding modes of non-covalent inhibitors [3, 48 - 52]. However, docking ligands that bind covalently to the receptor have been complicated by the reaction between the ligand and the receptor, which must be considered [53].

Docking Methods in Software

With this protein-ligand docking process, *in silico* analysis of drug design can be performed [54]. Thus, the cost and time required for the drug design process can be saved. Today, many software is used for molecular docking. This software is divided into 3 types according to the docking method [54]:

Fixed Docking

During docking, its molecular position is changed without changing its spatial shape. No changes can be made to the molecule [55].

Flexible-Fixed Docking

Docking is done without damaging the structure of the receptor containing macromolecules. The structure of small molecules shows their chemical effects in flexible, that is, mobile bonds. It has a higher accuracy rate than fixed docking.

Flexible Docking

The structure of ligands and receptors is flexible during docking. The movable bonds of the molecules and the angles of these bonds are movable in the program as in the natural structure of the molecules, which leads to more realistic results. However, it has more system requirements in terms of software and hardware than other types of software; it uses the processor capacity at a higher rate [56, 57].

Types of Docking Calculations Algorithms

Molecular docking software may have one or more search algorithms for the detection of suitable regions of the macromolecule to be docked with the ligand. Some of these are:

Stepwise Structure Algorithm

Ligand is divided into fixed and flexible structures according to the presence of variable, that is, rotatable bonds. First, the fixed part is docked to the binding sites of the macromolecule. Then the flexible parts are gradually added to the appropriate positions of the fixed part [58].

Monte Carlo Sampling Algorithm

By changing the 3-dimensional structure of the ligand, docking attempts are made to the active regions of the macromolecule at randomly determined angles. The energy compatibility in the docking regions of these docking processes is scored. These scoring results are compared among each other to determine the binding site with the best compatibility. In addition, it is stated that it is not an efficient method for docking large molecules [59, 60].

Genetic Algorithm

Successive generations are created with a logic similar to evolution. By changing the compatibility parameters of the molecule in each generation, the bonds with the lowest energy are eliminated at the end of the generation trials and the strong bonding pattern for docking is determined [61, 62].

Lamarckian Genetic Algorithm

Although it works in the same algorithm as the genetic algorithm, in addition to this, the information of the macromolecule regions that have been tried to bind with ligand in each generation, with less binding energy, is passed on to the next generation. These regions are also eliminated for next generation trials. This feature causes the algorithm to be faster and more efficient than the genetic algorithm [63, 64].

Biplane Space Sampling

Docking pockets are created by detecting the binding sites of the macromolecule. The docking pockets are docked at a speed of tens of thousands per second, leaving a 10-15 degree deviation in the rotatable bond angles in the ligand structure.

Shape Matching Algorithm

The algorithm predicts possible structures of binding pockets that can bind to the ligand. In line with these predictions, the appropriate binding sites on the macromolecule and the ligand are evaluated by docking with various alignments. The most compatible docking is detected [65].

Molecular Docking Software

Today, many programs are widely used for molecular docking. These programs:

1. DOCK [66]

2. FlexX [48]

3. GOLD [46]

4. Glide [67]

5. AutoDock [68]

6. Surflex [69]

7. AutoDock Vina [45]

8. Cdocker [70]

9. MOE-Dock [71]

10. FRED [1]

The comparison of the search algorithm, molecular docking software type, and access/license type criteria of these software is shown in Table **1**.

Table 1. Examples of molecular docking software.

Software	Search Algorithm	Molecular Docking Software Type	License Type
AutoDock	Genetic Algorithm & Lamarckian Genetic Algorithm	Flexible-Fixed Docking	Free
DOCK	Shape Matching Algorithm	Flexible Docking	Free
GOLD	Genetic Algorithm	Flexible Docking	Commercial
Glide	Monte Carlo Algorithm	Flexible Docking	Commercial
Maestro	Conformational-search algorithm	Flexible-Fixed Docking	Commercial
SwissDock	Biplane Space Sampling	Flexible-Fixed Docking	Free

The flexible docking software of Dock, GOLD and Glide software causes more realistic results than the flexible-fixed docking software of AutoDock and FlexX software. In addition, it has less accuracy, less system requirements and less CPU fatigue [72]. Lamarckian genetic algorithm, which is AutoDock's search algorithm used for ligand insertion, causes more accurate results than many other algorithms. Generalized formula for a protein-ligand (docking) interaction energy calculation proposed by Böhm in equations 1-4:

$$\Delta G_{binding} = \Delta G_0 + \Delta G_{bb} \sum h - bonds\ f\ (\Delta R, \Delta \alpha) + \Delta G_{bb} \sum ionic\ f(\Delta R, \Delta \alpha) +$$
$$\Delta G_{lipo}|A_{lipo}| + \Delta G_{rot}NROT \tag{1}$$

$$f(\Delta R, \Delta \alpha) = f1(\Delta R)\ f\ 2(\Delta \alpha) \tag{2}$$

$$f1(\Delta R) = \begin{cases} 1 & \Delta R \le 0.2\ A^0 \\ 1-(\Delta R - 0.2)/0.4 & \Delta R \le 0.6\ A^0 \\ 0, & \Delta R > 0.2\ A^0 \end{cases} \tag{3}$$

$$f2(\Delta \alpha) = \begin{cases} 1 & \Delta R \le 30^0 \\ 1-(\Delta R - 30)/50 & \Delta R \le 80^0 \\ 0, & \Delta R > 80^0 \end{cases} \tag{4}$$

There are many nesting software available in the market that include different embedding and scoring algorithms according to the system in question. For example, the target protein may contain heteroatoms *(i.e.,* metal atoms) in its structure, which must be considered in interaction maps. Or the target could be a DNA helix that requires more attention, containing phosphate groups responsible for hydrogen bonds and the overall shape of the helices. Researchers must carefully approach the system under study in order to choose the right algorithm for the calculations. A suggested formula for High Speed virtual Scanning Algorithm from Docking Algorithms:

$$\Delta G_{bind} = C_{lipo-lipo} \sum f(r_{lr}) + C_{hbond-neut-neut} \sum g(\Delta r) h(\Delta \alpha) +$$
$$C_{hbond-neut-chg} \sum g(\Delta r) h(\Delta \alpha) + C_{hbond-chg-chg} + \sum g(\Delta r)(\Delta \alpha) +$$
$$C_{max-metal-ion} \sum f(r_{lr}) + C_{rotb} H_{rotb} + C_{polar-phob} V_{pola-phob} + C_{coul} E_{coul} +$$
$$C_{vdW} E_{vdW} + \text{solvation term}$$

(5)

The interaction of molecules with proteins is an important factor that determines the activities of molecules. It should be well known that the more the molecule interacts with proteins, the higher its activity.

As it is generally known, it is known that Key-lock model molecules can enter the appropriate space and have high activities. Therefore, interaction of molecules with the appropriate region of the proteins is very important for molecular docking calculations, as shown in Fig. (1).

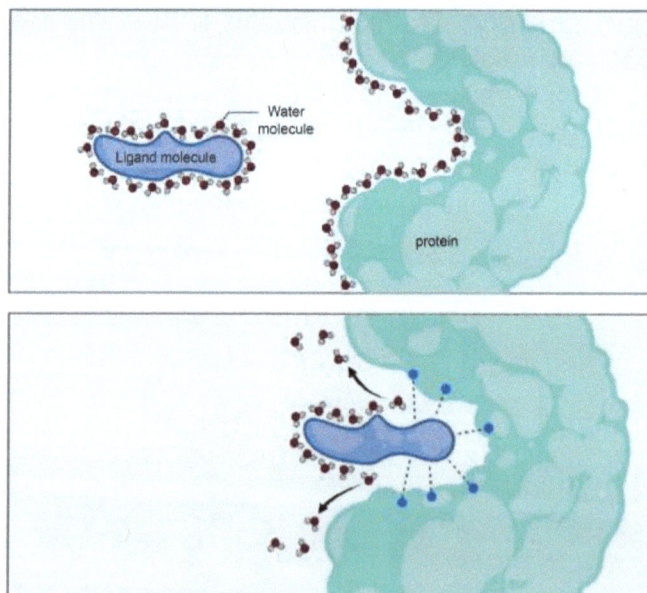

Fig. (1). Demonstration of the interaction between ligand and protein.

In order to examine the molecular docking calculations in more detail, the activities of 10 commonly known plants against cancer proteins were investigated with molecular docking calculations.

Artemisia sieversiana

Growing in the temperate regions of Europe, Asia, and North America, *Artemisia sieversiana* has attracted a lot of attention due to its richness of bioactive molecules [72]. Various phytochemicals have been found in different parts of the plant: sterols, coumarins and flavonoids, and flavonoids, terpenoids, and sterols, guaian-type sesquiterpenes, 3 germacrane-type sesquiterpenes, 1 muurolan-type sesquiterpene, and 1 diterpenoid [73]. The main compounds identified in the essential oils of the aerial parts are: borneol, geranyl butyrate, camphor and 1,8-cineol [73]. This plant is ethnomedically used to treat colds, diarrhoea, infections, fever, hysteria and jaundice [74, 75].

Various biological activities of *A. sieversiana* have been reported, spanning anti-inflammatory, antitumor- and antinematode activities [76]. Alcoholic extracts of *A. sieversiana* have displayed significant antioxidant activity [77]. In addition, the antifungal and antibacterial activity of polar and apolar extracts of *A. sieversiana* has been demonstrated against various microbial strains [78]. It has been reported that *A. sieversiana* has therapeutic effects on gastrointestinal disorders. Studies have also documented the herb's potential for the treatment of gastrointestinal ailments and pain [79]. Its methanol extracts have also been shown to have striking anthelmintic activity against the gastrointestinal nematode *Haemonchus contortus* [80]. In addition, extracts of the aerial parts inhibited the growth of SMMC-7721 hepatocarcinoma cells [72, 81] and showed a powerful cytotoxic activity against three different colorectal cancer cell lines: HCT-15, HT-29 and COLO [82].

The leaves and flowers contain varying concentrations of artemisinin [83]. It has been reported that artemisinin has a strong cytotoxic effect against 55 different human colorectal cancer cell lines [84]. Its derivatives have been found to induce apoptosis and inhibit angiogenesis in Fig. (**2**) [85]. In these activities, endoperoxide, part of artemisinin, may be responsible. Activation of the endoperoxide bond in the endosome by ferrous iron or reduced heme leads to the release of carbon-centered radicals. These radicals can target the mitochondrial electron transport chain and cause lysosomal damage leading to ROS generation, a series of caspase induction, cytochrome c release, DNA damage, and finally, apoptosis [86].

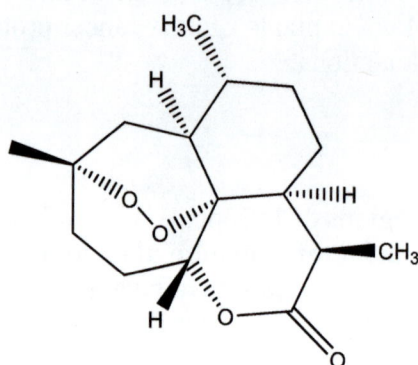

Fig. (2). Chemical structure of artemisinin.

In molecular docking calculations, it is possible to examine the interactions of molecules with various cancer proteins. The interaction of artemisinin molecule found in *A. sieversiana* plant extract with liver cancer protein (PDB ID: 2H80) [87] is given in Fig. (3).

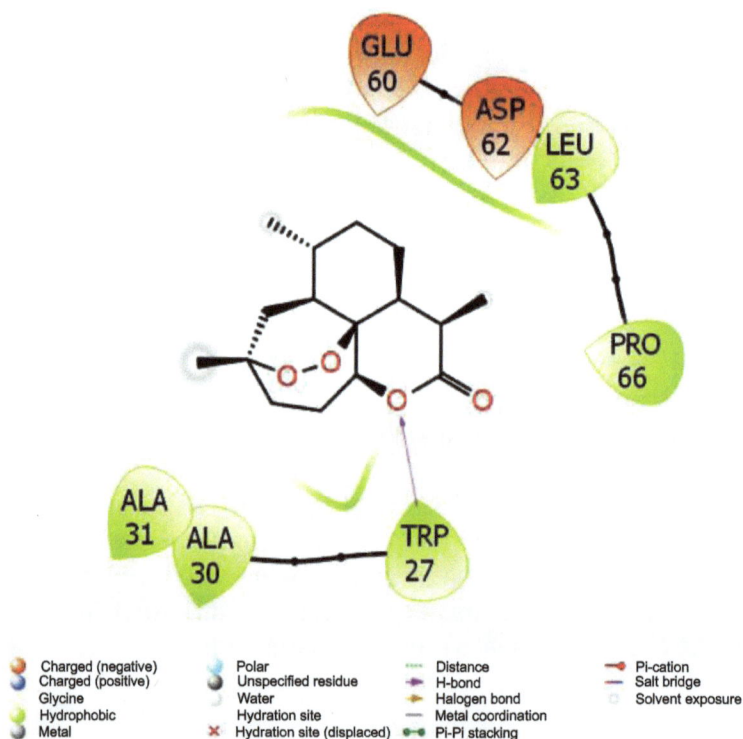

Fig. (3). Molecular docking interaction images of artemisinin molecule.

As seen in Fig. (**3**), the oxygen atom, which is a member of the hexadecimal ring in the artemisinin molecule, formed a hydrogen bond interaction with the TRP 27 protein.

Rosmarinus officinalis

Rosmarinus officinalis (rosemary) is an herb, abundant in the Mediterranean region. The plant is used as a spice and flavouring agent, making it the most important component of the Mediterranean diet [88]. As a medicinal plant, it has many biological effects: antioxidant, antithrombotic, antimicrobial, diuretic, antifungal, antidiabetic, hepatoprotective, anti-inflammatory, and anticancer [89]. There are many bioactive compounds belonging to the phenolic family in rosemary extracts; these are flavonoids (quercetin, rutin, kaempferol), terpenoids and acids (RA, caffeoylquinic acids, caffeic acid, quinic acid) [88]. The highest antioxidant properties of rosemary come from the phenolic diterpene extracts. The commonest rosemary diterpenes are carnosol, rosmanol, carnosic acid (CA), isorosmanol, and 7-methyl-epirosmanol, and epirosmanol. Carnosol and CA account for 90% of the antioxidant activity of the plant [90].

Carnosic acid, a natural benzenediol, is a betanediterpene and constitutes 1.5-2.5% mass of dried rosemary leaves. It has the typical o-diphenol structure and can therefore be oxidized to carnosol as it is unstable in solution [88]. As one of the strongest antioxidants in rosemary, it was found to have potent anticancer activity at different inhibitory concentrations against various cancer cell lines derived from human CRC, prostate, lung, breast, leukemia and liver malignant tissues [91]. It increases the activation of detoxifying enzyme genes such as glutathione peroxidase 8 (GPX8), glutathione peroxidase 3 (GPX3), monoamine oxidase B (MAOB) and aldehyde dehydrogenase 3 family at the transcriptional level [88]. Moreover, activation of ROS metabolism and alteration of some genes involved in oxidative degradation pathways have been observed recently during CA treatment [92].

Carnosol has antitumor, antioxidant and anti-inflammatory properties. It has been proposed as an anticancer drug in some *in vitro* studies because it targets several signalling pathways and therefore plays an important role in the inhibition of the growth and survival of cancer cells [88]. In addition to impeding tumorigenesis, carnosol has also been shown to have antimetastatic effects on murine melanoma cell lines [93]. Studies have shown that carnosol-mediated apoptotic cell death was observed in human colon cancer HCT116 cells in a dose- and time-dependent manner [94]. It has also been found that carnosol increased the antiproliferative activity of the phytochemicals capsaicin, quercetin and RA, and had a synergistic effect when used together with curcumin. The combination of carnosol and curcumin enhanced the antiproliferative effects of many different chemotherapeutic agents [92, 95, 96]. The docking picture that occurs as a result of this interaction is given in Fig. (**4**).

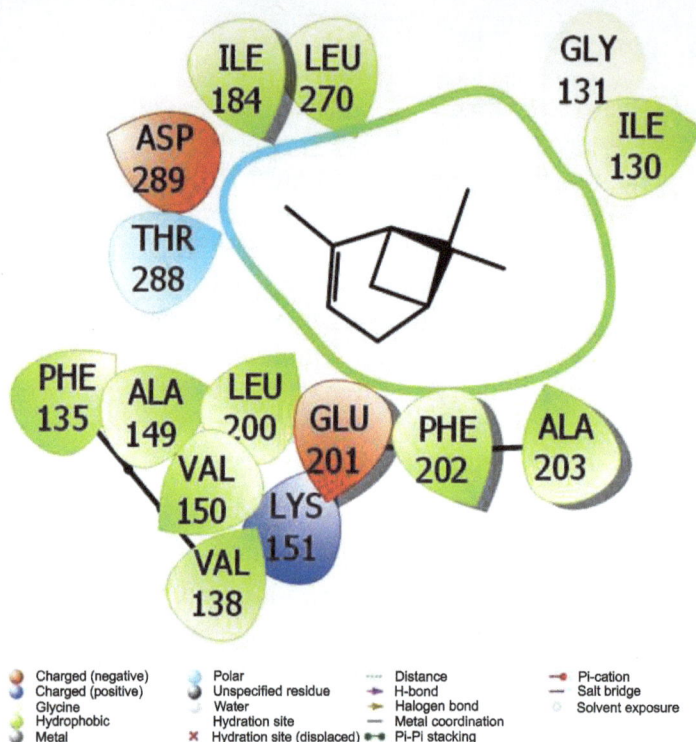

Fig. (4). Demonstration of the interaction of α-pinene molecule with colon cancer protein.

According to the results of the GC-MS analysis performed to determine the chemical molecules in the *Rosmarinus officinalis* plant, it is seen that the α-pinene molecule has the highest percentage in the plant, with 51.2%. MLK4 kinase (PDB ID: 4UYA) [97], one of the colon cancer receptors of this molecule, is affected.

Allium sativum

Allium sativum is a cultivated plant, garlic, used as food. Studies have shown that this plant could have a wide range of nutritional and medicinal functions. Its sulphur compounds, methyl allyl trisulfide, allicin, and diallyl trisulfide, had antithrombotic, anticancer and antibacterial effects, respectively [98]. /The plant also has many beneficial health benefits, such as antimicrobial, hypolipidemic, antiarthritic, hypoglycemic, antitumor and antithrombotic activities. Various garlic preparations have been shown to be chemopreventive. The two main compounds of garlic, S-allylcysteine and S-allylmercapto-L-cysteine, have the highest radical scavenging properties. In addition, S-allylcysteine has been found to reduce the growth of chemically induced and transplantable tumours in animals [99, 100]. The docking picture that occurs as a result of this interaction is given in Fig. (**5**).

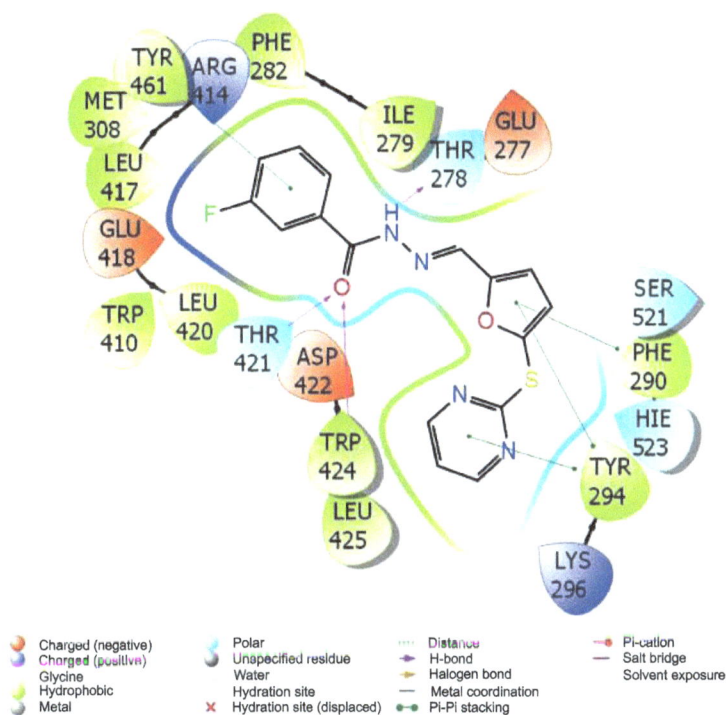

Fig. (5). Demonstration of the interaction of Allyl mercaptan molecule with lung cancer protein.

The plant *Allium sativum* has been used in many studies. GC-MC analysis was performed to determine the chemicals in this *Allium sativum* plant. As a result of this analysis, the interaction of the Allyl mercaptan molecule, the molecule with the highest mass, and the lung cancer protein (PDB ID: 5ZMA) [101] is given in Fig. (**5**).

Zingiber officinale

Belonging to the Zingiberaceae family, *Zingiber officinale* is commonly known as ginger or "inguru" (in Sinhalese) or "ingi" (in Tamil). It is widely used as a kitchen ingredient in Sri Lankan cuisine [102]. The plant is widely used among the cancer inhibitors of traditional medicine for the treatment of liver-, esophagus-, and gastrointestinal cancers. It is mostly used in polyherbal preparations. This herb contains many active ingredients; gingerol, shogaol, zingeron and paradol [103]. *Z. officinale* exerts its anticancer effects in several ways. For example, *Z. officinale* extract was found to significantly reduce NFκB expression in liver cancer in rats by inhibiting proinflammatory TNFα [104]. NFκB is a transcription factor, involved in biological processes such as inflammation, cell growth, and survival. It is also a cytokine that exerts its biological functions through NFκB activation [105]. Another study has shown a significant reduction in azoxymethane-induced colon cancer when male F344 rats. Studies have proven that gingerol can be a highly effective chemotherapeutic and chemopreventive agent in the treatment of colorectal cancer, both *in vitro* and *in vivo* in mice [106]. It can also be used as an antibacterial agent as well as for the treatment of nausea, diarrhea, and inflammation [107]. The docking picture that occurs as a result of this interaction is given in Fig. (**6**).

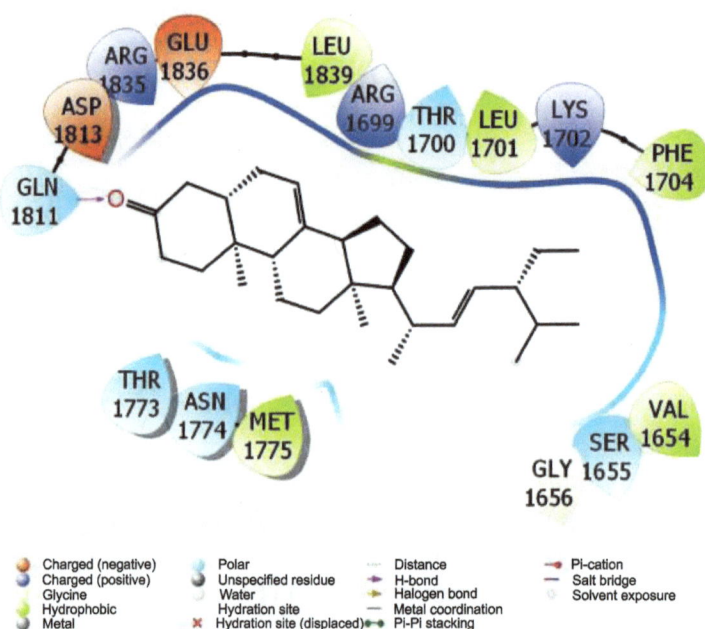

Fig. (6). Demonstration of the interaction of spinasterone molecule with breast cancer protein.

The *Zingiber officinale* plant has been used in many studies. GC-MC analysis was performed to determine the chemicals in this Allium sativum plant. As a result of this analysis, the interaction of the Spinasterone molecule, which has the highest mass, with the breast cancer protein (PDB ID: 1JNX) [108] is given in Fig. (**6**).

CONCLUSION

Molecular docking calculations are of great help in examining the interactions of molecules with various proteins, such as cancer, enzymes, viruses, and bacteria. With these calculations, it is obvious that great savings are made in both time and expense. In the study, after showing the basis of molecular docking calculations, the molecules with the highest percentage by mass were selected among the chemicals found in various medicinal and aromatic plants and their interactions with various cancer proteins were made.

REFERENCES

[1]   Pagadala NS, Syed K, Tuszynski J. Software for molecular docking: A review. Biophys Rev 2017; 9(2): 91-102.
        [http://dx.doi.org/10.1007/s12551-016-0247-1] [PMID: 28510083]

[2]   Guedes IA, de Magalhães CS, Dardenne LE. Receptor–ligand molecular docking. Biophys Rev 2014; 6(1): 75-87.
        [http://dx.doi.org/10.1007/s12551-013-0130-2] [PMID: 28509958]

[3]   Yuriev E, Ramsland PA. Latest developments in molecular docking: 2010-2011 in review. J Mol Recognit 2013; 26(5): 215-39.
        [http://dx.doi.org/10.1002/jmr.2266] [PMID: 23526775]

[4]   Gschwend DA, Good AC, Kuntz ID. Molecular docking towards drug discovery. J Mol Recognit 1996; 9(2): 175-86.
        [http://dx.doi.org/10.1002/(SICI)1099-1352(199603)9:2<175::AID-JMR260>3.0.CO;2-D]   [PMID: 8877811]

[5]   Singh AK, Rana HK, Singh V, Chand Yadav T, Varadwaj P, Pandey AK. Evaluation of antidiabetic activity of dietary phenolic compound chlorogenic acid in streptozotocin induced diabetic rats: Molecular docking, molecular dynamics, *in silico* toxicity, *in vitro* and *in vivo* studies. Comput Biol Med 2021; 134. 104462.
        [http://dx.doi.org/10.1016/j.compbiomed.2021.104462] [PMID: 34148008]

[6]   Ghosh S, Chetia D, Gogoi N, Rudrapal M. Design, molecular docking, drug-likeness, and molecular dynamics studies of 1,2,4-trioxane derivatives as novel *Plasmodium falciparum* falcipain-2 (FP-2) inhibitors. BioTechnologia 2021; 102(3): 257-75.
        [http://dx.doi.org/10.5114/bta.2021.108722] [PMID: 36606151]

[7]   Xie L, Li J, Xie L, Bourne PE. Drug discovery using chemical systems biology: Identification of the protein-ligand binding network to explain the side effects of CETP inhibitors. PLoS Comput Biol 2009; 5(5): e1000387.
        [http://dx.doi.org/10.1371/journal.pcbi.1000387] [PMID: 19436720]

[8]   Kandagalla S, Rimac H, Potemkin VA, Grishina MA. Complementarity principle in terms of electron density for the study of EGFR complexes. Future Med Chem 2021; 13(10): 863-75.
        [http://dx.doi.org/10.4155/fmc-2020-0265] [PMID: 33847171]

[9]   Luo L, Zhong A, Wang Q, Zheng T. Structure-based pharmacophore modeling, virtual screening, molecular docking, ADMET, and molecular dynamics (MD) simulation of potential inhibitors of PD-L1 from the library of marine natural products. Mar Drugs 2021; 20(1): 29.
        [http://dx.doi.org/10.3390/md20010029] [PMID: 35049884]

[10]   Marshall GR. Computer-aided drug design. Annu Rev Pharmacol Toxicol 1987; 27(1): 193-213.
        [http://dx.doi.org/10.1146/annurev.pa.27.040187.001205] [PMID: 3555315]

[11]   Macalino SJY, Gosu V, Hong S, Choi S. Role of computer-aided drug design in modern drug

discovery. Arch Pharm Res 2015; 38(9): 1686-701.
[http://dx.doi.org/10.1007/s12272-015-0640-5] [PMID: 26208641]

[12]   Hassan Baig M, Ahmad K, Roy S, *et al.* Computer aided drug design: Success and limitations. Curr Pharm Des 2016; 22(5): 572-81.
[http://dx.doi.org/10.2174/1381612822666151125000550] [PMID: 26601966]

[13]   Oliveira Viana JD, Scotti MT, Scotti L. Molecular docking studies in multitarget antitubercular drug discovery. Multi-Target Drug Design Using Chem-Bioinformatic Approaches. New York, NY: Humana Press 2018; pp. 107-54.
[http://dx.doi.org/10.1007/7653_2018_28]

[14]   Bacilieri M, Moro S. Ligand-based drug design methodologies in drug discovery process: An overview. Curr Drug Discov Technol 2006; 3(3): 155-65.
[http://dx.doi.org/10.2174/157016306780136781] [PMID: 17311561]

[15]   Merz KM Jr, Ringe D, Reynolds CH, Eds. Drug design: Structure-and ligand-based approaches. Cambridge University Press 2010.
[http://dx.doi.org/10.1017/CBO9780511730412]

[16]   Acharya C, Coop A, Polli JE, Mackerell AD Jr. Recent advances in ligand-based drug design: Relevance and utility of the conformationally sampled pharmacophore approach. Curr Comput Aided Drug Des 2011; 7(1): 10-22.
[http://dx.doi.org/10.2174/157340911793743547] [PMID: 20807187]

[17]   Bohl CE, Chang C, Mohler ML, *et al.* A ligand-based approach to identify quantitative structure-activity relationships for the androgen receptor. J Med Chem 2004; 47(15): 3765-76.
[http://dx.doi.org/10.1021/jm0499007] [PMID: 15239655]

[18]   Klebe G, Böhm HJ. Energetic and entropic factors determining binding affinity in protein-ligand complexes. J Recept Signal Transduct Res 1997; 17(1-3): 459-73.
[http://dx.doi.org/10.3109/10799899709036621] [PMID: 9029508]

[19]   Schueler FW, Keasling HH. The polymerization of pharmacophoric moieties and its effect upon biologic activity. I. Polymeric quaternary ammonium salts. J Am Pharm Assoc 1956; 45(12): 792-6.
[http://dx.doi.org/10.1002/jps.3030451207] [PMID: 13376374]

[20]   Schueler FW. The interaction of statistical and coulombic factors in the characterization of pharmacophoric moieties. Arch Int Pharmacodyn Ther 1953; 95(3-4): 376-97.
[PMID: 13125687]

[21]   Güner OF, Bowen JP. Setting the record straight: The origin of the pharmacophore concept. J Chem Inf Model 2014; 54(5): 1269-83.
[http://dx.doi.org/10.1021/ci5000533] [PMID: 24745881]

[22]   Bolton EE, Wang Y, Thiessen PA, Bryant SH. PubChem: integrated platform of small molecules and biological activities. Annual reports in computational chemistry 2008; 4: 217-41.

[23]   Lavecchia A, Cerchia C. *In silico* methods to address polypharmacology: Current status, applications and future perspectives. Drug Discov Today 2016; 21(2): 288-98.
[http://dx.doi.org/10.1016/j.drudis.2015.12.007] [PMID: 26743596]

[24]   Meng XY, Zhang HX, Mezei M, Cui M. Molecular docking: A powerful approach for structure-based drug discovery. Curr Comput Aided Drug Des 2011; 7(2): 146-57.
[http://dx.doi.org/10.2174/157340911795677602] [PMID: 21534921]

[25]   Kroemer RT. Structure-based drug design: Docking and scoring. Curr Protein Pept Sci 2007; 8(4): 312-28.
[http://dx.doi.org/10.2174/138920307781369382] [PMID: 17696866]

[26]   Kumar S, Kumar S. Chapter 6: Molecular docking: A structure-based approach for drug repurposing. In: Roy K. *In Silico* Drug Design: Repurposing Techniques and Methodologies. Academic Press 2019; pp. 161-89.

[http://dx.doi.org/10.1016/B978-0-12-816125-8.00006-7]

[27]   Reményi A, Good MC, Lim WA. Docking interactions in protein kinase and phosphatase networks. Curr Opin Struct Biol 2006; 16(6): 676-85.
[http://dx.doi.org/10.1016/j.sbi.2006.10.008] [PMID: 17079133]

[28]   Coupez B, Lewis RA. Docking and scoring-theoretically easy, practically impossible? Curr Med Chem 2006; 13(25): 2995-3003.
[http://dx.doi.org/10.2174/092986706778521797] [PMID: 17073642]

[29]   Jain AN. Scoring functions for protein-ligand docking. Curr Protein Pept Sci 2006; 7(5): 407-20.
[http://dx.doi.org/10.2174/138920306778559395] [PMID: 17073693]

[30]   Sarkı G, Tüzün B, Ünlüer D, Kantekin H. Synthesis, characterization, chemical and biological activities of 4-(4-methoxyphenethyl)-5- benzyl-2-hydroxy-2H-1,2,4-triazole-3(4H)-one phthalocyanine derivatives. Inorg Chim Acta 2023; 545: 121113.
[http://dx.doi.org/10.1016/j.ica.2022.121113]

[31]   Robertson JG. Enzymes as a special class of therapeutic target: Clinical drugs and modes of action. Curr Opin Struct Biol 2007; 17(6): 674-9.
[http://dx.doi.org/10.1016/j.sbi.2007.08.008] [PMID: 17884461]

[32]   Robertson JG. Mechanistic basis of enzyme-targeted drugs. Biochemistry 2005; 44(15): 5561-71.
[http://dx.doi.org/10.1021/bi050247e] [PMID: 15823014]

[33]   Doane T, Burda C. Nanoparticle mediated non-covalent drug delivery. Adv Drug Deliv Rev 2013; 65(5): 607-21.
[http://dx.doi.org/10.1016/j.addr.2012.05.012] [PMID: 22664231]

[34]   Jain NK, Gupta U. Application of dendrimer–drug complexation in the enhancement of drug solubility and bioavailability. Expert Opin Drug Metab Toxicol 2008; 4(8): 1035-52.
[http://dx.doi.org/10.1517/17425255.4.8.1035] [PMID: 18680439]

[35]   Kalgutkar AS, Dalvie DK. Drug discovery for a new generation of covalent drugs. Expert Opin Drug Discov 2012; 7(7): 561-81.
[http://dx.doi.org/10.1517/17460441.2012.688744] [PMID: 22607458]

[36]   Nassar AEF, Lopez-Anaya A. Strategies for dealing with reactive intermediates in drug discovery and development. Curr Opin Drug Discov Devel 2004; 7(1): 126-36.
[PMID: 14982156]

[37]   Pommier Y. Drugging topoisomerases: Lessons and challenges. ACS Chem Biol 2013; 8(1): 82-95.
[http://dx.doi.org/10.1021/cb300648v] [PMID: 23259582]

[38]   Zhou S, Chan E, Duan W, Huang M, Chen YZ. Drug bioactivation, covalent binding to target proteins and toxicity relevance. Drug Metab Rev 2005; 37(1): 41-213.
[http://dx.doi.org/10.1081/DMR-200028812] [PMID: 15747500]

[39]   Singh J, Petter RC, Baillie TA, Whitty A. The resurgence of covalent drugs. Nat Rev Drug Discov 2011; 10(4): 307-17.
[http://dx.doi.org/10.1038/nrd3410] [PMID: 21455239]

[40]   Yuriev E, Agostino M, Ramsland PA. Challenges and advances in computational docking: 2009 in review. J Mol Recognit 2011; 24(2): 149-64.
[http://dx.doi.org/10.1002/jmr.1077] [PMID: 21360606]

[41]   Mura C, McAnany CE. An introduction to biomolecular simulations and docking. Mol Simul 2014; 40(10-11): 732-64.
[http://dx.doi.org/10.1080/08927022.2014.935372]

[42]   Tantar AA, Conilleau S, Parent B, *et al.* Docking and biomolecular simulations on computer grids: Status and trends. Curr Comput Aided Drug Des 2008; 4(3): 235-49.
[http://dx.doi.org/10.2174/157340908785747438]

[43] Morris GM, Huey R, Lindstrom W, *et al.* Autodock4 and autodocktools4: Automated docking with selective receptor flexibility. J Comput Chem 2009; 30(16): 2785-91.
[http://dx.doi.org/10.1002/jcc.21256] [PMID: 19399780]

[44] Trott O, Olson AJ. AutoDock Vina: Improving the speed and accuracy of docking with a new scoring function, efficient optimization, and multithreading. J Comput Chem 2010; 31(2): 455-61.
[PMID: 19499576]

[45] Jones G, Willett P, Glen RC, Leach AR, Taylor R. Development and validation of a genetic algorithm for flexible docking. J Mol Biol 1997; 267(3): 727-48.
[http://dx.doi.org/10.1006/jmbi.1996.0897] [PMID: 9126849]

[46] Release S. 2021-3: Maestro. New York, NY: Schrödinger, LLC 2021.

[47] Rarey M, Kramer B, Lengauer T, Klebe G. A fast flexible docking method using an incremental construction algorithm. J Mol Biol 1996; 261(3): 470-89.
[http://dx.doi.org/10.1006/jmbi.1996.0477] [PMID: 8780787]

[48] Jacob RB, Andersen T, McDougal OM. Accessible high-throughput virtual screening molecular docking software for students and educators. PLoS Comput Biol 2012; 8(5): e1002499.
[http://dx.doi.org/10.1371/journal.pcbi.1002499] [PMID: 22693435]

[49] Fukunishi Y. Structural ensemble in computational drug screening. Expert Opin Drug Metab Toxicol 2010; 6(7): 835-49.
[http://dx.doi.org/10.1517/17425255.2010.486399] [PMID: 20465522]

[50] Hoffer L, Renaud JP, Horvath D. Fragment-based drug design: Computational & experimental state of the art. Comb Chem High Throughput Screen 2011; 14(6): 500-20.
[http://dx.doi.org/10.2174/138620711795767884] [PMID: 21521152]

[51] Konteatis ZD. *In silico* fragment-based drug design. Expert Opin Drug Discov 2010; 5(11): 1047-65.
[http://dx.doi.org/10.1517/17460441.2010.523697] [PMID: 22827744]

[52] Poustforoosh A, Faramarz S, Negahdaripour M, Tüzün B, Hashemipour H. Tracing the pathways and mechanisms involved in the anti-breast cancer activity of glycyrrhizin using bioinformatics tools and computational methods. J Biomol Struct Dyn 2023; 1-15.
[http://dx.doi.org/10.1080/07391102.2023.2196347] [PMID: 37042955]

[53] Chen YNP, Marnett LJ. Heme prosthetic group required for acetylation of prostaglandin H synthase by aspirin. FASEB J 1989; 3(11): 2294-7.
[http://dx.doi.org/10.1096/fasebj.3.11.2506093] [PMID: 2506093]

[54] Li X, Li Y, Cheng T, Liu Z, Wang R. Evaluation of the performance of four molecular docking programs on a diverse set of protein-ligand complexes. J Comput Chem 2010; 31(11): 2109-25.
[http://dx.doi.org/10.1002/jcc.21498] [PMID: 20127741]

[55] Ma L, Meng X, Liu Z, Du L. Suboptimal power-limited rendezvous with fixed docking direction and collision avoidance. J Guid Control Dyn 2013; 36(1): 229-39.
[http://dx.doi.org/10.2514/1.56449]

[56] Rosenfeld R, Vajda S, DeLisi C. Flexible docking and design. Annu Rev Biophys Biomol Struct 1995; 24(1): 677-700.
[http://dx.doi.org/10.1146/annurev.bb.24.060195.003333] [PMID: 7663131]

[57] Glen RC, Allen SC. Ligand-protein docking: Cancer research at the interface between biology and chemistry. Curr Med Chem 2003; 10(9): 763-77.
[http://dx.doi.org/10.2174/0929867033457809] [PMID: 12678780]

[58] Billings SA, Voon WSF. A prediction-error and stepwise-regression estimation algorithm for non-linear systems. Int J Control 1986; 44(3): 803-22.
[http://dx.doi.org/10.1080/00207178608933633]

[59] Kroese DP, Rubinstein RY. Monte Carlo methods. Wiley Interdiscip Rev Comput Stat 2012; 4(1): 48-

58.
[http://dx.doi.org/10.1002/wics.194]

[60]   Lee J. New monte carlo algorithm: Entropic sampling. Phys Rev Lett 1993; 71(2): 211-4.
[http://dx.doi.org/10.1103/PhysRevLett.71.211] [PMID: 10054892]

[61]   Mirjalili S. Genetic algorithm In: Evolutionary algorithms and neural networks. Cham: Springer 2019; pp. 43-55.
[http://dx.doi.org/10.1007/978-3-319-93025-1_4]

[62]   Kumar M, Husain D, Upreti N, Gupta D. Genetic algorithm: Review and application. SSRN 3529843.2010;

[63]   Fuhrmann J, Rurainski A, Lenhof HP, Neumann D. A new Lamarckian genetic algorithm for flexible ligand-receptor docking. J Comput Chem 2010; 31(9): 1911-8.
[http://dx.doi.org/10.1002/jcc.21478] [PMID: 20082382]

[64]   Wellock C, Ross BJ. An examination of Lamarckian genetic algorithms. In: Genetic and Evolutionary Computation Conference Late Breaking Papers 2001; 478-81.

[65]   Veltkamp RC. Shape matching: Similarity measures and algorithms Proceedings International Conference on Shape Modeling and Applications. 07-11 May, Genova, Italy, 2001, pp. 188-197.
[http://dx.doi.org/10.1109/SMA.2001.923389]

[66]   Venkatachalam CM, Jiang X, Oldfield T, Waldman M. LigandFit: A novel method for the shape-directed rapid docking of ligands to protein active sites. J Mol Graph Model 2003; 21(4): 289-307.
[http://dx.doi.org/10.1016/S1093-3263(02)00164-X] [PMID: 12479928]

[67]   Friesner RA, Banks JL, Murphy RB, *et al.* Glide: A new approach for rapid, accurate docking and scoring. 1. Method and assessment of docking accuracy. J Med Chem 2004; 47(7): 1739-49.
[http://dx.doi.org/10.1021/jm0306430] [PMID: 15027865]

[68]   Österberg F, Morris GM, Sanner MF, Olson AJ, Goodsell DS. Automated docking to multiple target structures: Incorporation of protein mobility and structural water heterogeneity in AutoDock. Proteins 2002; 46(1): 34-40.
[http://dx.doi.org/10.1002/prot.10028] [PMID: 11746701]

[69]   Jain AN. Surflex: Fully automatic flexible molecular docking using a molecular similarity-based search engine. J Med Chem 2003; 46(4): 499-511.
[http://dx.doi.org/10.1021/jm020406h] [PMID: 12570372]

[70]   Corbeil CR, Williams CI, Labute P. Variability in docking success rates due to dataset preparation. J Comput Aided Mol Des 2012; 26(6): 775-86.
[http://dx.doi.org/10.1007/s10822-012-9570-1] [PMID: 22566074]

[71]   McGann MR, Almond HR, Nicholls A, Grant JA, Brown FK. Gaussian docking functions. Biopolymers 2003; 68(1): 76-90.
[http://dx.doi.org/10.1002/bip.10207] [PMID: 12579581]

[72]   Benarba B, Pandiella A. Colorectal cancer and medicinal plants: Principle findings from recent studies. Biomed Pharmacother 2018; 107: 408-23.
[http://dx.doi.org/10.1016/j.biopha.2018.08.006] [PMID: 30099345]

[73]   Liu ZL, Liu QR, Chu SS, Jiang GH. Insecticidal activity and chemical composition of the essential oils of Artemisia lavandulaefolia and Artemisia sieversiana from China. Chem Biodivers 2010; 7(8): 2040-5.
[http://dx.doi.org/10.1002/cbdv.200900410] [PMID: 20730967]

[74]   Singh A, Singh PK. An ethnobotanical study of medicinal plants in chandauli district of uttar pradesh, India. J Ethnopharmacol 2009; 121(2): 324-9.
[http://dx.doi.org/10.1016/j.jep.2008.10.018] [PMID: 19022368]

[75]   Tayarani-Najaran Z, Sareban M, Gholami A, Emami SA, Mojarrab M. Cytotoxic and apoptotic effects

of different extracts of Artemisia turanica Krasch. on K562 and HL-60 cell lines. ScientificWorldJournal 2013; 2013: 1-6.
[http://dx.doi.org/10.1155/2013/628073] [PMID: 24288497]

[76] Chemesova I, Belenovskaya LM, Stukov AN. Anti-tumour activity of flavonoids from some Artemisia species. Rastit Resur 1987; 23(1): 100-3.

[77] Mangantbayaru K, Sravan K, Praveen A, *et al. In vitro* antioxidant studies on part of Origanum majoram linn and Artemesia sieversiana Ehrh. Pharmacogn Mag 2007; 3(9): 26-33.

[78] Malik JA, Wani AA. Ethnopharmacological properties of Artemisia genus used by the traditional healers of Kashmir. Indo Am JPharm Sci 2017; 4(8): 2738-43.

[79] Khan SW, Khatoon S. Ethnobotanical studies on some useful herbs of Haramosh and Bugrote valleys in Gilgit, northern areas of Pakistan. Pak J Bot 2008; 40(1): 43.

[80] Irum S, Ahmed H, Mirza B, *et al. In vitro* and *in vivo* anthelmintic activity of extracts from *Artemisia parviflora* and *A. sieversiana*. Helminthologia 2017; 54(3): 218-24.
[http://dx.doi.org/10.1515/helm-2017-0028]

[81] Zhang Q, Guo GN, Miao RD, Chen NY, Wang Q. Studies on the chemical constituents of Artemisia sieversiana and their anticancer activities. JLanzhou Unıv Nat Sci 2004; 40: 68-71.

[82] Tang J, Zhao JJ, Li ZH. Ethanol extract of *Artemisia sieversiana* exhibits anticancer effects and induces apoptosis through a mitochondrial pathway involving DNA damage in COLO-205 colon carcinoma cells. Bangladesh J Pharmacol 2015; 10(3): 518-23.
[http://dx.doi.org/10.3329/bjp.v10i3.23196]

[83] Mannan A, Ahmed I, Arshad W, *et al.* Survey of artemisinin production by diverse Artemisia species in northern Pakistan. Malar J 2010; 9(1): 310.
[http://dx.doi.org/10.1186/1475-2875-9-310] [PMID: 21047440]

[84] Ganguli A, Choudhury D, Datta S, Bhattacharya S, Chakrabarti G. Inhibition of autophagy by chloroquine potentiates synergistically anti-cancer property of artemisinin by promoting ROS dependent apoptosis. Biochimie 2014; 107(Pt B): 338-49.
[http://dx.doi.org/10.1016/j.biochi.2014.10.001] [PMID: 25308836]

[85] Li Z, Li Q, Wu J, Wang M, Yu J. Artemisinin and its derivatives as a repurposing anticancer agent: What else do we need to do? Molecules 2016; 21(10): 1331.
[http://dx.doi.org/10.3390/molecules21101331] [PMID: 27739410]

[86] Crespo-Ortiz MP, Wei MQ. Antitumor activity of artemisinin and its derivatives: From a well-known antimalarial agent to a potential anticancer drug. J Biomed Biotechnol 2012; 2012: 247597.
[http://dx.doi.org/10.1155/2012/247597]

[87] Li H, Sze K, Fung K. Validation of inter-helical orientation of the steril-alpha-motif of human deleted in liver cancer 2 by residual dipolar couplings. RCSB PDB 2008.
[http://dx.doi.org/10.2210/pdb2jw2/pdb]

[88] Petiwala SM, Johnson JJ. Diterpenes from rosemary (*Rosmarinus officinalis*): Defining their potential for anti-cancer activity. Cancer Lett 2015; 367(2): 93-102.
[http://dx.doi.org/10.1016/j.canlet.2015.07.005] [PMID: 26170168]

[89] Altinier G, Sosa S, Aquino RP, Mencherini T, Loggia RD, Tubaro A. Characterization of topical antiinflammatory compounds in *Rosmarinus officinalis* L. J Agric Food Chem 2007; 55(5): 1718-23.
[http://dx.doi.org/10.1021/jf062610+] [PMID: 17288440]

[90] Maldini M, Montoro P, Addis R, *et al.* A new approach to discriminate *Rosmarinus officinalis* L. plants with antioxidant activity, based on HPTLC fingerprint and targeted phenolic analysis combined with PCA. Ind Crops Prod 2016; 94: 665-72.
[http://dx.doi.org/10.1016/j.indcrop.2016.09.042]

[91] Yesil-Celiktas O, Sevimli C, Bedir E, Vardar-Sukan F. Inhibitory effects of rosemary extracts,

carnosic acid and rosmarinic acid on the growth of various human cancer cell lines. Plant Foods Hum Nutr 2010; 65(2): 158-63.
[http://dx.doi.org/10.1007/s11130-010-0166-4] [PMID: 20449663]

[92]   Martínez-Aledo N, Navas-Carrillo D, Orenes-Piñero E. Medicinal plants: Active compounds, properties and antiproliferative effects in colorectal cancer. Phytochem Rev 2020; 19(1): 123-37.
[http://dx.doi.org/10.1007/s11101-020-09660-1]

[93]   Huang SC, Ho CT, Lin-Shiau SY, Lin JK. Carnosol inhibits the invasion of B16/F10 mouse melanoma cells by suppressing metalloproteinase-9 through down-regulating nuclear factor-kappaB and c-Jun. Biochem Pharmacol 2005; 69(2): 221-32.
[http://dx.doi.org/10.1016/j.bcp.2004.09.019] [PMID: 15627474]

[94]   Park KW, Kundu J, Chae IG, *et al.* Carnosol induces apoptosis through generation of ROS and inactivation of STAT3 signaling in human colon cancer HCT116 cells. Int J Oncol 2014; 44(4): 1309-15.
[http://dx.doi.org/10.3892/ijo.2014.2281] [PMID: 24481553]

[95]   Vergara D, Simeone P, Bettini S, *et al.* Antitumor activity of the dietary diterpene carnosol against a panel of human cancer cell lines. Food Funct 2014; 5(6): 1261-9.
[http://dx.doi.org/10.1039/c4fo00023d] [PMID: 24733049]

[96]   Garzoli S, Laghezza Masci V, Franceschi S, Tiezzi A, Giacomello P, Ovidi E. Headspace/GC–MS analysis and investigation of antibacterial, antioxidant and cytotoxic activity of essential oils and hydrolates from *Rosmarinus officinalis* L. and *Lavandula angustifolia* Miller. Foods 2021; 10(8): 1768.
[http://dx.doi.org/10.3390/foods10081768] [PMID: 34441545]

[97]   Marusiak AA, Stephenson NL, Baik H, *et al.* Recurrent MLK4 loss-of-function mutations suppress JNK signaling to promote colon tumorigenesis. Cancer Res 2016; 76(3): 724-35.
[http://dx.doi.org/10.1158/0008-5472.CAN-15-0701-T] [PMID: 26637668]

[98]   Ariga T, Seki T. Antithrombotic and anticancer effects of garlic-derived sulfur compounds: A review. Biofactors 2006; 26(2): 93-103.
[http://dx.doi.org/10.1002/biof.5520260201] [PMID: 16823096]

[99]   Thomson M, Ali M. Garlic *Allium sativum*: A review of its potential use as an anti-cancer agent. Curr Cancer Drug Targets 2003; 3(1): 67-81.
[http://dx.doi.org/10.2174/1568009033333736] [PMID: 12570662]

[100] Sultana S, Asif HM, Nazar HMI, Akhtar N, Rehman JU, Rehman RU. Medicinal plants combating against cancer : A green anticancer approach. Asian Pac J Cancer Prev 2014; 15(11): 4385-94.
[http://dx.doi.org/10.7314/APJCP.2014.15.11.4385] [PMID: 24969858]

[101] Anantharajan J, Zhou H, Zhang L, *et al.* Structural and functional analyses of an allosteric EYA2 phosphatase inhibitor that has on-target effects in human lung cancer cells. Mol Cancer Ther 2019; 18(9): 1484-96.
[http://dx.doi.org/10.1158/1535-7163.MCT-18-1239] [PMID: 31285279]

[102] Williamson EM. Major Herbs of Ayurveda Churchill Livingstone Edimburgh. Elsevier Science Limited 2002.

[103] Rahmani AH, Shabrmi FM, Aly SM. Active ingredients of ginger as potential candidates in the prevention and treatment of diseases *via* modulation of biological activities. Int J Physiol Pathophysiol Pharmacol 2014; 6(2): 125-36.
[PMID: 25057339]

[104] Habib SHM, Makpol S, Hamid NAA, Das S, Ngah WZW, Yusof YAM. Ginger extract *(Zingiber officinale)* has anti-cancer and anti-inflammatory effects on ethionine-induced hepatoma rats. Clinics 2008; 63(6): 807-13.
[http://dx.doi.org/10.1590/S1807-59322008000600017] [PMID: 19061005]

[105] Wang X, Lin Y. Tumor necrosis factor and cancer, buddies or foes? Acta Pharmacol Sin 2008; 29(11): 1275-88.
[http://dx.doi.org/10.1111/j.1745-7254.2008.00889.x] [PMID: 18954521]

[106] Kim EC, Min JK, Kim TY, *et al.* [6]-Gingerol, a pungent ingredient of ginger, inhibits angiogenesis *in vitro* and *in vivo*. Biochem Biophys Res Commun 2005; 335(2): 300-8.
[http://dx.doi.org/10.1016/j.bbrc.2005.07.076] [PMID: 16081047]

[107] Kuruppu AI, Paranagama P, Goonasekara CL. Medicinal plants commonly used against cancer in traditional medicine formulae in Sri Lanka. Saudi Pharm J 2019; 27(4): 565-73.
[http://dx.doi.org/10.1016/j.jsps.2019.02.004] [PMID: 31061626]

[108] Williams RS, Green R, Glover JNM. Crystal structure of the BRCT repeat region from the breast cancer-associated protein BRCA1. Nat Struct Biol 2001; 8(10): 838-42.
[http://dx.doi.org/10.1038/nsb1001-838] [PMID: 11573086]

CHAPTER 6

FBDD & *De Novo* Drug Design

Anwesha Das[1,†], Arijit Nandi[2,†], Vijeta Kumari[3] and Mallika Alvala[4,*]

[1] *Department of Medicinal Chemistry, National Institute of Pharmaceutical Education and Research, Ahmedabad, Palaj, Gandhinagar 382355, Gujarat, India*

[2] *Department of Pharmacology, Dr. B.C. Roy College of Pharmacy and Allied Health Sciences, Durgapur-713206, West Bengal, India*

[3] *Laboratory of Natural Product Chemistry, Department of Pharmacy, Birla Institute of Technology and Science, Pilani (BITS Pilani), Pilani Campus, Pilani-333031, Rajasthan, India.*

[4] *MARS Training Academy, Hyderabad, India*

Abstract: Fragment-based drug or lead discovery (FBDD or FBLD) refers to as one of the most significant approaches in the domain of current research in the pharmaceutical industry as well as academia. It offers a number of advantages compared to the conventional drug discovery approach, which include – 1) It needs the lesser size of chemical databases for the development of fragments, 2) A wide spectrum of biophysical methodologies can be utilized for the selection of the best fit fragments against a particular receptor, and 3) It is far more simpler, feasible, and scalable in terms of the application when compared to the classical high-throughput screening methods, making it more popular day by day. For a fragment to become a drug candidate, they are analyzed and evaluated on the basis of numerous strategies and criteria, which are thoroughly explained in this chapter. One important term in the field of FBDD is *de novo* drug design (DNDD), which means the design and development of new ligand molecules or drug candidates from scratch using a wide range of *in silico* approaches and algorithmic tools, among which AI-based platforms are gaining large attraction. A principle segment of AI includes DRL that finds numerous applicabilities in the DNDD sector, such as the discovery of novel inhibitors of BACE1 enzyme, identification and optimization of new antagonists of DDR1 kinase enzyme, and development and design of ligand molecules specific to target adenosine A2A, *etc.* In this book chapter, several aspects of both FBDD and DNDD are briefly discussed.

Keywords: Artificial Intelligence, Autoencoder, Deep Learning, *De Novo* Drug Design, Drug Development, Drug Discovery, Evaluation Criteria, Expansion, Fragment-based Fragment to Lead, Hotspot analysis, *In silico*, Lead Optimization, Machine Learning, Molecular Docking, Optimization, Pharmacokinetic Properties, Property Prediction, Synthetic Accessibility.

---

[*] **Corresponding author Mallika Alvala:** MARS Training Academy, Hyderabad, India;
E-mail: mallikaalvala@yahoo.in
[†] *Authors with equal contribution.*

Igor José dos Santos Nascimento (Ed.)

INTRODUCTION

Since the last two decades, FBDD or FBLD has become one of the most triumphant methodologies in the area of early-stage drug development in the pharmaceutical industry as well as academia [1]. FBDD constitutes the screening of numerous molecules with lower molecular weights against clinically significant biological targets as these smaller fragments may fit into one or multiple binding sites of the protein and can act as potential beginning points in case of lead development. For fragment development, the physicochemical, pharmacokinetic, and toxic properties must be considered. One of the most popular methods, structure-based fragment screening, firstly employs a combination of multiple techniques, such as biophysical methods (thermophoresis, surface plasmon resonance [SPR], and differential scanning fluorimetry [DSF] *etc*.), later the employment of experimentations like X-ray crystallography or NMR, optimizes and structurally characterizes the fragments. Followed by that, further analytical stages like fragment growth also require the structural characterization of the screened hit fragments. The entire workflow of FBLD includes a massive high-throughput screening of all fragments that ultimately leads to the lead compound, and this approach is known as fragment-to-ligand optimization (F2L approach).

FBDD is referred to as one of the most attractive, effective, and popular approaches for chemical space exploration for perfectly fitting into the binding site of a biological target. While in the case of classical high-throughput screening (HTS), the screening of large libraries of complex molecules takes place against a target [2], in the case of FBDD, smaller libraries of lesser complex molecules that make fragments of larger drug-like molecules are usually screened against the target binding site for evaluating their binding efficiencies [3]. In spite of having lower potency than the larger drug-like compounds obtained *via* HTS, the fragments are considered potential starting points for designing larger drug-like molecules with higher affinity towards the target using the prior knowledge of the targets. This downside-up approach yields lead compounds with higher affinity and specificity, where a greater range of chemical space can be explored. Another advantage of FBDD includes that it requires lower expenses and lesser time for drug development through FBDD approach [4]. For example, Vemurafenib (Zelboraf™) is the first FBDD-derived drug that took only 6 years in all phases of the drug discovery pipeline before it went to FDA approval [5]. NMR can also be used in FBDD; for example, Bruker's Ligand Observed NMR is one of the most popular techniques for FBDD [6]. In the case of the computationally derived FBDD-approach, numerous tools can be employed for rationally designing a molecule. For example, AutoGrow4 is a genetic algorithm-based open source platform that can predict and design ligands computationally [7, 8]. Moreover, LigBuilder employs computational approaches to design ligands that can bind to multiple targets, multiple binding sites of a single target, or multiple conformations of a single target, thus forming a multi-target directed ligand (MTDL) [9]. This way, FBDD offers numerous attractive opportunities in the domain of drug discovery.

On the other hand, *De novo* drug design (DNDD) refers to the design of novel molecules that perfectly fit into a protein's binding site using several computational algorithms and approaches [10]. The meaning of the word "*De Novo*" is "starting from scratch or from the beginning", which implies that in DNDD, novel chemicals can be designed without any prior information of the starting point [11]. Among the several advantages of DNDD, such as larger chemical space exploration, new intellectual property containing compound design, time- and cost-effective development of novel chemical entities, and the strength of newer improved therapies as well as therapeutics, *etc.*, it shows one major disadvantage or challenge of synthesizability [12]. In this book chapter, several aspects of both FBDD and DNDD are briefly discussed.

TYPES OF DRUG DESIGN

DNDD can be defined as a drug designing methodology where new chemical entities (NCE) can be found from scratch from either the information related to the enzyme/receptor/biological target or its already known ligands having a strong inhibitory activity or good binding affinity towards the enzyme [13 - 25]. Needless to say, the main workflow behind the DNDD approach is - 1) A proper description and demonstration of the target's active binding site, 2) Pharmacophore modeling of the binding ligands, 3) Construction or generation of ligands by sampling, and 4) Evaluation of the constructed ligands. Principally, there are two types of DNDD approaches, namely, structure or receptor-based drug design (SBDD) and ligand-based drug design (LBDD).

Structure or Receptor-based Drug Design (SBDD)

SBDD is based on the three-dimensional structure (3D) of the biological target, where its structure is elucidated mainly by three methods, viz, electron microscopy, Nuclear magnetic resonance (NMR), and X-ray crystallography [26, 27]. Principally, SBDD starts with the determination of the receptor's active site. It is one of the most significant steps in SBDD as the reduction in the higher number of generated conformers and structures improves the specificity and selectivity towards the ligand. This specificity and tightness of the ligand binding at the receptor's active site are governed by the shape of the ligand molecule and its physical and chemical properties (non-covalent interactions, such as hydrophobic, hydrogen bonds, and electrostatic interactions) [12]. Among the number of methods involved in defining the active site, some are listed in Table **1**.

Table 1. Protein's active site predicting tool.

Name	Method	Description	Refs.
HSITE	Rule-based	Here, H-bond donors and acceptors are considered for predicting the active site.	[28]
LUDI and PRO_LIGAND	Rule-based	Here, hydrophobic contacts are considered for predicting the active site.	[29 - 31]
HIPPO	Rule-based	Here, covalent bonds and metal ion contacts are considered for predicting the active site.	[32]
-	Grid-based	Around the receptor active site, grid points are generated, and hydrophobic and H-bond interaction energies are calculated at each of the grid points of the fragments.	[33 - 35]
Multiple-copy simultaneous search (MCSS)	Docking	Favourable binding energy or docking score containing functional groups are minimized using force fields.	[36, 37]

Another methodology, homology modeling, can be adapted when the target structure is unavailable [38] that is based on the similarity in the amino acid sequences as well as the template structural quality. Here, the structure of the template protein and the similarity of the amino acid sequence between the target and template govern the homology modeled protein structure.

After performing all these methods, all generated structures are compared and evaluated by performing bounded free energy calculations. For this, several scoring functions are employed, like empirical and knowledge-based scoring functions, and force fields [39 - 45].

Ligand-based Drug Design (LBDD)

LBDD is performed when the 3D structural data of the target receptor is unavailable, but numerous data regarding the active binding ligands are available (3) from a structure-activity relationship (SAR)-related investigations, literature, or from ChEMBL-like molecular databases [46]. At first, a pharmacophore model is generated using the available known binders (either single or multiple), followed by plenty of novel ligand structures are designed by either performing a similarity search or generating a pseudo-target or pseudo-receptor [31, 46]. The pharmacophore model quality is an important tool in the case of LBDD, where it is highly dependent on and governed by the ligands' structural diversity. Another tool, *i.e.,* quantitative SAR (QSAR), is also an important model and plays a key role in LBDD [47]. Some of the LBDD tools are DOGS [48], SYNOPSIS [49], and TOPAS [50].

Sampling Methods in *De novo* Drug Design (DNDD)

Sampling of the generated structures can be performed using two methods, namely, atom-based and fragment-based methods [11, 12]. The atom-based method deals with the arbitrary placement of an atom in the receptor active that can be considered a base or seed to generate the remaining ligand structure. As a range of atoms along with their hybridization states are investigated in each step, a vast amount of structures along with a large amount of chemical space occupation are yielded. From these structures, the best possible fits can be found on the basis of their chemical and synthetic accessibility. One such atom-based sampling tool is "LEGEND" [34].

The second method, *i.e.,* fragment-based sampling method, involves the assembly of fragment-generated ligand structures, thus occupying a lesser chemical space, provides a wide range of diversity, chemical and synthetic accessibility, and sound pharmacokinetic properties (adsorption, distribution, metabolism, excretion, and toxicity [ADMET]) [11]. The assembly of fragments along with linker fragments can be obtained either experimentally or virtually [3], followed by which the molecular docking study is performed with the fragment at the active site. The docked fragment is considered as a base or seed to generate the remaining ligand structure [51 - 53]. After the fragment-based drug design (FBDD), the generated structures are filtered through a number of criterions such as drug likeness, ADMET properties, and Lipinski's rule of five, *etc* [54, 55]. Some of the fragment-based sampling algorithms include CONCERTS [40], LUDI [56], PRO_LIGAND [30], and SPROUT [57].

EVOLUTIONARY ALGORITHMS IN DNDD

In the case of DNDD, the evolutionary algorithms are widely used [11] that can be classified into several strategies like evolutionary programming, genetic programming, genetic algorithms, and evolutionary graphs and strategies. All of these are formed on the basis of optimization of the population using a variety of biological evolutions, *e.g.,* genetic recombination or crossover, mutation, reproduction, and selection *etc*. In the drug-designed domain, an arbitrarily generated chromosome-encoded population consists of generated ligand conformations and structures. The evolutionary algorithmic cycle is schematically represented in Fig. (**1**).

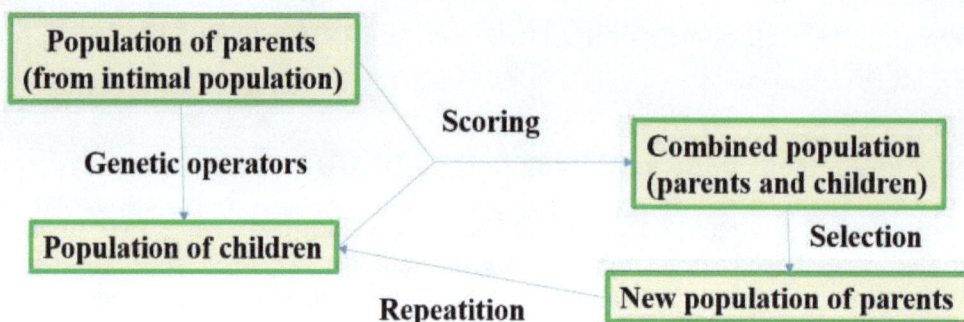

Fig. (1). Evolutionary algorithmic cycle.

Parents' new population can be used in the following cycle, and this process is repeated until all criterions of termination are fulfilled [58 - 60]. Examples of applications of DNDD are tabulated in Table **2**.

Table 2. Examples of DNDD.

Method	Example	Refs.
DNDD applications using genetic algorithm	LigBuilder, LEA, ADAPT, PEP, SYNOPSIS, LEA3D, GANDI, and ML GAN	[35, 49, 51, 61 - 65]
DNDD applications using evolutionary strategies	TOPAS, Flux (1), and Flux	[50, 66, 67]
DNDD applications using evolutionary graphs	MEGA and EvoMD	[68, 69]

ARTIFICIAL INTELLIGENCE (AI) IN DNDD

AI, along with its numerous subdivisions (Fig. **2**), can aid in revolutionizing the domain of drug discovery by previously detecting the possible outcomes in the same field [70], as it has been exceptionally successful in other domains too. These fields are - music [71], speech recognition [72], representations of videos [73], and formal languages, *etc.*

DEEP REINFORCEMENT LEARNING (DRL) IN DNDD

DRL is considered as the combination of both Deep Learning [DL] and Reinforcement Learning [RL] that plays a significant role in DNDD of the drug discovery domain [74]. In the DNDD segment, DRL has two components, viz, a generator and a DNDD agent, among which the first and second, respectively, utilize Artificial Neural Network (ANN) and RL, respectively. The generator or the generative model usually flows through multiple layers, *i.e.,* an input layers where either the molecular graphs or the SMILES [75] specific representation of a molecule, where characters represent atoms, and special characters represent connectivity [76] format of the ligand can be used, a series of intermediary hidden layers consisting of previously available data on the bioactive ligand molecules along with its specific receptor, and an output layer that is generated from repetitive training followed by reaching towards a decision [74]. The DNDD agents are considered as virtual robots and molecular interactors that can modify the molecules for their property improvement.

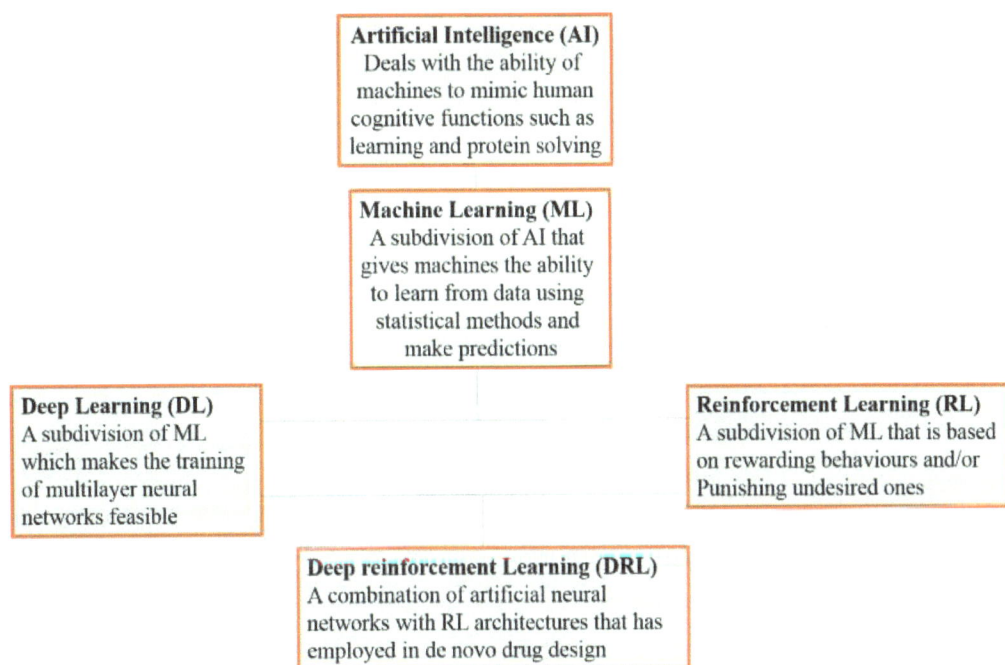

Artificial Intelligence (AI)
Deals with the ability of
machines to mimic human
cognitive functions such as
learning and protein solving

Machine Learning (ML)
A subdivision of AI that
gives machines the ability
to learn from data using
statistical methods and
make predictions

Deep Learning (DL)
A subdivision of ML
which makes the training
of multilayer neural
networks feasible

Reinforcement Learning (RL)
A subdivision of ML that is based
on rewarding behaviours and/or
Punishing undesired ones

Deep reinforcement Learning (DRL)
A combination of artificial neural
networks with RL architectures that has
employed in de novo drug design

Fig. (2). AI in DNDD.

Recurrent Neural Networks (RNN)

RNN can be defined as an architectural ANN that uses neuroral cyclic connections [77, 78], enabling the whole network to possess the present state's inner representation, so that it can recollect the previous steps' information present in several sequential data like SMILES format-represented molecules. The ligand molecules generated by RNN-driven DNDD are chemistry-driven, too, as it works in a step-by-step process sequentially.

The combination of RNN with RL has been applied extensively in DNDD [79 - 81], in which the first step includes a finely tuned and previously trained RNN utilizing several chemical databases generated small bioactive molecules. Usually, this training occurs *via* estimating the maximum amount of likelihood of the succeeding token of a target sequence, such as SMILES from the previous steps' provided token [82]. After training, it can be utilized for the generation of newer sequences [82]. The second step includes the generation of a DNDD agent on the basis of a mapping policy of each action probability state. The policy can be improved on the basis of these actions and receive rewards to elevate the amount of return. DNDD agent policy can be achieved *via* two approaches, 1) policy-based RL, where the policy is explicit for storing it in memory for the sake of learning, and 2) value-based RL, where the policy remains implicit, but the value function is kept. During the whole process, a clear endpoint or an episodic task is maintained for the generation of new ligand structures using the SMILES string. Some examples of RNN are provided in Table **3**.

Table 3. Examples of DRL.

Type	Name of Method	Description	Refs.
RNN	-	By this type, one model was trained to produce compounds that were sulfur-free by employing augmented episodic likelihood.	[82]
Stack-augmented RNN	Reinforcement Learning for Structural Evolution (ReLeaSE)	Design of chemical libraries consisting desirable biological and physicochemical properties. For this purpose, SMILES strings are taken as input.	[90]
Combination of RNN and RL	Fragment-based approach that is performed on the basis of actor-citric model	Using previously obtained bioactive lead compounds, novel chemical entities containing better properties are generated, and it is not required to search the whole chemical library.	[91]
Combination of RNN and TL (Transfer Learning)	Multiobjective evolutionary DNDD	As per the main principle of TL, at first, compounds are generated on the basis of RNN model, and later according to TL, the best compounds are selected and retrained, as TL can efficiently improve model accuracy.	[92, 93]
LSTM (Long short-term memory) RNN	DL	At first, LSTM-RNN models can be employed to develop a highly accurate SMILES strings containing chemical library, followed by that TL is employed to generate and screen chemicals that are structurally analogous to previously known lead bioactive compounds against the receptor target. Finally, the developed model is utilized in FBDD. By applying this generative model, the fragments are allowed to grow as starting points from the known active fragment to the known to lead compounds.	[94]

(Table 3) cont.....

Type	Name of Method	Description	Refs.
Combination of RNN with molecular descriptors	-	Here, the encoder is replaced by QSAR properties, RNN's conditional seed is replaced by the domains of the chemicals like the active compounds of the receptor.	[95]
RNN	Data-driven DNDD	Here, the RNN is trained on SMILES strings so that similar property- valid SMILES- containing compounds can be generated.	[80, 96]
Comparison of bidirectional RNN with unidirectional RNN	Bidirectional RNN	Computer-derived molecules containing generative SMILES strings are developed by the comparison of bidirectional RNN (combination of chemicobiological relevance, diversity of scaffolds, and novelty) with forward unidirectional RNN.	[97]
CNN	DeepScaffold	By employing CNN with molecular structural 2D graphs, a broad range of definitive scaffolds can be developed. For example, cyclic structure-containing scaffolds, specific side chains containing scaffolds, and Bemis-Murcko scaffolds. This method can efficiently perform chemical rule generalization in bond and atom addition on a specified scaffold, and can be successfully utilized in drug discovery.	[98]
Combination of CNN and RL	DeepGraphMolGen	It is a multipurpose approach where compounds are developed computationally and are represented as 2D graphical structures, because of being considered as a more natural representation compared to smile strings. Also here, the molecular properties are predicted, and on the basis of this graphical model, a novel framework for DNDD is suggested. Molecular development based on 2D graphical generation can be done by employing low-priced architecture of graphical convolution and a simpler decoder in computational terms.	[99, 100]
Combination of GAN and RL	Reinforced Adversarial Neural Computer (RANC)	It has improved the capability of molecular generation in DNDD, because it has added an explicit memory bank to solve the major challenges found commonly in generators. It develops structures similar to the SMILES length and main descriptors' distributions.	[101]
Combination of GAN and RL	Adversarial Threshold Neural Computer (ATNC)	It acts as a differentiable generator, which contains an adversarial threshold that can be served as a filter between the generator and the unbiased fraction of rewards along with the discriminator.	[102]
GAN	Internal Diversity Clustering (IDC)	Similar as above, but it can generate more diversified compounds, where the unbiased fraction of rewards is also newer.	[102]

(Table 3) cont.....

Type	Name of Method	Description	Refs.
Combination of AE and GAN	LatentGAN	It can generate 1) the molecules based on the actual drug randomly, and 2) molecules based on the target with positive results. Moreover, this method indicates that it is complementary towards the RNN-based approach.	[103]

Convolutional Neural Network (CNN)

CNN can be defined as an ANN that contains different layers for altering, convolution, and pooling for the automatic extraction of the features [83 - 85]. CNN was enormously applied in the field of processing images successfully [86]. Here, CNN is trained with several input features without the basis of the feature vector's absolute location [83]. Some examples of CNN are provided in Table **3**.

Generative Adversarial Network (GAN)

GAN can be defined as a specific type of neural network in which simultaneous training occurs on two networks, among which one network has a focus to generate images, while the other focuses on discrimination [87 - 89]. Here, the generator is usually responsible for true data (example) distribution capturing so that new data can be generated. On the other hand, the discriminator is responsible for discriminating the new generated data from the previously available examples accurately. It is categorized as a binary classifier. GANs find a wide range of applications in the field of medical image generation such as text-to-image, image-to-image translations, and super-resolution [87]. Some examples of GAN in the DNDD domain are tabulated in Table **3**.

Autoencoder (AE)

Variational Autoencoder (VAE)

A VAE is a stochastic variational inference and learning algorithm that is extensively used to represent high-dimensional complex data *via* a low-dimensional latent space learned in an unsupervised manner using encoders and decoders [104]. *De novo* drug design approaches using VAE include the development of a method to convert discrete representations of molecules to a multidimensional continuous representation [105]. In this study, a DNN was trained on hundreds of thousands of existing chemical structures to construct three coupled functions: an encoder, a decoder, and a predictor. The encoder converts the discrete representation of a molecule into a real-valued continuous vector, and the decoder converts these continuous vectors back into discrete molecular representations. The predictor estimates chemical properties from the latent continuous vector representation of the molecules. This model allowed efficient exploration of the chemical space through the development of optimized chemical structures. Another two examples are listed in Table **4**.

Sequence-to-Sequence Autoencoder (seq2seq AE)

A seq2seq AE is an artificial network architecture that maps an input sequence to a fixed-sized vector in the latent space using a gated recurrent unit (GRU) [106] or an LSTM network [107], and then maps the vector to a target sequence in another GRU or LSTM network [108]. Thus, the latent vector is an intermediate representation containing the "meaning" of the input sequence. In the case of DNDD, the input and output sequences are both SMILES strings [109]. Another examples are listed in Table **4**.

Adversarial Autoencoder (AAE)

An AAE is a probabilistic autoencoder that uses a GAN to perform variational inference by matching the aggregated posterior of the hidden code vector of an autoencoder with an arbitrary prior distribution [110]. Another examples are listed in Table **4**.

Table 4. Examples of AEs.

Type	Name of Method	Description	Refs.
Combination of CNN, RNN, and VAE	-	Here, SMILES strings of the novel molecules are generated using seed or base compounds' pharmacophoric characteristics, and 3D shapes. By employing this approach, uncovered portions of the chemical space can be explored in the search for drug-like properties containing newer functional groups as well as scaffolds.	[111]
Conditional VAE	-	Here, based on a conditional vector, novel molecules are designed and developed that can exhibit multiple targeting properties with very less error (10%). These properties include o/w partition coefficient (LogP), molecular weight, number of H-bond acceptors (HBA), and H-bond donors (HBD), and Topological surface area (TPSA). Using this approach, only the LogP with the alteration in other properties can be selectively and specifically controlled, and one specified property can be increased outside the training dataset range.	[112]

(Table 4) cont.....

Type	Name of Method	Description	Refs.
seq2seq AE	-	Here, desirable property-containing molecules can be developed by the optimization and identification of the scores of similarity and numerous chemical properties. The validation of all these optimizations and predictions is performed by employing the compound's fingerprint-based 2D predictors. The applications of this approach include – 1) Novel BACE1 inhibitor development, and 2) Generation of newer alternatives of eight preexisting candidate drugs such as Acalabrutinib, Ceritinib, Debrafenib, Enzalutamide, Idelalisib, Macimorelin, Panobinostat, and Ribociclib.	[109]
Combination of seq2seq hetero-AE, RNN, and LSTM	-	In this method, from a similar canonical SMILES formula, multiple SMILES strings are generated that exhibit a high similarity with the distance of the latent space, and show the cyclic molecular similarity.	[113]
AAE	-	This approach contains both the generator and a discriminator. Utilizing this method, desirable property-containing novel molecules can be generated using a training dataset that contains highly active antitumor compounds having similar molecular fingerprints.	[114]
AAE	druGan	Similar to above, but along with molecular structural fingerprints, additional properties such as solubility are also considered here. By this approach, distinct structures of chemicals can be developed rather than similar structures.	[115]

PARTICLE SWARM OPTIMIZATION (PSO) FOR DNDD

Particle swarm optimization is an evolutionary algorithm technique for screening small molecules having all desirable drug-like properties. Along with *in silico* prediction model, it has *an in silico* optimization algorithm too that optimizes the molecules with respect to the set properties of drug likeliness. Here, the query molecule (particles) is oriented and optimised in a sequential way to align with a preferred spot in a large search space. During their movement in an aligned chemical set, these molecules share information about the optimization rate. The velocity of these particles is a random function that is uniformly distributed and calculated. PSO utilizes the in-built sigmoidal transformation and equation to calculate the position of the molecule in the search space and evaluate its drug desirability according to the all *in silico* models [116]. Table (**5**) illustrates the various applications of PSO in the field of drug design.

PARAMETERS OF EVALUATION

The development of novel drug candidates initiates with drug designing, which is followed by its molecular optimization, synthesis, and biological evaluation.

Since the biological potency and effect on the target of newly designed molecules remain unpredicted, hence, it remains uncertain that every designed molecule gets screened in above-mentioned resource- intensive steps. It is, therefore, essential to focus on molecules generated *via* promising *de novo* techniques. The ideal properties of the best lead compound are – 1) It should be completely stable, novel, and should possess synthetic feasibility, 2) It should show balanced characteristics in terms of similarity with the previously available drug candidates, and in terms of difference so that it doesn't become entirely unpredictable, and 3) It should show high target affinity, pharmacokinetic, and drug-likeness properties.

Table 5. Applications of PSO in DNDD.

Type	Name of Method	Description	Refs.
PSO	COLIBREE	It is a combinatorial FB- and DNDD algorithm, the novel structures are developed using predefined chemical scaffold fragment-containing components of the molecules, and linkers. The fragments are generated from the previously available bioactive compounds by pseudo-retrosynthetic approach. The evaluation of the newly synthesized molecules is performed by checking their fitness score and how they are similar to the reference standards. In the entire process, PSO itself gains knowledge subsequently, which can help it in choosing linkers and newer fragments before reaching into the final compound.	[117]
Combination of PSO and DNN	-	In this methodology, the best drug molecule is optimized from a library of 75 million compounds on the basis of predicted targets that can perfectly fit into these leads, knowledge of SAR, and ADMET properties. The application of PSO is basically finding these chemicals' the compressed latent space to identify the best fit molecule. One of the significant merits of this method includes that a number of objective functions can be applied simultaneously during the identification and optimization. However, this approach should be optimized along with the modelable chemical constraints in the area of suitable applicability.	[118]

Diversity and Novelty

Computation models generate a large population of chemical compounds. However, not every generated compound appears to be inevitable and unique. Characteristics of the generated compounds may meet the reference standards or may be very dissimilar to the reference standard. The desirability completely depends on the selected method or model. When these methods are considered promising for screening drug-like molecules from a population of thousands, it becomes mandatory for the technical efficiency of these deep generative models. Certain technical failures like a model collapse in GANs [119] may happen where generated samples lack variety or blurry samples produced by AE- based models where all generated samples resemble to the mean value. There should be a validation guideline for computer assisted drug development to specify the work flow; such as the Organisation of Economic Cooperation and Development (OECD), which governs the validation of QSAR Models. Since there is no such authorized body to set the acceptable criteria of these evaluation matrices, the selection of a valid evaluation thresholds for screening thousands of chemical compounds remains subjective and domain-specific. Examples are listed in Table **6**.

Desired Properties

Results of all *in silico* prediction methods are based on the results of the evaluation parameters, which ascertain the desirability of the drug. However, there are other "side" properties that, during the generating process, evolve in a substantially unconstrained manner. Moreover, these side properties may represent the difference between the success and failure of the development stages following the design and synthesis of a candidate drug molecule. Examples are listed in Table **6**.

Synthetic Feasibility

Another concern is the actual capability of synthesizing the most promising *de novo* generated compounds for further evaluation and optimization [28, 32, 120 - 123]. Left unconstrained, generative models may propose overly complex or even impossible-to produce compounds. The generative process can be biased by penalizing the complexity of the molecules, but at the expense of reduced efficacy [124]. Examples of several evaluation functions are listed in Table **6**.

Table 6. List of criteria on the basis of the molecules to be evaluated.

Criterion	Name of the Methodology Utilized	Description	Refs.
Diversity and novelty	Levenshtein distance	It is utilized to check whether the SMILES string of the developed compound is similar to that of the reference compound's SMILES string.	[125]
Diversity and novelty	Dice distance and Tanimoto distance	These can be utilized in the case of substructural fingerprint representation of the developed compounds, for example, MACCS (Molecular access system), and ECFP (Extended-connectivity circular fingerprints).	[125]
Diversity and novelty	Convolutional kernel, and random walk kernel	These types of graphical kernels can be used in the case of graphical representation of the generated compounds.	[125]

(Table 6) cont.....

Criterion	Name of the Methodology Utilized	Description	Refs.
Desirable properties	Unexpected binding affinities towards the target complexes, and lesser drug-likeness	These undesired side properties can be utilized to filter out the lesser fit compounds, which may lead to cellular toxicity, generate adverse effects, and decrease the efficacy of the drug candidate, that finally causes the selection of the best fit lead candidate.	[126]
Desirable properties	Lipinski's filter, pharmacokinetic properties like Blood-brain barrier (BBB) permeation, binding efficiency and affinity towards the transporters, and solubility	Prior to the synthesis of the developed lead compound, these properties of the generated compounds should be evaluated as a mechanism to prioritize the best fit compounds.	[54, 125, 127]
Synthetic Feasibility	SA (Synthetic accessibility) score	In this methodology, some features of the molecules are considered, such as unconventional characteristics of the molecular structures, for example, unusual size of the compounds, complexity in terms of sterical features, non-classical fused rings, and large rings, *etc.*	[122]
Synthetic Feasibility	SC (Synthetic complexity) score	This methodology is trained by employing Reaxys database-generated twelve million reaction-based corpus to make sure the synthetic products show complexity than their reactants to foist a pairwise non-equal constraint.	[128]
Synthetic Feasibility	SPROUT	In this algorithm, previously known fragment database-based molecules are assembled, while every fragment is assigned with distinct penalty scores.	[57]
Synthetic Feasibility	MoleculeCHEF	In this approach, principle reactants along with a deep generative model consisting of reaction prediction, are employed to develop a simulative reaction series-generated target molecule.	[129]

BRIDGING TOXICOGENOMICS AND MOLECULAR DESIGN

Toxicogenomics is a field of study that links the safety assessment of chemicals to the underlying biological mechanisms [130, 131]. One important aspect tackled by toxicogenomics is the characterization of the mechanism-of-action (MOA) of a compound, represented as the set of all molecular alterations induced by the exposure of an organism (human) to it. Elucidation of the MOA allows an understanding of the chain of biological events (such as immune system activation, changes in metabolism, and effects on the cell cycle) triggered by a specific chemical (drug) exposure, which will lead to a phenotypic endpoint (for example, toxicity). Merging the cheminformatic and toxicogenomic methods, in combination with DL techniques, would facilitate and speed up the development of novel approaches where chemicals are designed *de novo* to exert specific molecular alterations and phenotypic effects. Most of the approaches proposed to date are chemocentric, but new methodologies that bridge toxicogenomics and molecular design are starting to emerge. One significant example is listed in Table 7.

Table 7. Examples of Bridging toxicogenomics and molecular design.

Methodology	Description	Refs.
GAN-based DL	It is a conditional model consisting of the data of gene expression, where both the representation as well as the desired properties of the generated compounds are introduced into the generative and training phases, so that it can learn a latent and decent space of representation. As a result, the model becomes so predictive and reconstructive that it can predict the altered transcriptomics that is pretty much similar to the same needed for the input.	[132]
Combination of GAN-based DL with conventional similarity search	The principle merit of employing this combinatorial methodology is that it does't only contain the initial reference set of chemical compounds against which, the measurement of gene expression occurred, but it also develops a set of novel molecules which tend to match the measurement of a query expression. However, this method's main demerit includes that further biological models such as cell cultures, should be optimized to validate the entire methodology.	[132]

DNDD FOR COVID-19

The coronavirus SARS-CoV-2 is responsible for the ongoing COVID-19 pandemic. The novel nature of this virus urgently requires the development of efficient drug repositioning and *de novo* drug design approaches. The scientific community has been actively working in this field and some of the well-known AI-based methods for drug design have been applied to generate new compounds [133 - 135]. Examples are listed in Table **8**.

Table 8. Case studies of DNDD for COVID-19.

Type	Name	Description	Refs.
DL-based method	Deep docking	In this SBDD approach, one thousand best fit ligands were found from the one billion compound-containing ZINC15 database against the target main protease (M^{Pro}) Severe acute respiratory syndrome coronavirus 2 (SARS-CoV-2).	[136]
VAE-based method	CogMol	In this computational SBDD approach, both molecular representations in SMILES format, and target proteins that cause the regression of the binding affinity are trained to develop the best compounds. By this approach, these molecules were optimized as the inhibitor of three spike proteins, namely, Plasmepsin-2, Plasmepsin-4, and ACE-2.	[137]

(Table 8) cont.....

Type	Name	Description	Refs.
Combination of FBDD approach and an advanced methodology of DL	Q-learning network	Using this methodology, 3CLPro or M^{Pro} inhibitory molecules were optimized and developed.	[138]
Combination of Wasserstein GAN (WGAN) and virtual screening approach	MolAICal	It is a three dimensional strategy to design and develop drug molecules on the basis of the target proteins. Using this approach, novel drug candidates have been developed from a fragment database consisting of FDA-approved candidates targeting two distinct proteins, namely, glucagon receptor GCGR (a membrane protein of SARS-CoV-2) and M^{Pro} (a non-membrane protein of SARS-CoV-2).	[139]

BUILDING COMMUNITY AND REGULATORY ACCEPTANCE OF DL-METHOD FOR DNDD

These COVID-19 examples demonstrate the power of DL methods for DNDD and are likely to further accelerate the drug discovery pipeline and the repurposing of existing drugs to alternative pathologies in the coming decade. However, since the development of DL-based DNDD approaches is still at an early stage, experimental validation of its effectiveness in drug discovery is crucial for the continuous improvement of these methods and to support their widespread uptake into medicinal chemistry practice and drug regulation. Few approaches are listed in Table **9**.

FBDD

Fragment Libraries

The selection and development of the fragment libraries is a pivotal step in the FBDD. Numerous databases are available with their unique characteristics of data representation. Diamond-SGC-iNEXT (DSiP) Poised library is an example of such type of fragment library [144]. It consists of 768 fragments and all compounds are available for purchase in Enamine as it is aligned with the Enamine REAL Database. However, other websites are available for the pharmacokinetics property optimization of existing phytoconstituents [145]. The fragment libraries satisfy the rule of 3, soluble in dimethyl sulfoxide (DMSO), or phosphate buffered saline. The number of fragments in the library is usually <1000 fragments, which is a user friendly feature available here, compared to the large volume databases [146].

Table 9. Approaches towards Building community and regulatory acceptance of DL-method for DNDD.

Approach	Fact	Refs.
Regulatory approach	1. Heads of Medical Agencies (HMA) and EMA (European Medicines Agency) suggest that the codes used for big data should be more simplified and clear. 2. All algorithms should go through postmarketing surveillance just as drugs do. 3. Whole documented reports for datasets and models should be provided, for example, for models consisting of QSAR, QMRF (QSAR model report format) should be formatted as per the OECD guidelines. These steps of the validation of the AI-based approaches can not only fight the COVID-19 pandemic, but may aid in the measurement and identification of further development of AI.	[140, 141]
Innovative open model	By employing this approach, methodologies and tools can be shared so that the promiscuity of the DL-based models can be obtained. As the entire data released can't be feasible due to privacy terms and conditions between the test and training dataset, merely the training one can be shared *via* the construction of a boundary tree so that it can predict the model meticulously. Across the boundary tree, datapoint crossing may make this model more believing and understanding towards the users. In spite of several challenges like the requirement for reliability and prudence regarding the patients' informations during the patients' treatment, this model is currently being applied in several situations for problem solving. Further rules and regulations regarding the validation of the model, defining the inputs and outputs, and mapping of data should also be incorporated.	[142, 143]

Fragment Expansion Strategy

After completion of the hit selection, the next step is the expansion of the fragments for better binding affinity and activity [1]. Several ways are available, one is taking the advice of a medicinal chemists and another way is to define vectors based on the stereochemistry of the structures. After that, there will be a screening for market available molecules from the large databases for a 1-3 heavy atom contain. The fragment enrichment is continued until the generation of high affinity molecules. The use of *the in silico* method using the fragments available in databases (*i.e.,* Zinc, Chembl) can potentiate this by ranking them based on predetermine docking parameters [146]. In the final stage, the nearer hits available within the same site can be linked [4]. The *in vitro* as well as *in vivo* experiments should satisfy the process of expansion.

Fragment Optimization Strategy

Followed by fragment expansion or hit identification, then comes fragment optimization, where both binding sites as well as the ligands' structural properties are considered. These optimization strategies have two main advantages. They are: – 1) Molecular fragments show more efficiency than the drug-like molecules, 2) The structures of molecular fragments can be optimized in a more efficient way compared to the drug-like molecules to possess a more favourable ADMET profile in the drug discovery and development pipeline, and these two are considered as the basis of the principle of the molecular fragment using. The possible cause can be explained by the following phenomenon - while the functional groups of the drug-like molecules may lead to poor protein binding or disruption of protein-ligand interaction, molecular fragments possess much easier protein binding as well as protein-ligand interactions of better quality, followed by a larger number of generated hits compared to that generated by the process of high-throughput screening (HTS) (Fig. **3**) [147, 148]. The reason behind this is that the complex molecules are inversely proportional to the probability of target-binding [149]. The other advantages include – 1) Fast progression towards hits, as instead of time-consuming chemical synthesis, simple-in-structure fragments can be easily purchased from several commercial chemical databases such as MolPort, ChemBridge, and ZINC15 *etc.* 2) FBDD shows better efficiency in the sampling of chemical space [147, 148]. 3) FBDD requires lesser expenses to be implemented compared to drug-like molecules because of the smaller size of the fragment libraries with respect to the HTS library [150].

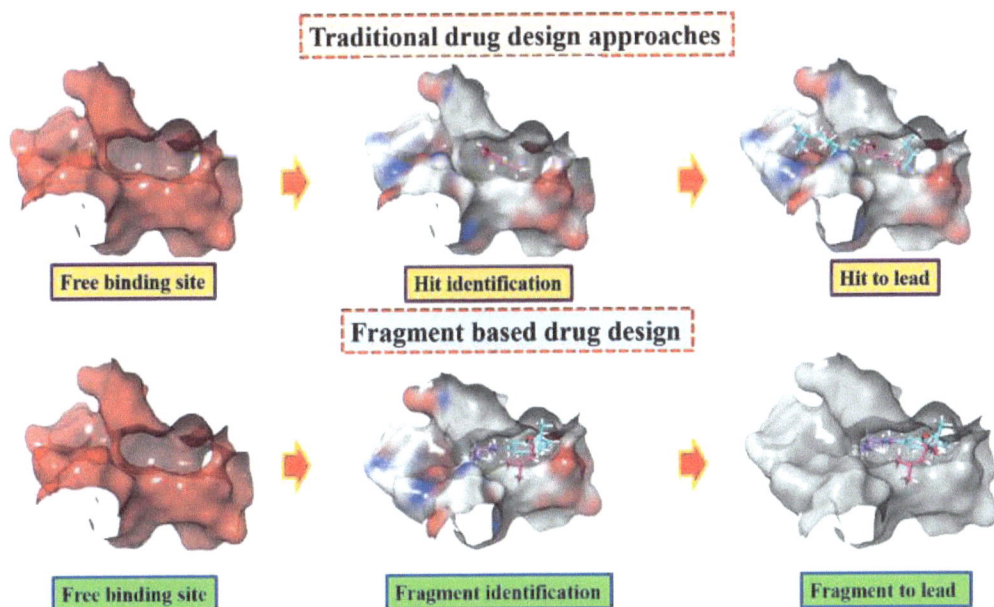

Fig. (3). Comparison between Conventional drug design and FBDD.

For optimization of the fragments, *i.e.,* for identifying the synthesizable vectors of the ligand, the evaluation of the fragment-protein binding site interactions should be properly and carefully carried out. In this regard, the structural data obtained from X-ray crystallography depicts only the snapshot of the system under investigation, however, the holistic environment of the structural dynamics of the proteins [151, 152] is known to affect smaller and weaker ligands as fragments without any alteration of the ligand-binding sites of the proteins [153, 154].

For these above-mentioned reasons, fragments are needed to be optimized or identified. For this purpose, orthogonal and complementary methods, such as thermodynamic data can be adopted [155]. Although mostly biophysical [156 - 159], nowadays, these methods can also be applied using biochemical approaches [148, 160, 161] in fragment-based lead discovery (FBLD). However, the biophysical approaches have a couple of advantages – 1) They can directly measure the protein-ligand binding, 2), They don't require any previous information of the function of the protein, and 3) They can detect smaller ligands with lower binding affinity [155]. Other than these two methods, the ligand efficiency (LE) and its related parameters play a significant role in going through the repetitive cycle of fragment optimization to identify the follow-up ligands' quality. These parameters include –

A. Binding Efficiency Index (BEI) – pKi/Molecular weight [162]

B. LE – Difference in energy/Heavy atom count [4, 163, 164]

C. Ligand Efficiency-Dependent Lipophlicity (LELP) – LogP/LE [4]

D. Lipophilic Efficiency (LipE/LLE) – pIC_{50} – cLogP [165]

E. Percentage Efficiency Index (PEI) – Percentage inhibition/ Molecular weight [4, 166]

F. Surface-binding Efficiency Index (SEI) – pIC_{50}/Topological polar surface area [162]

One of the principal reasons behind the generation of the higher number of hits using the FBDD approach compared to the library of drug-like molecules is that a protein's structural complexity makes smaller and lesser complex molecules more prone to bind, making them more efficient [149]. As a result, lesser complex fragments possess higher binding efficiency towards the protein, finally leading to an optimized small molecule. There can be mainly three strategies for fragment optimization (Fig. **4**).

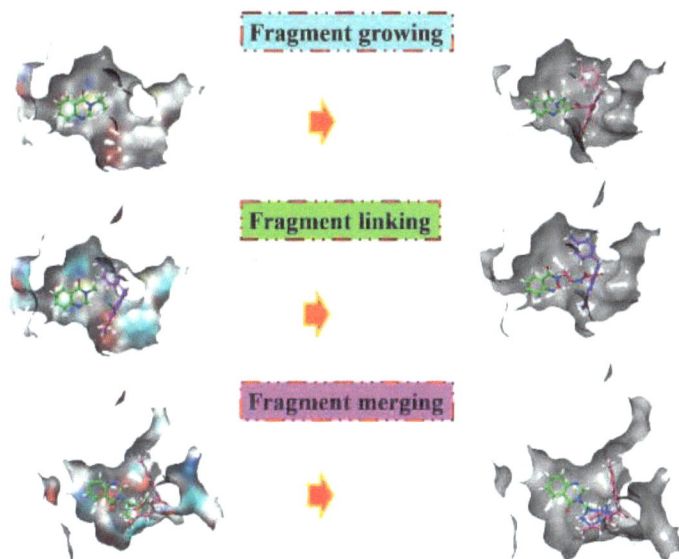

Fig. (4). Strategies for fragment optimization.

Fragment Growing

It is one of the most frequently and commonly used strategies, where the fragments are optimized for increasing their size by adding groups. This method is identical to the conventional method of modification of the compounds used in the hit identification approach through HTS (Table 10)

Table 10. Examples of fragment optimization strategy.

Type	Example	Refs.
Growing	Growing of phenyl fragment to a rigid moiety, *i.e.,* naphthyl gave two additional pi-pi interactions and an improved binding affinity towards the enzyme, B-specific-1-3-galactosyltransferase (PDB ID: 3U0X) from 800 μM to 271 μM. On the other hand, increased flexibility might worsen the binding affinity because of the retributed entropy. In terms of fragment growing approach, structural information should always be considered.	[172]
Merging	When some fragments' superimposing moieties bound to the protein responsible for the repression of transcription of the monooxygenase enzyme that was dependent on flavin using the fragment merging approach, the potency of inhibitors of the same enzyme got increased by two-fold.	[173]

Fragment Linking

It is one of the most conceptually simple methods, where two non-competitive fragments are linked through a chemical spacer or linker, and these two fragments fit in two separate subpockets of the same protein binding site. In spite of being attractive due to having the capacity of improving the ligand potency, still, the linker design with perfect flexibility as well as optimal and suitable orientation without hampering the original binding modes of the two different fragments makes it one of the most difficult fragment optimization strategies. Restricting the linked small molecule to two distinct fragments conformationally through variation of the linker's degree of rigidity may be one of the strategies of fragment optimization in case of linker identification [167]. Chung *et al.* reported that changing of oxime linkers to mono- and di-mines affects the rigidity of the final molecule product as well as protein binding. Moreover, in the case of fragment linking, the addition of rotatable bonds to the spacer system [52, 168] also often leads to unfavourable ADMET properties such as poorer permeability [169].

Fragment Merging

In this strategy, two different fragments partly bind to the common regions of the protein, making these fragments partially competitive towards each other w. r. t. to the protein binding site. In these cases, the common or overlapping portions of the fragments make a nucleus where two uncommon portions come. Fragment merging is considered a simpler and easier approach than fragment linking as it doesn't require the suitable design of linkers to join two separate fragments together [170, 171]. However, it possesses one major disadvantage of depending on the structural data of high quality for going to the later processes of fragment optimization [4]. As a result, fragment merging is considered highly related to the molecular hybridization approach in the domain of medicinal chemistry for novel molecule design through improving the drug potency *via* fusing the structures of other active compounds (Table **10**).

In silico Strategies for Fragment-to-ligand Optimization

Hotspot Analysis and Pocket Druggability Prediction

The knowledge of binding affinity and residue analysis plays a crucial role in the drug design process, but in the case of drug targeting PPIs, it becomes challenging to spot where exactly the drug binds. To overcome such challenges, hotspot analysis is an emerging strategy. Hotspots are generally termed as a small subset of residues present in a macromolecule's interface and are major contributors to binding affinity. There are different platforms for identification of hotspot surfaces of the target protein, one of the very commonly used among them is FTMap webserver [174]. It contains an algorithm inbuilt with 16 different organic probes of various shapes, sizes, and polarity to find a favorable position for each probe. Each of these probes remains clustered on the protein surface and the regions where these clusters overlap are called the consensus sites or hotspot areas on the protein surface. These hotspot areas are ranked based upon the number of numbers of overlapping probe clusters found in a particular area. The highly ranked hotspot site is where the fragment hit bind. FTMAPS have been well acknowledged to screen hotspot areas for various protein oncogenic B-RAF kinase, the target of the first marketed drug with fragment-based drug design, vemurafenib. During fragment screening, the fragment hits can bind to different sites of the protein. If the binding site is not well defined, researchers can use the pocket druggability prediction to move forward in F2L with the most druggable site able to accommodate ligands orally bioavailable. There are many available methods for predicting pocket druggability and these are well-described and reviewed elsewhere [175] (Fig. **5**).

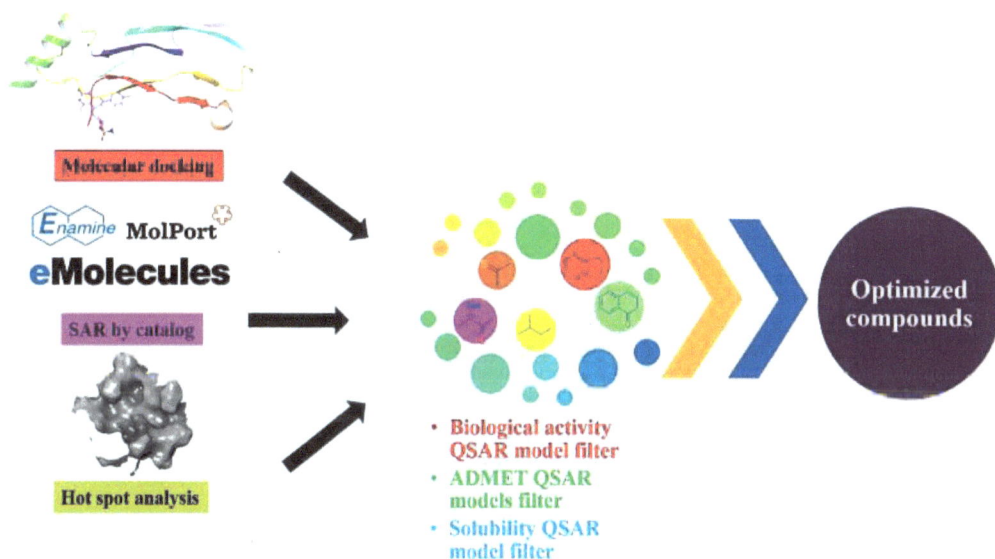

Fig. (5). Virtual screening workflow.

SAR Catalogue

It is one of the most efficient and fastest approaches for fragment-to--to-lead optimization. This approach is based on filtering the fragments from in-house database on the basis of ligand-based pharmacophore, shape-based, structure base, and fingerprint-based similarity. Several databases like ZINC, MolPort, and eMolecules are very often preferred for this purpose which contain in-built details about the commercial availability of the compound too. There are some target-based databases, *i.e.,* Chemdrive, ChemBridge, and Enamine, which contain a set of different fragments focusing a particular targets. Thus, using SAR -b--catalogue approach, we can avail information about the series of fragments that can be optimized for a particular target [175].

Molecular Docking

Molecular docking is algorithm-based computational approach used to predict the position, orientation, and binding scores of various ligands to the protein. This approach is used with other techniques to convert potential fragments into higher affinity ligands. Molecular docking and SAR by catalogue approach can be used in combination as a successful technique to select compounds that maintain the fragment hit binding mode while the binding energy is optimized. Thus, applying molecular docking, large compound datasets are efficiently assessed, and a small subset of the most promising compounds can be selected by binding modes and scores for experimental testing. The SAR catalogue is limited to cover only a finite number of chemical compounds that are commercially available, in this case, it is easy to generate virtual catalogues that resemble to a potential active fragments, which is synthetically feasible. Further, using the molecular docking approach, these virtually designed compounds can be evaluated for their binding affinity and experimental activity [175].

Machine Learning and Deep Learning

Structure based approaches optimizes fragments of high-affinity ligands by analyzing their electronic and steric constraints at the binding site, however the designed compounds may sometimes have poor synthetic feasibility or poor pharmacokinetic (absorption, distribution, metabolism and excretion) properties or toxic properties. These days several machine learning models are available having inbuilt advanced hardware like Graphical Processing Unit (GPU). Its large storage capacity syncs all theoretical models of ML to broaden its practical application in drug discovery. Quantitative structure-activity relationship (QSAR) model is a machine learning approach which applies the mathematical correlation between the molecular features of the compound (molecular weight, size, density, no. of rotatable bonds) and its therapeutic properties, *i.e.,* active, inactive, toxic, and nontoxic. Hence, a machine learning based approach can be utilised for the prediction and evaluation of different parameters of a potential fragment like solubility, biological activity, ADMET parameter, and synthetic feasibility [175].

DNDD

De novo design identifies a new chemical entity from scratch. This approach generates new fragments based on structural knowledge about the binding site (120). From scratch is referred to the phenomenon that these novel chemical entities are either formed by fragment growing or linking [51, 176], and due to these methods, *in silico* approach plays a significant role in FBDD [176].

Novel Molecules Generating Software for The Binding Pocket of Protein's Binding Site

The DNDD software design modified derivatives or analogues of previously known fragments by employing the experimentally or computationally derived binding modes of the fragments that were performed previously. Some of these software are described in the following table (Table **11**).

Table 11. List of software available for DNDD.

Type	Name	Explanation	Refs.
Novel molecules generating software for the binding pocket of protein's binding site	LUDI	Via the calculation of several sites of interaction present in the particular protein, and mapping fragment connection strategies can be employed utilizing linkers by empirical function.	[29]
Novel molecules generating softwares for the binding pocket of protein's binding site	Evolutionary algorithms	They are used when a large library of chemicals are considered.	[179]
Novel molecules generating softwares for the binding pocket of protein's binding site	GANDI	It uses the fragment connecting strategy using several linkers or bridges of the fragments that have already been docked by the aid of Tabu search and genetic algorithm. In essence, this methodology combines meas6rements of similarity along with force-field calculations.	[51]
Novel molecules generating softwares for the binding pocket of protein's binding site	BREED	In this fragment merging approach of Schrödinger, Inc., two small molecules or ligands' coordinated are overlapped as well as the superimposition of the three dimensionally, followed by the combination of these fragments which is known as shuffling of fragments.	[180]

(Table 11) cont.....

Type	Name	Explanation	Refs.
Novel molecules generating software for the binding pocket of protein's binding site	LigBuilder	It is a genetic algorithm-based methodology that utilizes fragment linking and growing strategies to develop compounds from the fragments which are made of organic chemicals. Its updated version, *i.e.,* the second one, can also determine the synthetic feasibility on the basis of the retrosynthetic approach along with a library of numerous chemical routes.	[35, 181]
Novel molecules generating softwares for the binding pocket of protein's binding site	Autogrow	It is also a genetic algorithm-based approach that utilizes a fragment growing strategy on the main moiety. At first, a single fragment is developed and docked against a particular receptor on the basis of which its best conformation is selected for the further development of the series of fragments. In its updated version, the drug-likeness score, as well as the synthetic feasibility, is also taken into account.	[8, 182]
Novel molecules generating softwares for the binding pocket of protein's binding site	ADAPT	It is also a genetic algorithm-based fragment growing approach that filters the best fits on the basis of the scores obtained in molecular docking as well as the ligand interactions with the proteins. Here, at first, the molecules are repetitively developed to fulfil the value of the target.	[62, 179]
Prediction of pharmacokinetic property	vNN	It is an open-source online platform for predicting the ADMET properties of compounds that are built on the basis of k-NN or k-nearest neighbour, where the structural similarities of the molecules are considered.	[183]
Prediction of pharmacokinetic property	Pred-hERG	This application is available online for the prediction of inhibitors and non-inhibitors of hERG channels. It can also identify significant anti-arrhythmic drugs. Its latest available (http://labmol.com.br/predherg/) version (4.2) works using the ChEMBL database version 23, which works on the basis of a random forest model (RF), making it more robust.	[184 - 187]
Prediction of pharmacokinetic property	admetSAR 2.0	This online approach is made on the basis of three distinct algorithms, namely, kNN, SVM, and RF, where a total of twenty-seven endpoints along with an ecotoxicity constraint are utilized. A scaffold hopping optimization tool, named ADMETopt is also included in it.	[178, 188]
Prediction of pharmacokinetic property	ADMETlab	With the help of this online available tool, the evaluation of pharmacokinetic and toxic properties are evaluated, and similarity scores (substructure searching) and drug likeness scores are predicted by utilizing five different algorithms, viz, DT (decision tree), SVM, native bias (NB), RF, and RP (recursive partitioning). Here, a total of thirty-one endpoints are employed.	[177]

(Table 11) cont.....

Type	Name	Explanation	Refs.
Prediction of pharmacokinetic property	SwissADME	It is one of the most promising and most utilized freely available online servers for pharmacokinetic and physicochemical property prediction along with the evaluation of drug likeness, too. Its latest version also provides the score of synthetic accessibility.	[189]
Prediction of pharmacokinetic property	QikProp	It is another most promising and most utilized module supplied by Schrödinger, Inc. for quick prediction of pharmacokinetic and physicochemical properties.	[190]
Prediction of synthesizability	SYLVIA	It predicts the synthetic accessibility score of an organic chemical on the basis of three structural variables, namely, complexity of stereochemistry, complexity of ring moieties, and complexity of the graphical representations of the generated molecules; and two variables consisting of database and library of reactions and reactants, namely, substructure of the reaction center and similitude of the reactants.	[191]
Prediction of synthesizability	-	This method follows the fragment combining strategy along with the consideration of the penalty of the complexity of the compounds. From the previously available chemicals' synthetic routes and their complexities, a method for the prediction of synthesizability is developed.	[122]
Prediction of synthesizability	-	Two synthesizability predicting methodologies are developed by employing SVMs on different descriptors of molecules. Among them, the first one includes retrosynthesis-based SVM (RSsvm) in which retrosynthetically derived molecules are identified and their reactants as well as the synthetic routes, are optimized. On the other hand, in the second approach, *i.e.,* the direct synthesis-based SVM (DRSVM), the forward synthetic routes for the compounds should be optimized.	[192]
Prediction of synthesizability	-	On the basis of marketed libraries of chemicals and their descriptors, a novel approach for the prediction of synthesizability is developed, where the symmetrical atoms' numbers, the complexity of the molecular graphs, the possibility for any molecules' structural analogues presence, and any molecules' chiral centers' amounts are considered.	[193]
Prediction and awareness of synthesizability	Lead+Op, and LeadOp+R	These algorithms are examples of fragment growing strategies, where at first, the synthetic routes linked to the beginning fragment are searched for. The routes can yield the conformers of synthetic products virtually, among which the best one should be selected, and from it, the other reactants are developed.	[194]
Strategy for DNDD	AUTO T and T	It is a one of a fragment merging strategy with the help of which the fragments are converted to full-length ligand molecules.	-

Pharmacokinetic Property Prediction of The Novel Compounds

Chemical and geometric constraints obtained from the native ligand of the target protein are considered as the primary constraints, while synthetic accessibility (SA) and ADMET properties are considered the secondary constraints in the case of FBDD. However, for the primary constraints, the internal constraints of the lead small molecule are required to be constructed, and for this reason, several of the compounds fail in clinical trials [177]. There are plenty of software and online tools/webservers available for predicting the ADMET properties and synthetic accessibility. However, many of them show disadvantages because of being expensive and a narrow coverage of chemical space [178]. The current predicting tools are mostly ML-based methods, such as tree-based approach, random forest (RF) method, and support vector machine (SVM)-based method, *etc*. (Table **11**).

Prediction of Synthesizability with The Novel Compounds

Several lead molecules obtained by DNDD process are not synthetically feasible (Dey and Caflisch, 2008), and hence, methods for determining the synthetic accessibility (SA) are being developed. This method employs the optimized lead molecule's complexity or its retrosynthetic pathway, where the entire synthetic route of the leads is required to be processed [122]. Among several SA predicting tools, some are listed in Table (**11**).

Synthesizability-aware Methods

As most of the lead compounds generated *via* FBDD-DNDD approach are challenging to synthesize, some of the tools have added methods to calculate the SA score (Table **11**).

Case Studies

In this section, some of the last 5 years' drugs developed *via* the FBDD approach are tabulated in Table (**12**).

PROTAC AND MOLECULAR GLUE

Inhibitors and recruiters of ubiquitin ligase (E3 ligases) are utilized for targeted protein degradation, including proteolysis-targeting chimeras (PROTACs) [200]. Ubiquitination governs most cellular activities, including protein breakdown, homeostasis, cell cycle regulation, and immunological signalling. PROTACs are heterobifunctional small molecules made up of two ligands linked together by a linker: one recruits and binds a protein of interest (POI), while the other recruits and binds an E3 ubiquitin ligase [201]. The PROTACs simultaneous binding of the POI and ligase produces ubiquitylation of the POI and its subsequent breakdown by the ubiquitin-proteasome system, after which the PROTAC is recycled to target another copy of the POI. PROTACs are distinguished from traditional inhibitors by their catalytic-type mode of action and event-driven pharmacology [202]. Traditional inhibitors have a one-to-one connection with the POI and their pharmacology is driven by stoichiometry and, in most cases, interacts with a catalytic site. FBDD can discover binding hot spots in E3 ligases pockets more effectively than typical small molecule screens. After then, fragments may be joined together to form molecules with increased binding affinities. Covalent FBDD has lately prompted the development of new ligands for E3 ligases. Covalent fragment screening assists in finding ligand shallow E3 ligase PPI sites. Drug-like leads with adequate physicochemical qualities have yet to be documented, and most studies involve reactive electrophilic warheads. PPIs control all biological functions. Small compounds promote protein proximity to modify PPIs [203]. Molecular glue degraders, monovalent chemicals, organize contacts between a target protein and an E3 ubiquitin ligase, causing proteasomal degradation of the former. Molecular glue degraders may destroy unligandable proteins by manipulating PPIs, expanding the targetable proteome in new ways [204]. However, insufficient knowledge of regulating parameters makes molecular glue design challenging [205]. Different types of strategies can be implemented in the designing of small molecules for the PROTAC and molecular glue with the help of the AI, and DRI.

Table 12. Case studies.

Drug Type	Disease	Target	Approach	Refs.
Anti-mycobacterial agents	Tuberculosis (TB)	*Mycobacterium tuberculosis* (Mtb), and *Mycobacterium abscessus* (*Mab*) strains. In particular, *Mtb* hydrolase that is responsible for the metabolism of cholesterol	Soaking	[195]
Anti-mycobacterial agents	TB	*Mtb* pantothenate synthetase	Growing and linking	[167, 196]
Anti-mycobacterial agents	Principally, cystic fibrosis and leprosy	*Mab* tRNA methyltransferase, and *Mycobacterium leprae*	Merging	[197]
Anti-Dengue agents	Dengue viral disease	Dengue virus (DENV) NS5 mRNA Methyltransferase and NS3 helicase	-	[147]
Anti-Dengue agents	Dengue viral disease	DENV NS5 mRNA Methyltransferase S-Adenosyl-L-methionine	Linking	[147]

(Table 12) cont.....

Drug Type	Disease	Target	Approach	Refs.
Anticancer agents	Cancer	Human mutT homolog 1	Combination of SAR by catalogue, docking, and wet lab experiments	[198]
Anticholinergic agents	Alzheimer's Disease	Human acetylcholinesterase	Cavity and Build tools of LigBuilder, QikProp, Glide, and Induced fit docking tools of Schrödinger	[199]

CONCLUSION

FBDD along with DNDD has become one of the most significant and popular strategies in the area of drug design and development in the pharmaceutical industry as well as academia as it possesses a number of attractive advantages such as its scalable and simpler implementation, lesser need for large chemical space, and the possibility to employ a vast range of biophysical methodologies in case of screening against targets. This approach offers more consistent and efficient ligand optimization during the screening of hit-to-lead identification compared to the conventional HTS approach of designing a larger, drug-like molecule. A range of *in silico* methodologies (prediction of ADMET, binding site analysis, and determination of synthetic accessibility) have been quite useful and can be combined and integrated with experimental wet lab techniques for F2L identification and optimization, so that all fragments can show higher affinity towards the biological target. In the case of DNDD, several advancements in AI-based approaches such as ML- and DL-based methods have been quite helpful in F2L optimization, and finally reaching into the lead compound.

REFERENCES

[1]   Lamoree B, Hubbard RE. Current perspectives in fragment-based lead discovery (FBLD). Essays Biochem 2017; 61(5): 453-64.
      [http://dx.doi.org/10.1042/EBC20170028] [PMID: 29118093]

[2]   Hall RJ, Mortenson PN, Murray CW. Efficient exploration of chemical space by fragment-based screening. Prog Biophys Mol Biol 2014; 116(2-3): 82-91.
      [http://dx.doi.org/10.1016/j.pbiomolbio.2014.09.007] [PMID: 25268064]

[3]   Thomas SE, Mendes V, Kim SY, *et al.* Structural biology and the design of new therapeutics: From HIV and cancer to mycobacterial infections. J Mol Biol 2017; 429(17): 2677-93.
      [http://dx.doi.org/10.1016/j.jmb.2017.06.014] [PMID: 28648615]

[4]   Davis BJ, Roughley SD. Fragment-based lead discovery.Ann Rep Med Chem. Elsevier 2017; 50: pp. 371-439.
      [http://dx.doi.org/10.1016/bs.armc.2017.07.002]

[5]   Erlanson DA, Fesik SW, Hubbard RE, Jahnke W, Jhoti H. Twenty years on: The impact of fragments on drug discovery. Nat Rev Drug Discov 2016; 15(9): 605-19.
[http://dx.doi.org/10.1038/nrd.2016.109] [PMID: 27417849]

[6]   Fragment   screening   by   ligand   observed   nmr | Bruker.   Available   from:
https://www.bruker.com/en/resources/library/application-note-
-mr/fragment-screening-by-ligand-observed-nmr.html (cited 2022 Dec 20).

[7]   Spiegel JO, Durrant JD. AutoGrow4: An open-source genetic algorithm for *de novo* drug design and lead optimization. J Cheminform 2020; 12(1): 25.
[http://dx.doi.org/10.1186/s13321-020-00429-4] [PMID: 33431021]

[8]   Durrant JD, Amaro RE, McCammon JA. AutoGrow: A novel algorithm for protein inhibitor design. Chem Biol Drug Des 2009; 73(2): 168-78.
[http://dx.doi.org/10.1111/j.1747-0285.2008.00761.x] [PMID: 19207419]

[9]   Yuan Y, Pei J, Lai L. LigBuilder V3: A multi-target *de novo* drug design approach. Front Chem 2020; 8(February): 142.
[http://dx.doi.org/10.3389/fchem.2020.00142] [PMID: 32181242]

[10]   Schneider P, Schneider G. *De novo* design at the edge of chaos. J Med Chem 2016; 59(9): 4077-86.
[http://dx.doi.org/10.1021/acs.jmedchem.5b01849] [PMID: 26881908]

[11]   Devi RV, Sathya SS, Coumar MS. Evolutionary algorithms for *de novo* drug design : A survey. Appl Soft Comput 2015; 27: 543-52. [Internet].
[http://dx.doi.org/10.1016/j.asoc.2014.09.042]

[12]   Schneider G, Fechner U. Computer-based *de novo* design of drug-like molecules. Nat Rev Drug Discov 2005; 4(8): 649-63.
[http://dx.doi.org/10.1038/nrd1799] [PMID: 16056391]

[13]   Nicolaou C, Kannas C, Loizidou E. Multi-objective optimization methods in *de novo* drug design. Mini Rev Med Chem 2012; 12(10): 979-87.
[http://dx.doi.org/10.2174/138955712802762284] [PMID: 22420573]

[14]   Nicolaou CA, Brown N. Multi-objective optimization methods in drug design. Drug Discov Today Technol 2013; 10(3): e427-35.
[http://dx.doi.org/10.1016/j.ddtec.2013.02.001] [PMID: 24050140]

[15]   dos Santos Nascimento IJ, da Silva Rodrigues ÉE, da Silva MF, de Araújo-Júnior JX, de Moura RO. Advances in computational methods to discover new NS2B-NS3 inhibitors useful against dengue and zika viruses. Curr Top Med Chem 2022; 22(29): 2435-62.
[http://dx.doi.org/10.2174/1568026623666221122121330] [PMID: 36415099]

[16]   Perez-Castillo Y, Sánchez-Rodríguez A, Tejera E, *et al.* A desirability-based multi objective approach for the virtual screening discovery of broad-spectrum anti-gastric cancer agents. PLoS One 2018; 13(2): e0192176.
[http://dx.doi.org/10.1371/journal.pone.0192176] [PMID: 29420638]

[17]   Sánchez-Rodríguez A, Pérez-Castillo Y, Schürer SC, *et al.* From flamingo dance to (desirable) drug discovery: A nature-inspired approach. Drug Discov Today 2017; 22(10): 1489-502.
[http://dx.doi.org/10.1016/j.drudis.2017.05.008] [PMID: 28624633]

[18]   Nascimento IJ dos S, Mendonça de Aquino T, Ferreira da Silva-Júnior E. Molecular dynamics applied to discover antiviral agents. Disco Antiv Agent 2022; 7: 62-131.

[19]   dos Santos Nascimento IJ, da Silva Santos-Júnior PF, de Araújo-Júnior JX, da Silva-Júnior EF. Strategies in medicinal chemistry to discover new hit compounds against ebola virus: Challenges and perspectives in drug discovery. Mini Rev Med Chem 2022; 22(22): 2896-924.
[http://dx.doi.org/10.2174/1389557522666220404085858] [PMID: 35379146]

[20]   Cruz-Monteagudo M, Borges F, Cordeiro MNDS. Desirability-based multiobjective optimization for

global QSAR studies: Application to the design of novel NSAIDs with improved analgesic, antiinflammatory, and ulcerogenic profiles. J Comput Chem 2008; 29(14): 2445-59.
[http://dx.doi.org/10.1002/jcc.20994] [PMID: 18452123]

[21] Chattaraj B, Nandi A, Das A, *et al.* Inhibitory activity of *Enhydra fluctuans* Lour. on calcium oxalate crystallisation through *in silico* and *in vitro* studies. Front Pharmacol 2023; 13(January): 982419.
[http://dx.doi.org/10.3389/fphar.2022.982419] [PMID: 36744215]

[22] Nandi A, Das A, Dey YN, Roy KK. The abundant phytocannabinoids in rheumatoid arthritis: Therapeutic targets and molecular processes identified using integrated bioinformatics and network pharmacology. Life 2023; 13(3): 700.
[http://dx.doi.org/10.3390/life13030700] [PMID: 36983855]

[23] Chattaraj B, Nandi A, Das A, *et al. Enhydra fluctuans* Lour. aqueous extract inhibited the growth of calcium phosphate crystals: An *in vitro* study. Food Chem Adv 2023; 2: 100287.
[http://dx.doi.org/10.1016/j.focha.2023.100287]

[24] Chattaraj B, Khanal P, Nandi A, *et al.* Network pharmacology and molecular modelling study of *Enhydra fluctuans* for the prediction of the molecular mechanisms involved in the amelioration of nephrolithiasis. J Biomol Struct Dyn 2023; 1-11.
[http://dx.doi.org/10.1080/07391102.2023.2189476] [PMID: 36914227]

[25] Madan A, Garg M, Satija G, *et al.* SAR based review on diverse heterocyclic compounds with various potential molecular targets in the fight against COVID-19: A medicinal chemist perspective. Curr Top Med Chem 2023; 23.
[PMID: 36703601]

[26] Berman HM, Westbrook J, Feng Z, *et al.* The protein data bank. Nucleic Acids Res 2000; 28(1): 235-42.
[http://dx.doi.org/10.1093/nar/28.1.235] [PMID: 10592235]

[27] Burley SK, Berman HM, Bhikadiya C, *et al.* RCSB protein data bank: Biological macromolecular structures enabling research and education in fundamental biology, biomedicine, biotechnology and energy. Nucleic Acids Res 2019; 47(D1): D464-74.
[http://dx.doi.org/10.1093/nar/gky1004] [PMID: 30357411]

[28] Danziger DJ, Dean PM. Automated site-directed drug design: A general algorithm for knowledge acquisition about hydrogen-bonding regions at protein surfaces. Proc R Soc Lond B Biol Sci 1989; 236(1283): 101-13.
[http://dx.doi.org/10.1098/rspb.1989.0015] [PMID: 2565575]

[29] Böhm HJ. LUDI: Rule-based automatic design of new substituents for enzyme inhibitor leads. J Comput Aided Mol Des 1992; 6(6): 593-606.
[http://dx.doi.org/10.1007/BF00126217] [PMID: 1291628]

[30] Clark DE, Frenkel D, Levy SA, *et al.* PRO_LIGAND: An approach to *de novo* molecular design. 1. Application to the design of organic molecules. J Comput Aided Mol Des 1995; 9(1): 13-32.
[http://dx.doi.org/10.1007/BF00117275] [PMID: 7751867]

[31] Waszkowycz B, Clark DE, Frenkel D, *et al.* PRO_LIGAND: An approach to *de novo* molecular design. 2. Design of novel molecules from molecular field analysis (MFA) models and pharmacophores. J Med Chem 1994; 37(23): 3994-4002.
[http://dx.doi.org/10.1021/jm00049a019] [PMID: 7966160]

[32] Gillet VJ, Myatt G, Zsoldos Z, Johnson AP. SPROUT, HIPPO and CAESA: Tools for *de novo* structure generation and estimation of synthetic accessibility. Perspect Drug Discov Des 1995; 3(1): 34-50.
[http://dx.doi.org/10.1007/BF02174466]

[33] Bohacek RS, McMartin C. Multiple highly diverse structures complementary to enzyme binding sites: Results of extensive application of a *de novo* design method incorporating combinatorial growth. J Am Chem Soc 1994; 116(13): 5560-71.

[http://dx.doi.org/10.1021/ja00092a006]

[34]   Nishibata Y, Itai A. Automatic creation of drug candidate structures based on receptor structure. Starting point for artificial lead generation. Tetrahedron 1991; 47(43): 8985-90.
[http://dx.doi.org/10.1016/S0040-4020(01)86503-0]

[35]   Wang R, Gao Y, Lai L. LigBuilder: A multi-purpose program for structure-based drug design. J Mol Model 2000; 6(7-8): 498-516.
[http://dx.doi.org/10.1007/s0089400060498]

[36]   Miranker A, Karplus M. Functionality maps of binding sites: A multiple copy simultaneous search method. Proteins 1991; 11(1): 29-34.
[http://dx.doi.org/10.1002/prot.340110104] [PMID: 1961699]

[37]   Eisen MB, Wiley DC, Karplus M, Hubbard RE. HOOK: A program for finding novel molecular architectures that satisfy the chemical and steric requirements of a macromolecule binding site. Proteins 1994; 19(3): 199-221.
[http://dx.doi.org/10.1002/prot.340190305] [PMID: 7937734]

[38]   Muhammed MT, Aki-Yalcin E. Homology modeling in drug discovery: Overview, current applications, and future perspectives. Chem Biol Drug Des 2019; 93(1): 12-20.
[http://dx.doi.org/10.1111/cbdd.13388] [PMID: 30187647]

[39]   Luo Z, Wang R, Lai L. RASSE: A new method for structure-based drug design. J Chem Inf Comput Sci 1996; 36(6): 1187-94.
[http://dx.doi.org/10.1021/ci950277w] [PMID: 8941995]

[40]   Pearlman DA, Murcko MA. CONCERTS: Dynamic connection of fragments as an approach to *de novo* ligand design. J Med Chem 1996; 39(8): 1651-63.
[http://dx.doi.org/10.1021/jm950792l] [PMID: 8648605]

[41]   Patel S, Das A, Meshram P, *et al.* Pyruvate kinase M2 in chronic inflammations: A potpourri of crucial protein–protein interactions. Cell Biol Toxicol 2021; 37(5): 653-78.
[http://dx.doi.org/10.1007/s10565-021-09605-0] [PMID: 33864549]

[42]   Chowdhury A, Patel S, Sharma A, Das A, Meshram P, Shard A. A perspective on environmentally benign protocols of thiazole synthesis. Chem Heterocycl Compd 2020; 56(4): 455-63.
[http://dx.doi.org/10.1007/s10593-020-02680-x]

[43]   Zhu J, Fan H, Liu H, Shi Y. Structure-based ligand design for flexible proteins: Application of new F-DycoBlock. J Comput Aided Mol Des 2001; 15(11): 979-96.
[http://dx.doi.org/10.1023/A:1014817911249] [PMID: 11989626]

[44]   Makhal PN, Nandi A, Kaki VR. Insights into the recent synthetic advances of organoselenium compounds. ChemistrySelect 2021; 6(4): 663-79.
[http://dx.doi.org/10.1002/slct.202004029]

[45]   Zhu J, Yu H, Fan H, Liu H, Shi Y. Design of new selective inhibitors of cyclooxygenase-2 by dynamic assembly of molecular building blocks. J Comput Aided Mol Des 2001; 15(5): 447-63.
[http://dx.doi.org/10.1023/A:1011114307711] [PMID: 11394738]

[46]   Gaulton A, Hersey A, Nowotka M, *et al.* The ChEMBL database in 2017. Nucleic Acids Res 2017; 45(D1): D945-54.
[http://dx.doi.org/10.1093/nar/gkw1074] [PMID: 27899562]

[47]   Afantitis A, Melagraki G, Koutentis PA, Sarimveis H, Kollias G. Ligand - based virtual screening procedure for the prediction and the identification of novel β-amyloid aggregation inhibitors using kohonen maps and counterpropagation artificial neural networks. Eur J Med Chem 2011; 46(2): 497-508.
[http://dx.doi.org/10.1016/j.ejmech.2010.11.029] [PMID: 21167625]

[48]   Hartenfeller M, Zettl H, Walter M, *et al.* DOGS: Reaction-driven *de novo* design of bioactive compounds. PLOS Comput Biol 2012; 8(2): e1002380.

[http://dx.doi.org/10.1371/journal.pcbi.1002380] [PMID: 22359493]

[49]   Vinkers HM, de Jonge MR, Daeyaert FFD, *et al.* SYNOPSIS: Synthesize and optimize system *in silico.* J Med Chem 2003; 46(13): 2765-73.
[http://dx.doi.org/10.1021/jm030809x] [PMID: 12801239]

[50]   Schneider G, Lee ML, Stahl M, Schneider P. *De novo* design of molecular architectures by evolutionary assembly of drug-derived building blocks. J Comput Aided Mol Des 2000; 14(5): 487-94.
[http://dx.doi.org/10.1023/A:1008184403558] [PMID: 10896320]

[51]   Dey F, Caflisch A. Fragment-based *de novo* ligand design by multi-objective evolutionary optimization. Supporting Information. J Chem Inf Model 2008; 48(3): 679-90.
[http://dx.doi.org/10.1021/ci700424b] [PMID: 18307332]

[52]   Ichihara O, Barker J, Law RJ, Whittaker M. Compound design by fragment-linking. Mol Inform 2011; 30(4): 298-306.
[http://dx.doi.org/10.1002/minf.201000174] [PMID: 27466947]

[53]   Schneider G. Future *de novo* drug design. Mol Inform 2014; 33(6-7): 397-402.
[http://dx.doi.org/10.1002/minf.201400034] [PMID: 27485977]

[54]   Lipinski CA, Lombardo F, Dominy BW, Feeney PJ. Experimental and computational approaches to estimate solubility and permeability in drug discovery and development settings. Adv Drug Deliv Rev 2012; 64 (Suppl.): 4-17.
[http://dx.doi.org/10.1016/j.addr.2012.09.019] [PMID: 11259830]

[55]   Aronov A. Predictive in silico modeling for hERG channel blockers. Drug Discov Today 2005; 10(2): 149-55.
[http://dx.doi.org/10.1016/S1359-6446(04)03278-7] [PMID: 15718164]

[56]   Böhm HJ. The computer program LUDI: A new method for the *de novo* design of enzyme inhibitors. J Comput Aided Mol Des 1992; 6(1): 61-78.
[http://dx.doi.org/10.1007/BF00124387] [PMID: 1583540]

[57]   Gillet VJ, Newell W, Mata P, *et al.* SPROUT: Recent developments in the *de novo* design of molecules. J Chem Inf Comput Sci 1994; 34(1): 207-17.
[http://dx.doi.org/10.1021/ci00017a027] [PMID: 8144711]

[58]   McGarrah DB, Judson RS. Analysis of the genetic algorithm method of molecular conformation determination. J Comput Chem 1993; 14(11): 1385-95.
[http://dx.doi.org/10.1002/jcc.540141115]

[59]   Clark DE, Westhead DR. Evolutionary algorithms in computer-aided molecular design. J Comput Aided Mol Des 1996; 10(4): 337-58.
[http://dx.doi.org/10.1007/BF00124503] [PMID: 8877705]

[60]   Masek BB, Baker DS, Dorfman RJ, *et al.* Multistep reaction based *de novo* drug design: Generating synthetically feasible design ideas. J Chem Inf Model 2016; 56(4): 605-20.
[http://dx.doi.org/10.1021/acs.jcim.5b00697] [PMID: 27031173]

[61]   Douguet D, Thoreau E, Grassy G. A genetic algorithm for the automated generation of small organic molecules: Drug design using an evolutionary algorithm. J Comput Aided Mol Des 2000; 14(5): 449-66.
[http://dx.doi.org/10.1023/A:1008108423895] [PMID: 10896317]

[62]   Pegg SCH, Haresco JJ, Kuntz ID. A genetic algorithm for structure-based *de novo* design. J Comput Aided Mol Des 2001; 15(10): 911-33.
[http://dx.doi.org/10.1023/A:1014389729000] [PMID: 11918076]

[63]   Budin N, Ahmed S, Majeux N, Caflisch A. An evolutionary approach for structure-based design of natural and non-natural peptidic ligands. Comb Chem High Throughput Screen 2001; 4(8): 661-73.
[http://dx.doi.org/10.2174/1386207013330652] [PMID: 11812261]

[64] Douguet D, Munier-Lehmann H, Labesse G, Pochet S. LEA3D: A computer-aided ligand design for structure-based drug design. J Med Chem 2005; 48(7): 2457-68.
[http://dx.doi.org/10.1021/jm0492296] [PMID: 15801836]

[65] Barigye SJ, García de la Vega JM, Perez-Castillo Y. Generative adversarial networks (GANs) based synthetic sampling for predictive modeling. Mol Inform 2020; 39(10): 2000086.
[http://dx.doi.org/10.1002/minf.202000086] [PMID: 32558335]

[66] Fechner U, Schneider G. Flux (1): A virtual synthesis scheme for fragment-based *de novo* design. J Chem Inf Model 2006; 46(2): 699-707.
[http://dx.doi.org/10.1021/ci0503560] [PMID: 16563000]

[67] Schüller A, Suhartono M, Fechner U, *et al.* The concept of template-based *de novo* design from drug-derived molecular fragments and its application to TAR RNA. J Comput Aided Mol Des 2008; 22(2): 59-68.
[http://dx.doi.org/10.1007/s10822-007-9157-4] [PMID: 18064402]

[68] Nicolaou CA, Apostolakis J, Pattichis CS. *De novo* drug design using multiobjective evolutionary graphs. J Chem Inf Model 2009; 49(2): 295-307.
[http://dx.doi.org/10.1021/ci800308h] [PMID: 19434831]

[69] Wong SSY, Weimin Luo , Chan KCC, Evo MD. EvoMD: An algorithm for evolutionary molecular design. IEEE/ACM Trans Comput Biol Bioinformatics 2011; 8(4): 987-1003.
[http://dx.doi.org/10.1109/TCBB.2010.100] [PMID: 20876937]

[70] Klambauer G, Hochreiter S, Rarey M. Machine learning in drug discovery. J Chem Inf Model 2019; 59(3): 945-6.
[http://dx.doi.org/10.1021/acs.jcim.9b00136] [PMID: 30905159]

[71] Eck D, Schmidhuber J. Finding temporal structure in music: Blues improvisation with LSTM recurrent networks. neural networks for signal processing. Proceedings of the IEEE Workshop,. 2002;2002-Janua:747–56.

[72] Kawakami K. Supervised sequence labelling with recurrent neural networks. Technical University of Munich 2008.

[73] Srivastava N, Mansimov E, Salakhutdinov R. Unsupervised learning of video representations using LSTMs. 32nd International Conference on Machine Learning, ICML 2015,. 843-52.

[74] Liu X, Ye K, van Vlijmen HWT, IJzerman AP, van Westen GJP. An exploration strategy improves the diversity of *de novo* ligands using deep reinforcement learning: A case for the adenosine A_{2A} receptor. J Cheminform 2019; 11(1): 35.
[http://dx.doi.org/10.1186/s13321-019-0355-6] [PMID: 31127405]

[75] David L, Thakkar A, Mercado R, Engkvist O. Molecular representations in AI-driven drug discovery: A review and practical guide. J Cheminform 2020; 12(1): 56.
[http://dx.doi.org/10.1186/s13321-020-00460-5] [PMID: 33431035]

[76] Weininger D. SMILES, a chemical language and information system. 1. Introduction to methodology and encoding rules. J Chem Inf Comput Sci 1988; 28(1): 31-6.
[http://dx.doi.org/10.1021/ci00057a005]

[77] Lipton ZC, Berkowitz J, Elkan C. A critical review of recurrent neural networks for sequence learning. arXiv 2015; 1506: 00019v4. Available from: http://arxiv.org/abs/1506.00019

[78] Pineda FJ. Generalization of back-propagation to recurrent neural networks. Phys Rev Lett 1987; 59(19): 2229-32.
[http://dx.doi.org/10.1103/PhysRevLett.59.2229] [PMID: 10035458]

[79] Merk D, Friedrich L, Grisoni F, Schneider G. *De Novo* design of bioactive small molecules by artificial intelligence. Mol Inform 2018; 37(1-2): 1700153.
[http://dx.doi.org/10.1002/minf.201700153] [PMID: 29319225]

[80]   Segler MHS, Kogej T, Tyrchan C, Waller MP. Generating focused molecule libraries for drug discovery with recurrent neural networks. ACS Cent Sci 2018; 4(1): 120-31.
[http://dx.doi.org/10.1021/acscentsci.7b00512] [PMID: 29392184]

[81]   Maragakis P, Nisonoff H, Cole B, Shaw DE. A deep-learning view of chemical space designed to facilitate drug discovery. J Chem Inf Model 2020; 60(10): 4487-96.
[http://dx.doi.org/10.1021/acs.jcim.0c00321] [PMID: 32697578]

[82]   Olivecrona M, Blaschke T, Engkvist O, Chen H. Molecular de-novo design through deep reinforcement learning. J Cheminform 2017; 9(1): 48.
[http://dx.doi.org/10.1186/s13321-017-0235-x] [PMID: 29086083]

[83]   Rifaioglu AS, Nalbat E, Atalay V, Martin MJ, Cetin-Atalay R, Doğan T. DEEPScreen: high performance drug–target interaction prediction with convolutional neural networks using 2-D structural compound representations. Chem Sci 2020; 11(9): 2531-57.
[http://dx.doi.org/10.1039/C9SC03414E] [PMID: 33209251]

[84]   Lecun Y, Bottou L, Bengio Y, Haffner P. Gradient-based learning applied to document recognition. Proc IEEE 1998; 86(11): 2278-324.
[http://dx.doi.org/10.1109/5.726791]

[85]   Sun M, Zhao S, Gilvary C, Elemento O, Zhou J, Wang F. Graph convolutional networks for computational drug development and discovery. Brief Bioinform 2020; 21(3): 919-35.
[http://dx.doi.org/10.1093/bib/bbz042] [PMID: 31155636]

[86]   LeCun Y, Bengio Y, Hinton G. Deep learning. Nature 2015; 521(7553): 436-44.
[http://dx.doi.org/10.1038/nature14539] [PMID: 26017442]

[87]   Yi X, Walia E, Babyn P. Generative adversarial network in medical imaging: A review. Med Image Anal 2019; 58: 101552.
[http://dx.doi.org/10.1016/j.media.2019.101552] [PMID: 31521965]

[88]   Gui J, Sun Z, Wen Y, Tao D, Ye J. A Review on Generative Adversarial Networks: Algorithms, Theory, and Applications. IEEE Trans Knowl Data Eng 2021; 14(8): 1-28.

[89]   Vanhaelen Q, Lin YC, Zhavoronkov A. The advent of generative chemistry. ACS Med Chem Lett 2020; 11(8): 1496-505.
[http://dx.doi.org/10.1021/acsmedchemlett.0c00088] [PMID: 32832015]

[90]   Popova M, Isayev O, Tropsha A. Deep reinforcement learning for *de novo* drug design. Sci Adv 2018; 4(7): eaap7885.
[http://dx.doi.org/10.1126/sciadv.aap7885] [PMID: 30050984]

[91]   Ståhl N, Falkman G, Karlsson A, Mathiason G, Boström J. Deep reinforcement learning for multiparameter optimization in *de novo* drug design. J Chem Inf Model 2019; 59(7): 3166-76.
[http://dx.doi.org/10.1021/acs.jcim.9b00325] [PMID: 31273995]

[92]   Yasonik J. Multiobjective *de novo* drug design with recurrent neural networks and nondominated sorting. J Cheminform 2020; 12(1): 14.
[http://dx.doi.org/10.1186/s13321-020-00419-6] [PMID: 33430996]

[93]   Pan SJ, Yang Q. A survey on transfer learning. IEEE Trans Knowl Data Eng 2010; 22(10): 1345-59.
[http://dx.doi.org/10.1109/TKDE.2009.191]

[94]   Gupta A, Müller AT, Huisman BJH, Fuchs JA, Schneider P, Schneider G. Generative recurrent networks for *de novo* drug design. Mol Inform 2018; 37(1-2): 1700111.
[http://dx.doi.org/10.1002/minf.201700111]

[95]   Kotsias PC, Arús-Pous J, Chen H, Engkvist O, Tyrchan C, Bjerrum EJ. Direct steering of *de novo* molecular generation with descriptor conditional recurrent neural networks. Nat Mach Intell 2020; 2(5): 254-65.
[http://dx.doi.org/10.1038/s42256-020-0174-5]

[96]   Kusner MJ, Paige B, Hemández-Lobato JM. Grammar variational autoencoder. 34th International Conference on Machine Learning, ICML 2017,. 3072-84.

[97]   Grisoni F, Moret M, Lingwood R, Schneider G. Bidirectional molecule generation with recurrent neural networks. J Chem Inf Model 2020; 60(3): 1175-83.
[http://dx.doi.org/10.1021/acs.jcim.9b00943] [PMID: 31904964]

[98]   Li Y, Hu J, Wang Y, Zhou J, Zhang L, Liu Z. DeepScaffold: A comprehensive tool for scaffold-based *de novo* drug discovery using deep learning. J Chem Inf Model 2020; 60(1): 77-91.
[http://dx.doi.org/10.1021/acs.jcim.9b00727] [PMID: 31809029]

[99]   Khemchandani Y, O'Hagan S, Samanta S, *et al.* DeepGraphMolGen, a multi-objective, computational strategy for generating molecules with desirable properties: A graph convolution and reinforcement learning approach. J Cheminform 2020; 12(1): 53.
[http://dx.doi.org/10.1186/s13321-020-00454-3] [PMID: 33431037]

[100] Li Y, Zhang L, Liu Z. Multi-objective *de novo* drug design with conditional graph generative model. J Cheminform 2018; 10(1): 33.
[http://dx.doi.org/10.1186/s13321-018-0287-6] [PMID: 30043127]

[101] Putin E, Asadulaev A, Ivanenkov Y, *et al.* Reinforced adversarial neural computer for *de novo* molecular design. J Chem Inf Model 2018; 58(6): 1194-204.
[http://dx.doi.org/10.1021/acs.jcim.7b00690] [PMID: 29762023]

[102] Putin E, Asadulaev A, Vanhaelen Q, *et al.* Adversarial threshold neural computer for molecular *de novo* design. Mol Pharm 2018; 15(10): 4386-97.
[http://dx.doi.org/10.1021/acs.molpharmaceut.7b01137] [PMID: 29569445]

[103] Prykhodko O, Johansson SV, Kotsias PC, *et al.* A *de novo* molecular generation method using latent vector based generative adversarial network. J Cheminform 2019; 11(1): 74.
[http://dx.doi.org/10.1186/s13321-019-0397-9] [PMID: 33430938]

[104] Girin L, Leglaive S, Bie X, *et al.* Dynamical variational autoencoders : A comprehensive review. ARXIV 2022; 2008: 12595.

[105] Gómez-Bombarelli R, Wei JN, Duvenaud D, *et al.* Automatic chemical design using a data-driven continuous representation of molecules. ACS Cent Sci 2018; 4(2): 268-76.
[http://dx.doi.org/10.1021/acscentsci.7b00572] [PMID: 29532027]

[106] Cho K, Van Merriënboer B, Gulcehre C, *et al.* Learning phrase representations using RNN encoder-decoder for statistical machine translation. Proceedings of the Conference 2014; 1724-34.
[http://dx.doi.org/10.3115/v1/D14-1179]

[107] Hochreiter S, Schmidhuber J. Long short-term memory. Neural Comput 1997; 9(8): 1735-80.
[http://dx.doi.org/10.1162/neco.1997.9.8.1735] [PMID: 9377276]

[108] Sutskever I, Vinyals O, Le QV. Sequence to sequence learning with neural networks. Adv Neural Inf Process Syst 2014; 4(January): 3104-12.

[109] Gao K, Nguyen DD, Tu M, Wei GW. Generative network complex for the automated generation of drug-like molecules. J Chem Inf Model 2020; 60(12): 5682-98.
[http://dx.doi.org/10.1021/acs.jcim.0c00599] [PMID: 32686938]

[110] Makhzani A, Shlens J, Jaitly N, Goodfellow I, Frey B. Adversarial autoencoders. arXiv 2015; 1511: 05644. Available from: http://arxiv.org/abs/1511.05644

[111] Skalic M, Jiménez J, Sabbadin D, De Fabritiis G. Shape-based generative modeling for *de novo* drug design. J Chem Inf Model 2019; 59(3): 1205-14.
[http://dx.doi.org/10.1021/acs.jcim.8b00706] [PMID: 30762364]

[112] Lim J, Ryu S, Kim JW, Kim WY. Molecular generative model based on conditional variational autoencoder for *de novo* molecular design. J Cheminform 2018; 10(1): 31.
[http://dx.doi.org/10.1186/s13321-018-0286-7] [PMID: 29995272]

[113] Bjerrum E, Sattarov B. Improving chemical autoencoder latent space and molecular *de novo* generation diversity with heteroencoders. Biomolecules 2018; 8(4): 131.
[http://dx.doi.org/10.3390/biom8040131] [PMID: 30380783]

[114] Kadurin A, Aliper A, Kazennov A, *et al.* The cornucopia of meaningful leads: Applying deep adversarial autoencoders for new molecule development in oncology. Oncotarget 2017; 8(7): 10883-90.
[http://dx.doi.org/10.18632/oncotarget.14073] [PMID: 28029644]

[115] Kadurin A, Nikolenko S, Khrabrov K, Aliper A, Zhavoronkov A. druGAN: An Advanced Generative Adversarial Autoencoder Model for *de Novo* Generation of New Molecules with Desired Molecular Properties *in Silico*. Mol Pharm 2017; 14(9): 3098-104.
[http://dx.doi.org/10.1021/acs.molpharmaceut.7b00346] [PMID: 28703000]

[116] Namasivayam V, Bajorath J. Multiobjective particle swarm optimization: Automated identification of structure-activity relationship-informative compounds with favorable physicochemical property distributions. J Chem Inf Model 2012; 52(11): 2848-55.
[http://dx.doi.org/10.1021/ci300402g] [PMID: 23039232]

[117] Hartenfeller M, Proschak E, Schüller A, Schneider G. Concept of combinatorial *de novo* design of drug-like molecules by particle swarm optimization. Chem Biol Drug Des 2008; 72(1): 16-26.
[http://dx.doi.org/10.1111/j.1747-0285.2008.00672.x] [PMID: 18564216]

[118] Winter R, Montanari F, Steffen A, Briem H, Noé F, Clevert DA. Efficient multi-objective molecular optimization in a continuous latent space. Chem Sci 2019; 10(34): 8016-24.
[http://dx.doi.org/10.1039/C9SC01928F] [PMID: 31853357]

[119] Metz L, Poole B, Pfau D, Sohl-Dickstein J. Unrolled generative adversarial networks. Adv Neural Inf Process Syst 2017; 1-25.

[120] Schneider G, Clark DE. Automated *de novo* drug design: Are we nearly there yet? Angew Chem Int Ed 2019; 58(32): 10792-803.
[http://dx.doi.org/10.1002/anie.201814681] [PMID: 30730601]

[121] Auti PS, Nandi A, Kumari V, Paul AT. Design, synthesis, biological evaluation and molecular modelling studies of oxoacetamide warhead containing indole-quinazolinone based novel hybrid analogues as potential pancreatic lipase inhibitors. New J Chem 2022; 46(24): 11648-61.
[http://dx.doi.org/10.1039/D2NJ01210C]

[122] Ertl P, Schuffenhauer A. Estimation of synthetic accessibility score of drug-like molecules based on molecular complexity and fragment contributions. J Cheminform 2009; 1(1): 8.
[http://dx.doi.org/10.1186/1758-2946-1-8] [PMID: 20298526]

[123] Afantitis A, Tsoumanis A, Melagraki G. Enalos suite of tools: Enhancing cheminformatics and nanoinformatics through KNIME. Curr Med Chem 2020; 27(38): 6523-35.
[http://dx.doi.org/10.2174/0929867327666200727114410] [PMID: 32718281]

[124] Gao W, Coley CW. The synthesizability of molecules proposed by generative models. J Chem Inf Model 2020; 60(12): 5714-23.
[http://dx.doi.org/10.1021/acs.jcim.0c00174] [PMID: 32250616]

[125] Rupp M, Schneider G. Graph kernels for molecular similarity. Mol Inform 2010; 29(4): 266-73.
[http://dx.doi.org/10.1002/minf.200900080] [PMID: 27463053]

[126] Bickerton GR, Paolini GV, Besnard J, Muresan S, Hopkins AL. Quantifying the chemical beauty of drugs. Nat Chem 2012; 4(2): 90-8.
[http://dx.doi.org/10.1038/nchem.1243] [PMID: 22270643]

[127] Hutter M. *In silico* prediction of drug properties. Curr Med Chem 2009; 16(2): 189-202.
[http://dx.doi.org/10.2174/092986709787002736] [PMID: 19149571]

[128] Coley CW, Rogers L, Green WH, Jensen KF. SCScore: Synthetic complexity learned from a reaction

corpus. J Chem Inf Model 2018; 58(2): 252-61.
[http://dx.doi.org/10.1021/acs.jcim.7b00622] [PMID: 29309147]

[129] Boda K, Johnson AP. Molecular complexity analysis of *de novo* designed ligands. J Med Chem 2006; 49(20): 5869-79.
[http://dx.doi.org/10.1021/jm050054p] [PMID: 17004702]

[130] Kinaret PAS, Serra A, Federico A, *et al.* Transcriptomics in toxicogenomics, part i: Experimental design, technologies, publicly available data, and regulatory aspects. Nanomaterials 2020; 10(4): 750.
[http://dx.doi.org/10.3390/nano10040750] [PMID: 32326418]

[131] Federico A, Serra A, Ha MK, *et al.* Transcriptomics in toxicogenomics, part ii: Preprocessing and differential expression analysis for high quality data. Nanomaterials 2020; 10(5): 903.
[http://dx.doi.org/10.3390/nano10050903] [PMID: 32397130]

[132] Méndez-Lucio O, Baillif B, Clevert DA, Rouquié D, Wichard J. *De novo* generation of hit-like molecules from gene expression signatures using artificial intelligence. Nat Commun 2020; 11(1): 10.
[http://dx.doi.org/10.1038/s41467-019-13807-w] [PMID: 31900408]

[133] Keshavarzi Arshadi A, Webb J, Salem M, *et al.* Artificial intelligence for COVID-19 drug discovery and vaccine development. Front Artif Intell 2020; 3(August): 65.
[http://dx.doi.org/10.3389/frai.2020.00065] [PMID: 33733182]

[134] Lalmuanawma S, Hussain J, Chhakchhuak L. Applications of machine learning and artificial intelligence for COVID-19 (SARS-CoV-2) pandemic: A review. Chaos Solit Fract 2020; 139: 110059.
[http://dx.doi.org/10.1016/j.chaos.2020.110059] [PMID: 32834612]

[135] Mohanty S, Harun AI Rashid M, Mridul M, Mohanty C, Swayamsiddha S, Swayamsiddha S. Application of artificial intelligence in COVID-19 drug repurposing. Diabetes Metab Syndr 2020; 14(5): 1027-31.
[http://dx.doi.org/10.1016/j.dsx.2020.06.068] [PMID: 32634717]

[136] Ton AT, Gentile F, Hsing M, Ban F, Cherkasov A. Rapid identification of potential inhibitors of SARS-CoV-2 main protease by deep docking of 1.3 billion compounds. Mol Inform 2020; 39(8): 2000028.
[http://dx.doi.org/10.1002/minf.202000028] [PMID: 32162456]

[137] Chenthamarakshan V, Das P, Hoffman SC, *et al.* CogMol: Target-specific and selective drug design for COVID-19 using deep generative models. Advances in Neural Information Processing Systems 2020; 2020: 1-13.

[138] Tang B, He F, Liu D, *et al.* AI-aided design of novel targeted covalent inhibitors against SARS-Co--2. Biomolecules 2022; 12(6): 746.
[http://dx.doi.org/10.3390/biom12060746] [PMID: 35740872]

[139] Bai Q, Tan S, Xu T, Liu H, Huang J, Yao X. MolAICal: A soft tool for 3D drug design of protein targets by artificial intelligence and classical algorithm. Brief Bioinform 2021; 22(3): bbaa161.
[http://dx.doi.org/10.1093/bib/bbaa161] [PMID: 32778891]

[140] OECD. Using artificial intelligence to help combat COVID-19. Oecd. 2020; 1-5. Available from: https://www.oecd.org/coronavirus/policy-responses/using-artificial-intelligence-to-help-comba--covid-19-ae4c5c21/

[141] Baruffaldi S, Van Beuzekom B, Dernis H, *et al.* dentifying and measuring developments in artificial intelligence : Making the impossible possible. OECD Science, Technology and Industry Working Papers 2020; 1-68. Available from: https://www.oecd-ilibrary.org/content/paper/5f65ff7--en%0Ahttps://dx.doi.org/10.1787/5f65ff7e-en

[142] Wu H, Wang C, Yin J, Lu K, Zhu L. Interpreting shared deep learning models *via* explicable boundary trees. arXiv 2017; 1709: 03730. Available from: http://arxiv.org/abs/1709.03730

[143] Zhao S, Talasila M, Jacobson G, Borcea C, Aftab SA, Murray JF. Packaging and sharing machine learning models *via* the acumos AI open platform. 17th IEEE International Conference on Machine

Learning and Applications, ICMLA 2018,. Orlando, FL, USA, December 17-20, 2018, pp,841–6.

[144] DSi-Poised Library - - Diamond Light Source. Available from: https://www.diamond.ac.uk/Instruments/Mx/Fragment-Screening/Fragment-Libraries/DSi-Poised-Library.html# (cited 2022 Dec 21)

[145] Schuffenhauer A, Ruedisser S, Marzinzik A, *et al*. Library design for fragment based screening. Curr Top Med Chem 2005; 5(8): 751-62.
[http://dx.doi.org/10.2174/1568026054637700] [PMID: 16101415]

[146] Trevizani R, Custódio FL, dos Santos KB, Dardenne LE. Critical features of fragment libraries for protein structure prediction. PLoS One 2017; 12(1): e0170131.
[http://dx.doi.org/10.1371/journal.pone.0170131] [PMID: 28085928]

[147] Coutard B, Decroly E, Li C, *et al*. Assessment of dengue virus helicase and methyltransferase as targets for fragment-based drug discovery. Antiviral Res 2014; 106(1): 61-70.
[http://dx.doi.org/10.1016/j.antiviral.2014.03.013] [PMID: 24704437]

[148] Mondal M, Groothuis DE, Hirsch AKH. Fragment growing exploiting dynamic combinatorial chemistry of inhibitors of the aspartic protease endothiapepsin. MedChemComm 2015; 6(7): 1267-71. [Internet].
[http://dx.doi.org/10.1039/C5MD00157A]

[149] Hann MM, Leach AR, Harper G. Molecular complexity and its impact on the probability of finding leads for drug discovery. J Chem Inf Comput Sci 2001; 41(3): 856-64.
[http://dx.doi.org/10.1021/ci000403i] [PMID: 11410068]

[150] Macarron R, Banks MN, Bojanic D, *et al*. Impact of high-throughput screening in biomedical research. Nat Rev Drug Discov 2011; 10(3): 188-95.
[http://dx.doi.org/10.1038/nrd3368] [PMID: 21358738]

[151] Henzler-Wildman K, Kern D. Dynamic personalities of proteins. Nature 2007; 450(7172): 964-72.
[http://dx.doi.org/10.1038/nature06522] [PMID: 18075575]

[152] Boehr DD, Nussinov R, Wright PE. The role of dynamic conformational ensembles in biomolecular recognition. Nat Chem Biol 2009; 5(11): 789-96.
[http://dx.doi.org/10.1038/nchembio.232] [PMID: 19841628]

[153] Matias PM, Donner P, Coelho R, *et al*. Structural evidence for ligand specificity in the binding domain of the human androgen receptor. Implications for pathogenic gene mutations. J Biol Chem 2000; 275(34): 26164-71.
[http://dx.doi.org/10.1074/jbc.M004571200] [PMID: 10840043]

[154] Seo MH, Park J, Kim E, Hohng S, Kim HS. Protein conformational dynamics dictate the binding affinity for a ligand. Nat Commun 2014; 5(1): 3724.
[http://dx.doi.org/10.1038/ncomms4724] [PMID: 24758940]

[155] Ciulli A. Biophysical screening for the discovery of small-molecule ligands. Methods Mol Biol 2013; 1008: 357-88.
[http://dx.doi.org/10.1007/978-1-62703-398-5_13]

[156] Shuker SB, Hajduk PJ, Meadows RP, Fesik SW. Discovering high-affinity ligands for proteins: SAR by NMR. Science 1996; 274(5292): 1531-4.
[http://dx.doi.org/10.1126/science.274.5292.1531] [PMID: 8929414]

[157] Lo MC, Aulabaugh A, Jin G, *et al*. Evaluation of fluorescence-based thermal shift assays for hit identification in drug discovery. Anal Biochem 2004; 332(1): 153-9.
[http://dx.doi.org/10.1016/j.ab.2004.04.031] [PMID: 15301960]

[158] Navratilova I, Hopkins AL. Fragment screening by surface plasmon resonance. ACS Med Chem Lett 2010; 1(1): 44-8.
[http://dx.doi.org/10.1021/ml900002k] [PMID: 24900174]

[159] Pedro L, Quinn R. Native mass spectrometry in fragment-based drug discovery. Molecules 2016; 21(8): 984.
[http://dx.doi.org/10.3390/molecules21080984] [PMID: 27483215]

[160] Godemann R, Madden J, Krämer J, *et al.* Fragment-based discovery of BACE1 inhibitors using functional assays. Biochemistry 2009; 48(45): 10743-51.
[http://dx.doi.org/10.1021/bi901061a] [PMID: 19799414]

[161] Boettcher A, Ruedisser S, Erbel P, *et al.* Fragment-based screening by biochemical assays: Systematic feasibility studies with trypsin and MMP12. SLAS Discov 2010; 15(9): 1029-41.
[http://dx.doi.org/10.1177/1087057110380455] [PMID: 20855559]

[162] Abadzapatero C, Metz J. Ligand efficiency indices as guideposts for drug discovery. Drug Discov Today 2005; 10(7): 464-9.
[http://dx.doi.org/10.1016/S1359-6446(05)03386-6] [PMID: 15809192]

[163] Hopkins AL, Groom CR, Alex A. Ligand efficiency: A useful metric for lead selection. Drug Discov Today 2004; 9(10): 430-1.
[http://dx.doi.org/10.1016/S1359-6446(04)03069-7] [PMID: 15109945]

[164] Nissink JWM. Simple size-independent measure of ligand efficiency. J Chem Inf Model 2009; 49(6): 1617-22.
[http://dx.doi.org/10.1021/ci900094m] [PMID: 19438171]

[165] Shultz MD. Setting expectations in molecular optimizations: Strengths and limitations of commonly used composite parameters. Bioorg Med Chem Lett 2013; 23(21): 5980-91.
[http://dx.doi.org/10.1016/j.bmcl.2013.08.029] [PMID: 24018190]

[166] Abad-Zapatero C. Are SAR tables obsolete? Drug Discov Today 2017; 22(2): 195-8.
[http://dx.doi.org/10.1016/j.drudis.2016.12.002] [PMID: 27993550]

[167] Chung S, Parker JB, Bianchet M, Amzel LM, Stivers JT. Impact of linker strain and flexibility in the design of a fragment-based inhibitor. Nat Chem Biol 2009; 5(6): 407-13.
[http://dx.doi.org/10.1038/nchembio.163] [PMID: 19396178]

[168] De Fusco C, Brear P, Iegre J, *et al.* A fragment-based approach leading to the discovery of a novel binding site and the selective CK2 inhibitor CAM4066. Bioorg Med Chem 2017, 25(13): 3471 82.
[http://dx.doi.org/10.1016/j.bmc.2017.04.037] [PMID: 28495381]

[169] Veber DF, Johnson SR, Cheng HY, Smith BR, Ward KW, Kopple KD. Molecular properties that influence the oral bioavailability of drug candidates. J Med Chem 2002; 45(12): 2615-23.
[http://dx.doi.org/10.1021/jm020017n] [PMID: 12036371]

[170] Xu X, Fang X, Wang J, Zhu H. Identification of novel ROS inducer by merging the fragments of piperlongumine and dicoumarol. Bioorg Med Chem Lett 2017; 27(5): 1325-8.
[http://dx.doi.org/10.1016/j.bmcl.2016.08.016] [PMID: 28159415]

[171] Miyake Y, Itoh Y, Hatanaka A, *et al.* Identification of novel lysine demethylase 5-selective inhibitors by inhibitor-based fragment merging strategy. Bioorg Med Chem 2019; 27(6): 1119-29.
[http://dx.doi.org/10.1016/j.bmc.2019.02.006] [PMID: 30745098]

[172] Strecker C, Peters H, Hackl T, Peters T, Meyer B. Fragment growing to design optimized inhibitors for human blood group B galactosyltransferase (GTB). ChemMedChem 2019; 14(14): 1336-42.
[http://dx.doi.org/10.1002/cmdc.201900296] [PMID: 31207161]

[173] Nikiforov PO, Surade S, Blaszczyk M, *et al.* A fragment merging approach towards the development of small molecule inhibitors of Mycobacterium tuberculosis EthR for use as ethionamide boosters. Org Biomol Chem 2016; 14(7): 2318-26.
[http://dx.doi.org/10.1039/C5OB02630J] [PMID: 26806381]

[174] Kozakov D, Grove LE, Hall DR, *et al.* The FTMap family of web servers for determining and characterizing ligand-binding hot spots of proteins. Nat Protoc 2015; 10(5): 733-55.

[http://dx.doi.org/10.1038/nprot.2015.043] [PMID: 25855957]

[175] de Souza Neto LR, Moreira-Filho JT, Neves BJ, *et al. In silico* strategies to support fragment-to-lead optimization in drug discovery. Front Chem 2020; 8(February): 93.
[http://dx.doi.org/10.3389/fchem.2020.00093] [PMID: 32133344]

[176] Kumar A, Voet A, Zhang KYJ. Fragment based drug design: Fom experimental to computational approaches. Curr Med Chem 2012; 19(30): 5128-47.
[http://dx.doi.org/10.2174/092986712803530467] [PMID: 22934764]

[177] Dong J, Wang NN, Yao ZJ, *et al.* ADMETlab: A platform for systematic ADMET evaluation based on a comprehensively collected ADMET database. J Cheminform 2018; 10(1): 29.
[http://dx.doi.org/10.1186/s13321-018-0283-x] [PMID: 29943074]

[178] Cheng F, Li W, Zhou Y, *et al.* admetSAR: A comprehensive source and free tool for assessment of chemical ADMET properties. J Chem Inf Model 2012; 52(11): 3099-105.
[http://dx.doi.org/10.1021/ci300367a] [PMID: 23092397]

[179] Reddy AS, Chen L, Zhang S. Structure-based de novo drug design.De novo Molecular Design. Wiley Online Library 2013; pp. 97-124.

[180] Pierce AC, Rao G, Bemis GW. BREED: Generating novel inhibitors through hybridization of known ligands. Application to CDK2, p38, and HIV protease. J Med Chem 2004; 47(11): 2768-75.
[http://dx.doi.org/10.1021/jm030543u] [PMID: 15139755]

[181] Yuan Y, Pei J, Lai L. LigBuilder 2: A practical *de novo* drug design approach. J Chem Inf Model 2011; 51(5): 1083-91.
[http://dx.doi.org/10.1021/ci100350u] [PMID: 21513346]

[182] Durrant JD, Lindert S, McCammon JA. AutoGrow 3.0: An improved algorithm for chemically tractable, semi-automated protein inhibitor design. J Mol Graph Model 2013; 44: 104-12.
[http://dx.doi.org/10.1016/j.jmgm.2013.05.006] [PMID: 23792207]

[183] Schyman P, Liu R, Desai V, Wallqvist A. vNN web server for ADMET predictions. Front Pharmacol 2017; 8(DEC): 889.
[http://dx.doi.org/10.3389/fphar.2017.00889] [PMID: 29255418]

[184] Braga RC, Alves VM, Silva MFB, *et al.* Pred-hERG: A novel web-accessible computational tool for predicting cardiac toxicity. Mol Inform 2015; 34(10): 698-701.
[http://dx.doi.org/10.1002/minf.201500040] [PMID: 27490970]

[185] Alves V, Braga R, Muratov E, Andrade C. Development of web and mobile applications for chemical toxicity prediction. J Braz Chem Soc 2018; 29(5): 982-8.
[http://dx.doi.org/10.21577/0103-5053.20180013]

[186] Mitcheson JS, Chen J, Lin M, Culberson C, Sanguinetti MC. A structural basis for drug-induced long QT syndrome. Proc Natl Acad Sci 2000; 97(22): 12329-33.
[http://dx.doi.org/10.1073/pnas.210244497] [PMID: 11005845]

[187] Willighagen EL, Waagmeester A, Spjuth O, *et al.* The ChEMBL database as linked open data. J Cheminform 2013; 5(1): 23.
[http://dx.doi.org/10.1186/1758-2946-5-23] [PMID: 23657106]

[188] Yang H, Chaofeng L, Lixia S, *et al.* AdmetSAR 2.0: Web-service for prediction and optimization of chemical ADMET properties. Bioinformatics 2017; 33(16): 1-7.
[http://dx.doi.org/10.1093/bioinformatics/btw552] [PMID: 28419194]

[189] Daina A, Michielin O, Zoete V. SwissADME: A free web tool to evaluate pharmacokinetics, drug-likeness and medicinal chemistry friendliness of small molecules. Sci Rep 2017; 7(1): 42717.
[http://dx.doi.org/10.1038/srep42717] [PMID: 28256516]

[190] QikProp | Schrödinger. Available from: https://www.schrodinger.com/products/qikprop (cited 2022 Aug 19).

[191] Boda K, Seidel T, Gasteiger J. Structure and reaction based evaluation of synthetic accessibility. J Comput Aided Mol Des 2007; 21(6): 311-25.
[http://dx.doi.org/10.1007/s10822-006-9099-2] [PMID: 17294248]

[192] Podolyan Y, Walters MA, Karypis G. Assessing synthetic accessibility of chemical compounds using machine learning methods. J Chem Inf Model 2010; 50(6): 979-91.
[http://dx.doi.org/10.1021/ci900301v] [PMID: 20536191]

[193] Fukunishi Y, Kurosawa T, Mikami Y, Nakamura H. Prediction of synthetic accessibility based on commercially available compound databases. J Chem Inf Model 2014; 54(12): 3259-67.
[http://dx.doi.org/10.1021/ci500568d] [PMID: 25420000]

[194] Lin FY, Esposito EX, Tseng YJ. LeadOp+R: Structure-based lead optimization with synthetic accessibility. Front Pharmacol 2018; 9(MAR): 96.
[http://dx.doi.org/10.3389/fphar.2018.00096] [PMID: 29556192]

[195] Ryan A, Polycarpou E, Lack NA, *et al.* Investigation of the mycobacterial enzyme HsaD as a potential novel target for anti-tubercular agents using a fragment-based drug design approach. Br J Pharmacol 2017; 174(14): 2209-24.
[http://dx.doi.org/10.1111/bph.13810] [PMID: 28380256]

[196] Silvestre HL, Blundell TL, Abell C, Ciulli A. Integrated biophysical approach to fragment screening and validation for fragment-based lead discovery. Proc Natl Acad Sci 2013; 110(32): 12984-9.
[http://dx.doi.org/10.1073/pnas.1304045110] [PMID: 23872845]

[197] Whitehouse AJ, Thomas SE, Brown KP, *et al.* Development of inhibitors against mycobacterium abscessus tRNA (m1G37) Methyltransferase (TrmD) using fragment-based approaches. J Med Chem 2019; 62(15): 7210-32.
[http://dx.doi.org/10.1021/acs.jmedchem.9b00809] [PMID: 31282680]

[198] Rudling A, Gustafsson R, Almlöf I, *et al.* Fragment-based discovery and optimization of enzyme inhibitors by docking of commercial chemical space. J Med Chem 2017; 60(19): 8160-9.
[http://dx.doi.org/10.1021/acs.jmedchem.7b01006] [PMID: 28929756]

[199] Pascoini AL, Federico LB, Arêas ALF, Verde BA, Freitas PG, Camps I. *In silico* development of new acetylcholinesterase inhibitors. J Biomol Struct Dyn 2019; 37(4): 1007-21.
[http://dx.doi.org/10.1080/07391102.2018.1447513] [PMID: 29607738]

[200] Schapira M, Calabrese MF, Bullock AN, Crews CM. Targeted protein degradation: Expanding the toolbox. Nat Rev Drug Discov 2019; 18(12): 949-63.
[http://dx.doi.org/10.1038/s41573-019-0047-y] [PMID: 31666732]

[201] Yang Z, Sun Y, Ni Z, *et al.* Merging PROTAC and molecular glue for degrading BTK and GSPT1 proteins concurrently. Cell Res 2021; 31(12): 1315-8.
[http://dx.doi.org/10.1038/s41422-021-00533-6] [PMID: 34417569]

[202] Michaelides IN, Collie GW. E3 ligases meet their match: Fragment-based approaches to discover new E3 ligands and to unravel E3 biology. J Med Chem 2023; 66(5): 3173-94.
[http://dx.doi.org/10.1021/acs.jmedchem.2c01882] [PMID: 36821822]

[203] Kennedy C, McPhie K, Rittinger K. Targeting the ubiquitin system by fragment-based drug discovery. Front Mol Biosci 2022; 9(September): 1019636.
[http://dx.doi.org/10.3389/fmolb.2022.1019636] [PMID: 36275626]

[204] Domostegui A, Nieto-Barrado L, Perez-Lopez C, Mayor-Ruiz C. Chasing molecular glue degraders: Screening approaches. Chem Soc Rev 2022; 51(13): 5498-517.
[http://dx.doi.org/10.1039/D2CS00197G] [PMID: 35723413]

[205] Békés M, Langley DR, Crews CM. PROTAC targeted protein degraders: The past is prologue. Nat Rev Drug Discov 2022; 21(3): 181-200.
[http://dx.doi.org/10.1038/s41573-021-00371-6] [PMID: 35042991]

CHAPTER 7

Molecular Simulation in Drug Design: An Overview of Molecular Dynamics Methods

Fernando D. Prieto-Martínez[1,*], Yelzyn Galván-Ciprés[2] and Blanca Colín-Lozano[3]

[1] *Instituto de Química, Universidad Nacional Autónoma de México, Ciudad de México, México*

[2] *Escuela Nacional de Ciencias Biológicas, Instituto Politécnico Nacional, Ciudad de México, México*

[3] *Facultad de Ciencias Químicas, Benemérita Universidad Autónoma de Puebla, Puebla, México*

Abstract: Molecular interaction is the basis for protein and cellular function. Careful inhibition or modulation of these is the main goal of therapeutic compounds. In the pharmaceutical field, this process is referred to as pharmacodynamics. Over the years, there have been several hypotheses attempting to describe this complex phenomenon. From a purely biophysical point of view, molecular interactions may be attributed to pairwise contributions such as charge angles, torsions, and overall energy. Thus, the computation of binding affinity is possible, at least in principle. Over the last half of the past century, molecular simulation was developed using a combination of physics, mathematics, and thermodynamics. Currently, these methods are known as structure-based drug design (SBDD) and it has become a staple of computer-aided drug design (CADD). In this chapter, we present an overview of the theory, current advances, and limitations of molecular dynamics simulations. We put a special focus on their application to virtual screening and drug development.

Keywords: Drug Design, Enhanced Sampling, Molecular Interaction, Molecular Simulation.

INTRODUCTION

Traditional methods for drug development often involve a multidisciplinary approach; the process usually begins by selecting what is known as a drug target and the consequent study of its biochemistry. Then comes the molecular design followed by organic synthesis, and subsequently, pre-clinical *in vitro*, *ex vivo*, and *in vivo* studies are carried out, when possible, depending on the task at hand. After

---

* **Corresponding author Fernando D. Prieto-Martínez:** Instituto de Química, Universidad Nacional Autónoma de México, Ciudad de México, México; E-mail: ferdpm4@hotmail.com

Igor José dos Santos Nascimento (Ed.)

gathering enough information, the research team can now decide which compounds can be considered as leads and a pharmacophore is then identifiable. Later, the ADMET (absorption, distribution, metabolism, elimination and toxicity) properties are optimized for clinical trials to finally market the best compound as a drug [1]. It should be noted, however, that this route represents a challenging, long, and expensive process that, on average, takes up to 15 years. Since 2019, the average cost of developing a drug can go from $161 to $4,540 million dollars, being anticancer drugs the most expensive to develop with a cost that goes between $944 and $4,540 million dollars [2].

Interestingly, around 90% of clinical drug development fails due to poor ADMET properties: absorption, solubility, permeability, efficacy, metabolism, excretion and high toxicity [3].

As a result of the above, there are several more novel strategies for discovering new drug candidates, such as: optimization of existing drugs, drug repurposing, systematic biological assays, use of available biological information, rational drug design and computer-aided drug design (CADD) also called *in silico* drug discovery methods.

The term *in silico* comes from Latin 'in silicon' and it refers to performed by using computers or *via* computer simulation. A mathematician Pedro Miramontes from the Universidad Nacional Autónoma de México (UNAM), who was the son of Luis Ernesto Miramontes Cárdenas, responsible for the synthesis of the active pharmaceutical ingredient (API) norethisterone of the first anticontraceptive pill, presented in his talk "DNA and RNA Physicochemical Constraints, Cellular Automata and Molecular Evolution" the term "*in silico*" to explain biological experiments performed *via* computer simulation [4].

CADD provides a complement to explain and predict biological activities and it comprises various methods, such as QSAR, virtual high-throughput screening, pharmacophore modeling, fragment-based screening, molecular docking and molecular dynamics simulations (MDS). It is worth mentioning that in recent years, these techniques have been put in the spotlight since they considerably reduce time and costs in all stages of drug development from the initial lead design to final stage clinical trials. Particularly in 2021, *in silico* drug discovery methods have gained popularity, as they have made it possible to optimize research work even remotely, which is a very useful tool in complex scenarios such as the COVID-19 pandemic.

Molecular recognition processes arise from pairwise interactions. Physical descriptions of these are possible using potential energy functions. In the literature, these functions are referred to as force fields, serving as angular stones in molecular mechanics and other related methods. In CADD, these approaches are grouped under structure-based (SBDD) approaches where computational resources are used to make numerical simulations of molecular phenomena. As of today, molecular docking has become the most prominent method for SBDD efforts, mostly due to its ease of implementation, flexibility, and overall prompt results.

Nevertheless, even with such positive attributes, there is no denying that molecular docking is prone to erratic or even aberrant results. Moreover, the technique has been trivialized in recent years. This has led to what we may call 'literature flooding', as evidenced in the trend for keyword docking (Fig. **1**) in recent years. Of course, the causes for this are multifactorial; still, a conserved tendency seems to be an overreliance on docking scores.

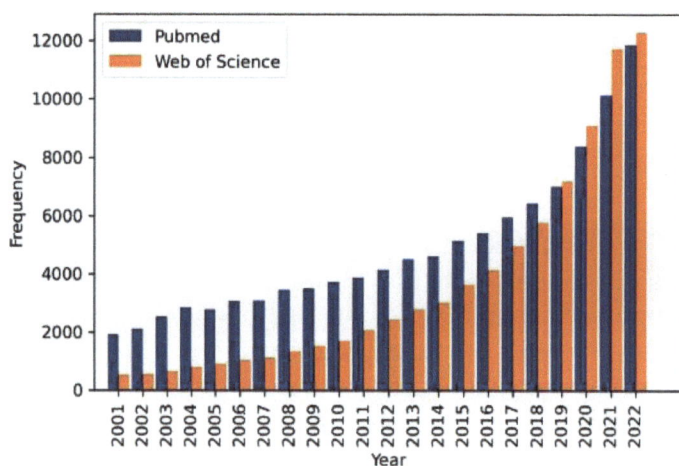

Fig. (1). Trends for "molecular docking" as a search query from two major academic search engines.

The truth of the matter is that the development of these tools has proven to be a complex task. Even with recent implementations of artificial intelligence, the development of universal scoring functions remains daunting. This raises the need for more exhaustive methods, such as MM-PBSA/GBSA and free energy perturbations (see Chapter 9).

For decades, one of the main problems was the computing power needed to solve the equations of motion for the *N*-atoms composing the system. In the beginning, a rather small system, *i.e.*, a couple of thousand atoms, could take months to simulate the movement for a couple of hundred picoseconds. Recently, thanks to technological advances, it has been possible to build powerful workstations that are on par with the last generation high computing clusters (HPC). In sharp contrast, on today's hardware, even a rather "discrete" workstation can increase up to six orders of magnitude for a similarly sized system.

Historical Background

The term molecular mechanics is hardly a novel concept. Nonetheless, its legacy and implications cannot be understated, and the proof of this is the laureates for the Nobel Prize in chemistry in 2013. These methods began their development as early as 1940, a noteworthy precedent being the work of Berni Alder, using a physical descriptions of hard spheres to approximate particle behavior. Even with these rudimentary implementations, early works of molecular simulations paved the way to unprecedented possibilities. Later on, came notable research papers from prominent authors, such as Michelle Parrinello, Martin Karplus, Arieh Warshel, Andrew McCammon, Herman Berendsen, and Peter Kollman, just to name a few.

As mentioned previously, the development of molecular dynamics goes back a few decades, and in Fig. (**2**), some of the more outstanding advances are highlighted. Starting in the late 1960s, the first systematic force fields were well underway by research groups led by important contributors such as Norman L. Allinger and Shneior Lifson, among others. Moving to the 1970s, we can see the creation of the Protein Data Bank (PDB) and the first simulation of protein folding, but it wasn't until the eighties when simulations of biomacromolecules in vacuum started to come out. However, the 90's was a big decade for molecular dynamics with the development and launching of the first versions of software like GROMACS and NAMD as well as an essential application of these computational tools for the biggest public health emergency at the time; HIV. This last achievement truly helped establish molecular dynamics as an important tool for studying protein dynamics without the associated hardships of working with easily degradable biomacromolecules like enzymes.

Recently, during the COVID-19 pandemic, molecular dynamics was certainly exploited, which can be easily seen by the notable increase in publications from 2020 - to date when searching for keywords like molecular dynamics, simulations and covid-19 in different search engines. During 2020, 735 different research papers were published under the categories 'molecular dynamics' and 'covid-19', in 2021, a little over twice the amount was published with a total of 1598 entries and lastly, to date in 2022, 1402 works were published. In contrast, on average, around 300 works are published a year using molecular dynamics for the research of diabetes and parasite infections.

With this, we conclude our brief context for molecular dynamics simulations. In the next section, we present some of the theory and inner workings behind the numerical simulation of biomolecules.

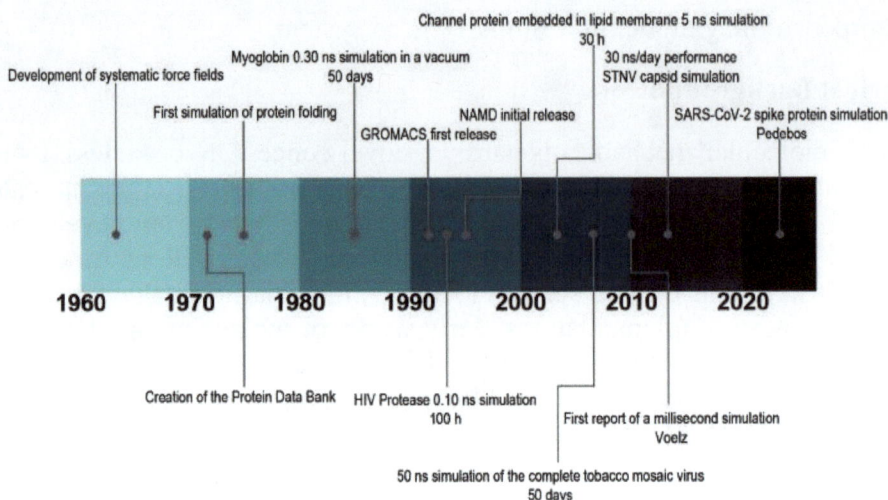

Fig. (2). A brief timeline of historical events regarding molecular modeling and dynamics.

THEORETICAL INTERLUDE

To the uninitiated, molecular simulation may hold a mysticism of sorts, as the mere premise sounds out of sci-fi books. In reality, most of it follows Occam's razor, as the mechanisms under the hood have been known for hundreds of years. Essentially, we are tackling a *N*-body problem; which can be described by Newton's laws of motion. Therefore, as long as we neglect the quantum nature of matter, we can use classical mechanics.

The Basics: Generating Equations of Motion

As of today, the laws of motion are well known to most people. However, their implications are not. It is useful to remember that the mathematics behind them only applies to simple models. Thus, the main drawback is that due to the system's complexity, an analytical solution is out of the question. To approach a solution, we do the next best thing which is using the variational principle.

Performing infinitesimally small variations of functions is the essence of calculus. For this case, to obtain a trajectory, we take two variables: a position (*x*) and time (*t*). From physics, we know that the relation between the two equals the overall velocity of a given particle. Now, if we propose small changes along both variables, we approach the limit of this relationship, which is the instantaneous velocity of the particle.

This implies a shift in the interpretation of Newton's laws, by transforming the expression for the second law of motion:

$$\vec{F} = \frac{d\vec{p}}{dt} = m\frac{d\vec{v}}{dt} = m\vec{a} \tag{1}$$

With a function known as the Lagrangian:

$$\mathcal{L} = K - U \tag{2}$$

Where K stands for the kinetic energy and U represents the potential energy. This simplifies the system of equations needed to describe the trajectory of the system. Yielding the principle of action, which is the optimum "path" as it minimizes the total energy. Another useful consequence of the transformation is that instead of forces we can express changes as means of configuration: a pair of positions and velocities.

Practically speaking, we just need to solve these equations sequentially to yield the trajectory approaching a numerical solution. The first proposal goes hand in hand with the Lagrangian interpretation of mechanics. The Euler-Lagrange or forward Euler method is a well-known algorithm to solve differential equations. For this particular case, it establishes the following:

$$\frac{dx}{dt} = f(x,t); x_t = x_{t-1} + f(x,t)\Delta t \therefore x = x_0 + v_0\Delta t \tag{3}$$

We pause for a moment to discuss an important detail. The selected interval of time (Δt) is known as the integration timestep and it has a direct effect on the expected result. The forward Euler method, like any numerical approach, possesses a local truncation error. In algorithm jargon, this is known as instability. In practice, these local errors may accumulate, leading to a serious failure in the integration. A common workaround is diminishing the timestep to a heuristic value.

Generally speaking, the timestep must have enough "resolution" to account for the fastest movement that must be described. For molecular dynamics simulations, this value is on the order of femtoseconds; as it relates to the vibrational frequencies of chemical bonds (between 10^{13} and 10^{14} Hz). Typical values for it are between 1 or 2 fs, but higher values (4-6 fs) are possible with more sophisticated integrators or by introducing a grouping scheme [5 - 9].

Another aspect worth mentioning from equation **3** is its resemblance to the kinematic equation. However, we can notice there is a missing term: acceleration. This also limits the implementation of the forward Euler method as a pertinent solution. To better appreciate this, we now write the said equation:

$$x = x_0 + v_0 t + \frac{1}{2} a t^2 \tag{4}$$

A more competent approach to accomplish integration is the algorithm proposed by Loup Verlet in 1967 [10]. We will provide a small overview of it as it is often the go-to implementation on most MD software. Briefly, we take the kinematic equation and use Taylor expansion about x (t ± Δt), arriving to:

$$x(t + \Delta t) = x(t) + v(t)\Delta t + \frac{1}{2} a(t)\Delta t^2 + \frac{1}{6} b(t)\Delta t^3 + \mathcal{O}\Delta t^4 \tag{5}$$

The expansion implies that to determine the value of x we need the current value of x, y, a, b, and so on. However, the b term is often nullified, hence the error corresponds to O (Δt^3). Next, we take advantage of the time-symmetric properties of classical mechanics. So, we include the value of x in the previous timestep and add both equations together:

$$x(t + \Delta t) = 2x(t) - x(t - \Delta t) + a(t)\Delta t^2 + \mathcal{O}(\Delta t^4) \tag{6}$$

This may seem rather convoluted, still, contrary to the forward Euler method, which approaches first-order derivatives, Verlet integration approaches second-order ones providing three main advantages:

• The velocity and jerk (b) terms have been effectively canceled, making the integration more accurate than the Taylor expansion.

• To compute the next value for x, all that is needed are the current and previous ones plus the acceleration.

• The error term now corresponds to O (Δt^4).

In summary, a molecular dynamics simulation consists of the following:

• Initial positions of atoms are assigned

• Initial velocities are assigned and the acceleration is calculated

• The integrator solves the equations of motion for Δt

• Positions are updated

• Iterate n times

A final note is that for molecular dynamics simulations, an additional consideration must be made. There is yet another derivation from classical mechanics that results are more useful. Suppose that the energy of a given system can be expressed as a function of position and momentum; therefore:

$$E(x,p) = \frac{p^2}{2m} + U(x) \tag{7}$$

Accounting for energy conservation, it can be established that changes in potential energy correspond to changes in kinetic energy. It follows that we can compute the said changes by differentiating by position and momentum; both of which are independent. In other words, we arrive at the following equation:

$$\frac{\partial E}{\partial x} = \frac{\partial U(x)}{\partial} \tag{8}$$

Which again, can be related to Newton's second law:

$$\vec{F} = -\nabla U_i \therefore F = -\frac{dp}{dt} \tag{9}$$

Based on this, we arrive at new equations of motion, which allow the time evolution of the system:

$$\frac{dp}{dt} = -\frac{\partial E}{\partial x} \, ; \, \frac{dx}{dt} = \frac{\partial E}{\partial x} \tag{10}$$

Just like Lagrangian mechanics, this new framework for classical mechanics offers notable advantages over the original formalism. The first one arises from the implementation of momentum, as this magnitude can be generalized better than velocity. In addition, it can be stated that we are expanding upon the basis set by Lagrange. In fact, these two approaches are indeed related by means of a mathematical transformation or Legendre transform:

$$f(x) \Rightarrow f^* \left(\frac{df}{dx}\right) \tag{11}$$

$$\mathcal{L}^* \left(q_i, \frac{\partial \mathcal{L}}{\partial q_i}\right) = \frac{\partial \mathcal{L}}{\partial q_i} - \mathcal{L} \Rightarrow \mathcal{L}^*(q_i, p_i) = p_i q_i - \mathcal{L} \tag{12}$$

Now, we name this new variable H or the Hamiltonian. For a multiple particle system, it comes from:

$$\mathcal{H}(q_i, p_i) = \sum_i p_i \dot{q}_i - \mathcal{L} \tag{13}$$

As a result of these foundations, the Hamiltonian is also related to the total energy of the system. Moreover, Hamilton's equations yield a rather abstract but useful concept known as phase space. While not really a space in the conventional sense, phase space is a manifold that represents flows of motion. The rationale behind this may not be evident now but suffice to say this concept will come in handy later on.

We have seen the basics of classical mechanics and some underpinnings of their implementation for molecular simulation, we must now see the other side of the coin. Recall that no matter which description, classical mechanics requires a potential energy function known as a force field, a concept adopted from physics; in computational chemistry it has some slight variations. In the following section, we discuss the main features and progress in molecular force fields.

Breaking Molecular Interactions Down to Physical Contributions: Enter Molecular Force Field

In physics, a force field is a vector field that accounts for non-contact forces acting on a body or particle, common examples are gravity or electric fields. Thus, for our given case, we need to identify the forces at play. In other words, the potentials are broken down to pairwise contributions accounting for:

• Lennard-Jones potential

• Electrostatics (long-range)

• Motion along chemical bonds (vibrations, rotation, and torsion)

A general form for pair i, j would be:

$$U = \sum_{i<j} \Sigma \epsilon_{ij} \left[\left(\frac{\sigma_{ij}^{12}}{r_{ij}} - \frac{\sigma_{ij}^6}{r_{ij}} \right) \right] + \sum_{ij} \Sigma \frac{q_i q_j}{4\pi\epsilon_0 r_{ij}} + \sum_{bonds} \frac{1}{2} k_b (r - r_0)^2$$
$$+ \sum_{angles} \frac{1}{2} k_b (\sigma - \sigma_0) + \sum_{dihedrals} k_\phi [1 + cos(n\phi - \delta)] \tag{14}$$

However, there is another layer of complexity as atoms have distinct values for these parameters. Therefore, force field development often uses both experimental and theoretical information to determine proper values for individual cases. This is known as parametrization and it is not short from being a form of art. This is also the reason for the inherent limitations of chemical force fields, as these can hardly be universal. For example, protein force fields have reached enough maturity; yet the parameters of these cannot be indistinctly applied to organic molecules. Mostly due to atom typing, a common result is duplicate information on extended forcefields [11]. Also, for small organic molecules, geometric descriptions are often more complex when compared to peptides. Recently, there have been notable efforts to develop robust force fields and/or potentials that aim to mitigate these issues.

A Primer on Thermodynamics and Statistical Mechanics

We have seen the mathematical aspects of simulation, but still a question remains: How can we ensure that these are representative of real molecular motion? To bridge this gap, we need to cover the basis of thermodynamics.

Thermodynamics describes the interaction between a system and its surroundings. A common notion in thermodynamics is equilibrium, usually meaning that all forces within it are in balance. Another perspective for this is that an equilibrium is reached when macroscopic variables remain constant, *i.e.*, they become time independent.

In principle, the equilibrium conditions can be displaced, nevertheless the system tends to resist these sudden changes and the said changes may occur in very different timescales. This has an important consequence as, contrary to physical phenomena, thermodynamics uses a rather flexible frame of time. Still, thermodynamic notions establish that time goes definitively in one direction, this is known as the arrow of time, and it has been related to another thermodynamic concept: entropy.

Of the many concepts thermodynamics introduces, entropy is perhaps one of the most cryptic ones at first. Commonly associated with disorder or chaos, it may be interpreted as the fate of an isolated system. While true, this is a rather limited perspective on the implications of entropy. So, what is entropy then? To get a better notion, consider the following experiment: suppose we have a particle in a given space; if we were to delimit the said space incrementally, each area would correspond to a macrostate, whereas each position in the space would be a microstate (Fig. **3**).

In principle, any given microstate is equally probable, but such is not the case for macrostates. Thus, it follows that the bigger the area, the higher the probability of the corresponding macrostate. Following a similar reasoning, Ludwig Boltzmann [12] formulated a relationship between entropy (S) and the number of microstates (Ω) accessible to an N-particle system:

$$S = k_B ln \, \Omega \tag{15}$$

From this equation, it is possible to make another crucial conceptualization: entropy is also information. To illustrate this, recall our thought experiment, to provide the particle's position, we could use a binary notation with a zero value for the particle's starting position. Therefore, for macrostate B we only need one bit of information, whereas for macrostate C we would need three bits and so on. Thus, for any additional degree of freedom needed to describe a system, it follows that the entropy has increased.

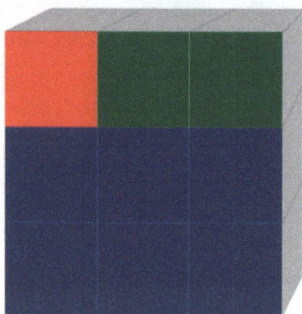

Fig. (3). Scheme representing micro and macrostates as delimited by the degrees of freedom needed for their description.

Now, we can understand why entropy is the absolute direction of the arrow of time. As for an isolated system, the entropy always increases; equilibrium is only reached when the entropy is at its maximum value, meaning that every microstate is equally probable. This serves as the foundation of statistical mechanics, as we use microscopical phenomena to describe macroscopic observables.

To better enunciate this, we need yet another concept: ensemble. This refers to a group of microstates belonging to a specific macrostate. In other words, a given macroscopic observable (A) is an ensemble average:

$$A = \langle a \rangle = \frac{1}{Z} \sum_{\lambda}^{Z} a(X_\lambda) \tag{16}$$

Where (a) is the ensemble average, Z is the number of members in the ensemble and a (X λ) is a microscopic phase space function. To fill in the missing link, recall that the phase space describes a flow of motion. Based on this, it is important to note that the trajectories in phase space are unique. As a consequence, trajectories in phase space do not intersect unless they are the same. For a high number of particles (such as an ensemble), this flow may be interpreted as a physical fluid. This abstraction is made to understand two facts:

• The density of this 'fluid' is a function of position and momentum

• The volume of this 'fluid' remains constant with time

This last statement is known as the Liouville theorem [13] and it has deep implications: Due to the nature of the phase space volume, the state of a system can be determined at any given point in time using Hamiltonian mechanics. Thus far, we have covered most of the basics, however, it is important to remember that all these foundations are related to an isolated system. This type of system is one that does not exchange either matter or energy with its surroundings. Thus, the macroscopic states are the number of particles (N), volume (V), and energy (E). In statistical mechanics, this ensemble is known as the microcanonical ensemble. Moreover, as discussed above, entropy is the thermodynamic state function of this ensemble:

$$S(N,V,E) = k_B ln\Omega(N,V,E) \tag{17}$$

To sum it up, this implies that the standard molecular dynamics simulation always samples the microcanonical ensemble. These conditions are theoretically useful but impractical, as biological systems are on a different playground altogether. To solve this problem, Maxwell relations can be used to transform the entropy into a new state variable corresponding to the desired macroscopic observables:

$$T = (\frac{\partial E}{\partial S})_{N,V}, \qquad P = -(\frac{\partial E}{\partial V})_{N,S}, \qquad \mu = (\frac{\partial E}{\partial N})_{V,S} \tag{18}$$

With these, the following ensembles can be described:

• Canonical ensemble (NVT)

• Isobaric-isothermal ensemble (NPT)

• Grand canonical ensemble (μVT)

As for the state functions, these are related to free energy; either Helmoltz, Gibbs, or both.

With this, we have provided a comprehensive framework for molecular simulations. Of course, the above discussions are far from exhaustive or formal. For interested readers, we suggest the following materials for [14 - 17].

In the following section, we discuss the importance of sampling and best practices for representative simulations.

THE OVERARCHING PROBLEM: SAMPLING

In the previous section, it was mentioned that with molecular dynamics we sample the conformations of a system, and these can be seen as members of a given ensemble as a flow in phase space. In addition, it was stated that macroscopic observables come from an average. This is easy to visualize in a time-independent manner; such is the case in Monte Carlo simulations, where samples are obtained iteratively from stochastic permutations of a selected degree of freedom in the system. This, in addition to a pruning criterion such as the Metropolis-Hastings algorithm, ensures proper sampling as long as the iteration number approaches a high value [18].

For molecular dynamics, analogue reasoning applies to ensure proper sampling; one must ensure that the system has explored all possible microstates. However, this is not possible in practice as the simulation time must approach infinity. To work around this limitation, a compromise must be made. It must be assumed that the phenomenon under study is ergodic in nature. This is known as the ergodic hypothesis:

$$\langle a \rangle = \int d\vec{q}^N dr^N A\left(r^N, \vec{q}^N\right) \rho\left(r^N, \vec{q}^N\right) \equiv \lim_{t \to \infty} \frac{1}{t} \int_0^t d\tau A\left(r(\tau), \vec{q}(\tau)\right) \tag{19}$$

In layman's terms, the ergodic hypothesis assumes that a temporal average equals the ensemble average. As long as the phenomenon is indeed ergodic and the simulation time approaches infinity, the obtained sample can be representative. Alas, this is yet to be proven formally and cannot be universally applied [19]. Nonetheless, all simulations assume ergodicity *a priori*.

From this, a question arises: is a longer simulation better for all cases? To answer this, let us consider a toy model: a single particle bound to an asymmetric potential: the double well. Since the potential is asymmetric, it serves as an energy barrier the particle must overcome. Thus, the two macrostates are clearly distinguishable based on the particle's position along the x axis.

Here, we introduce the canonical partition function, a statistical mechanics concept which describes the properties of a system in thermodynamic equilibrium. With it, the computation of probability is possible:

$$Z = \sum e^{-\beta E_i}, \beta = \frac{1}{k_B T} \therefore P(x) = \frac{1}{Z_x} e^{\frac{U(x)}{k_B T}} \tag{20}$$

And the potential for the double well ($U(x)$) comes from:

$$U(x) = E_0[(\frac{x}{a})^4 - (\frac{x}{a})^2] - \frac{b}{a}x \qquad (21)$$

Hence, if one could plot the possible values for *P(x)*, two normal distributions are found, each for distinct macrostates (Fig. **4**). Of course, this exercise is only theoretical as, in practice, we do not know the exact form of the potential function nor the value for Z.

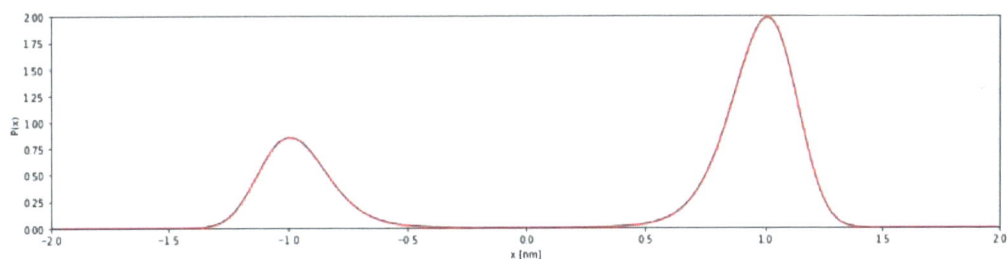

Fig. (4). Theoretical distribution of macrostates in an asymmetric double system. The plot is obtained from the canonical partition function.

Using the ergodic hypothesis, we can approximate this plot using the recovered positions in a simulated trajectory. A comparison of the probability density with simulation time is presented in Fig. (**5**).

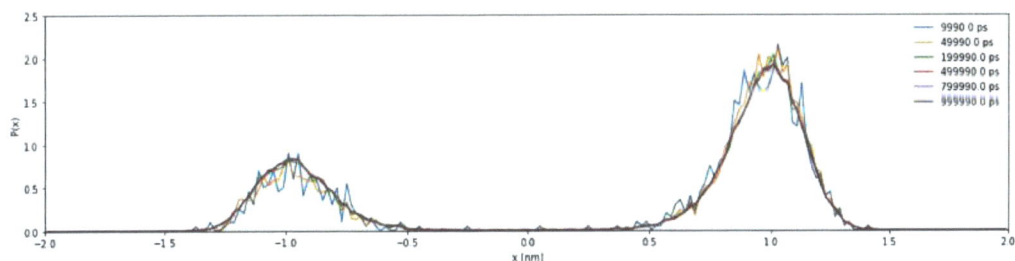

Fig. (5). Graphical representation of the ergodic hypothesis. Notice with an increasing simulation time, how the partition function is smoother and approximate to the theoretical estimate.

As seen, with increasing simulation time, the overall description approaches the theoretical distribution. However, even for this rather simple model, the convergence takes roughly 200 million integration steps. Notice also how the approximation may overestimate or underestimate the intended function. This is expected as the integration was numeric in nature; nonetheless, these subtle artifacts in the trajectory can have a notable influence on the interpretation of the simulation.

Considering a biomolecular system, the complexity notably increases. This implies that a proper sampling could need very long simulation times (in the millisecond scale or even longer). While it has been shown that the current force fields can be stable enough in said time scale [20]. The truth of the matter is that the time and resources needed for that kind of study are hardly available universally.

In light of this, a better practice is to obtain independent trajectories to establish trends and distinguish putative artifacts. In contrast, several approaches have been proposed to increase sampling. These are known as enhanced sampling methods and often introduce different biases in order to prevent the simulation from getting stuck in metastable states. Table (**1**) presents a brief description of some of the most common methods for enhanced sampling.

Table 1. Common methods for enhanced sampling.

Method	Description	Refs.
Umbrella sampling	The system is biased along the permutation of specific degrees of freedom. These are known as collective variables (CVs). CVs are chosen to model a structural transition or phenomenon.	[21]
Replica exchange	An array of replica systems run in parallel at different temperatures. After a given time, the replicas exchange temperatures. Following several iterations of this, the thermodynamics of the overall samples are calculated using weighted probability analysis.	[22]
Adaptive sampling	This method utilizes parallel simulations to identify states based on empiric criteria. These can be used to construct Markov state models (MSM) to measure the associated probability of states. This information can be used to spawn another batch of simulations iteratively.	[23]
Metadynamics	This method also uses CV biases to increase sampling. In this case. Gaussian functions are used to prevent the system from revisiting the same values for the given CVs. This bias is implanted in a history-dependent manner. Free energy landscape can be obtained by reversing the biased potential.	[24]
Accelerated molecular dynamics	Simulations are subjected to a boost potential whenever the energy reaches a threshold value and this potential may be applied integrally or partially. With this bias, sampling can be increased notably. Simulations of hundreds of nanoseconds are able to sample the free energy landscape robustly.	[25]
τRAMD	A recent protocol, based on aMD. This implementation is conceived as a means to study ligand residence times and unbinding paths. In this case, the ligand is subjected to the boost potential in a random direction. Whenever the distance between the ligand and a protein falls below a given threshold, thus, ligand unbinding events can be characterized with shorter simulations.	[26]

To gain a better grasp of the true power of enhanced sampling, let us return to the toy model. Now, consider a set of replica systems, each simulated at different temperatures. After a given time interval, an exchange between neighboring pairs is attempted. The acceptance criterion for the exchange is the well-known Metropolis-Hastings, which, for this case, comes from:

$$w\left(x_m^{[i]}\middle|x_n^{[j]}\right) = \begin{cases} 1, \text{if}\Delta \leq 0 \\ e^{-\Delta}, \text{if}\Delta > 0 \end{cases}$$

$$e^{[\beta_n - \beta_m](E(q^{[i]}) - E(q^{[j]}))}, \qquad \beta = \frac{1}{k_B T}$$

(22)

In essence, this implies that whenever the exchange is accepted, the simulation is able to explore novel regions of the free energy landscape as low temperature conformations slowly acquire more kinetic energy. Just for comparison purposes, let us contrast this replica exchange system with two independent classical simulations, each one with a different length. Now, if we plot the obtained potential energy, we will obtain something similar to Fig. (**6**).

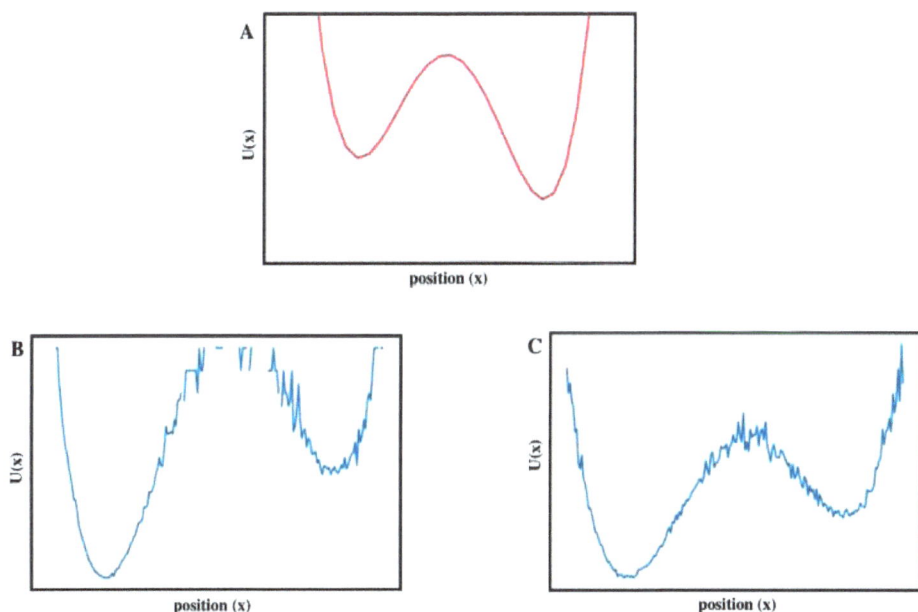

Fig. (6). Comparison of brute force sampling and enhanced protocols (replica exchange). **A)** The target potential. **B)** Surface recovered from 2 ns of conventional simulation. **C)** Surface recovered using a four-replica scheme, the total simulation time was 2 ns.

Indeed, it can be seen that the simulation explored the potential energy surface in its entirety. The original implementation of this protocol was proposed by Sugita and Okamoto [22]. Suffice it to say, that this example is just illustrative in nature. For protein simulations, the distribution of replicas is more complex. As temperature differences between replicas must be low to enforce the exchange. Moreover, the simulations must be analyzed with additional methods to ensure a correct assessment of the observed probabilities and thermodynamic observables [27].

Indeed, we have seen how sampling critically affects the quality and overall meaningfulness of simulation results. Nonetheless, there are additional limitations to consider in any MDS protocol. Thus, we will comment on some of these in the next section.

CURRENT LIMITATIONS OF MOLECULAR DYNAMICS

While powerful, molecular dynamics still have important setbacks that need to be discussed.

Due to the nature of how classical mechanics is used to simplify the calculations, no electron density is taken into consideration. Hence, any function regarding electrons is beyond the limits of this approach. The modelling of covalent processes or reaction intermediates is simply not feasible. For this, hybrid methods have been developed, currently known as QM/MM, where the systems are divided into two regions: the QM zone is calculated with a semiempirical potential; while the rest of the system is updated using conventional MM treatment. While promising, these methods are still rather limited and their implementation is hardly straightforward. In most instances, QM/MM frameworks still suffer from hardware limitations, limiting the accessible system's size and timescale [28].

In this regard, modelling of metal centers is another important subject where MD is still insufficient. In most instances, users must decide which component is more important to model: either geometry or electrostatics. However, as previously stated, these calculations do not account for the inductive effects. Regular cases such as zinc or magnesium centers have been extensively studied and parametrized [29 - 31]. However, other metal centers often need parametrization and validation using extensive quantum calculations [32].

Force fields have reached a notable maturity for protein modelling. Yet, there are some instances where further developments are due. For example, as of today, accurate modelling of small organic molecules is not solved [33, 34]. As current implementations are mostly derivatives of protein force fields (*i.e.*, GAFF or CGenFF). This also extends to other molecules, such as polymers or even glycosides. Which need a different approach altogether to accurately describe their flexibility. Notable efforts towards this are currently in the works. OpenFF, for instance, is a collective endeavor to develop the first force field for small molecules without atom typing [11, 35].

MOLECULAR DYNAMICS PRACTICE AT A GLANCE

Implementing molecular dynamics protocols can now be straightforward. Mostly due to the maturing of simulation software, prior cryptic usage has been streamlined with user-friendly steps or interfaces. Still, there are several critical and technical settings that are decisive in obtaining significant results. In the following section, we break down some of these.

Prior to Simulation

Before discussing software or setup, we take a brief moment to develop hardware settings and recommendations for computing architecture. As previously mentioned, molecular dynamics require extensive use of computational resources. Not quite as much as quantum methods, but to conduct real work or comprehensive calculations, a laptop is generally inadvisable.

This is not to say that MD exclusively needs HPC; it is true this used to be the case. Take, for instance, the supercomputer Anton, developed by the DESRES group, which made use of a special architecture to improve solely on MD performance [36]. As years went by, with the advent of GPU computing, it became possible to obtain significant amounts of data in reasonable time spans. To put this into perspective, let us consider the most common benchmark system: dihydrofolate reductase (DHFR), which comprises ~ 24 000 atoms. Fig. (7) shows the expected performance on this "small" system comparing CPU and GPU architectures.

As shown, the most efficient way to run simulations is using GPUs by far. It is also notable that consumer grade GPUs have achieved performance on par with "productivity" solutions. The main difference between the two, at least for MD simulations, is their performance while using double float precision. However, in practice, most MD engines make use of hybrid or "mixed" precision, which is very good on consumer grade hardware. Other features like ECC memory or VRAM capacity may be desired, but most consumer grade cards have enough memory to account for 1×10^{6} atoms. As for GPU models, cards produced by NVIDIA tend to be preferred, mostly due to the optimization and performance gained with CUDA computing, but OpenCL implementations are also possible, it all depends on selected software.

Finally, we would like to comment on the settings for increased performance. As stated in the theoretical interlude, integration timestep is critical to this. As of today, different proposals and algorithms have been implemented to allow for increased timesteps. Yet, we strongly advise against a focus solely on obtaining fast results. For most cases, a timestep of 4 fs can be used without much issue, however studies accounting for subtle changes such as diffusion or bulk liquid timestep should be kept at 2 fs.

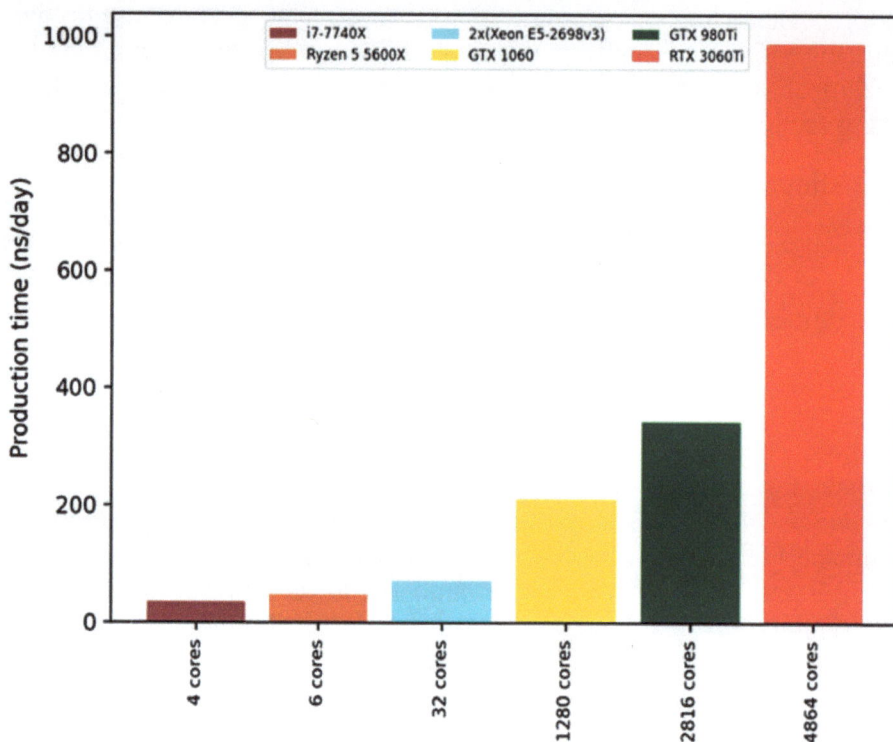

Fig. (7). Comparison of achieved performance using CPU and GPU hardware for molecular dynamics. The benchmark system was DHFR in a cubic box, integration timestep was 4 fs for all cases.

First Steps

In the drug discovery context, there are two common starting points for molecular simulation. The first one is a crystal structure, either locally or from repositories like the Protein Data Bank (PDB) or homology model. The second one is a protein-ligand complex obtained from molecular docking. While a PDB file offers a reasonable starting point for docking, it has several downsides for molecular dynamics.

One of the most underrated stages in terms of making sure that it is correctly executed, is the preparation of the files to be used by the system. Although the Protein Data Bank already comes with information regarding the resolution of the crystal (if this applies), and the presence or lack of mutations, among other details, it is important to note that, more common than not, there are missing loops and residues from the structure due to the many positions the atoms are able to adopt in such a place. Not only that, pdb files do not come with explicit non-polar hydrogens, which would complicate the recording of each coordinate. So, taking this into consideration, there are a few preparation steps that need to be taken, which we will be discussing in this section.

You have chosen an appropriate pdb file for the desired simulation, now, what? The first thing we need to look at is the amino acid sequence and compare it to the wild type, if there are missing residues, we need to reconstruct them. Next is adding explicit non-polar hydrogens. Once we have performed these corrections, it is necessary to save the new file and create the corresponding topology and coordinates files.

Next, we need to look at the simulation box, the usual shape used for the box is cubic, however, one can choose to use different options as long as they allow for an optimum size as well as minimum solvent. In this step, it is convenient to mention the so-called periodic boundary conditions. Briefly, setting this parameter allows for a space made up of a couple of angstroms, to be taken into consideration, that will prevent the protein from interacting with copies of itself, thus minimizing the edge and finite size effects of the box.

Moving on to solvation, with the box already established, for the simulation of biomolecules, the solvent used is water with the appropriate concentration of sodium, chloride and potassium ions to resemble the conditions inside or outside a cell. In terms of temperature and pressure, it is also necessary to take physiological conditions to set these values correctly. Finally, overall minimization of the system is required.

All these steps take time, but ensuring a proper system setup allows the user to obtain a result that is closer to reality and can more accurately predict what could happen experimentally. In short, the user must prioritize accuracy and validity over speed to obtain earnest results.

Commonly Used Force Fields

For more than a decade, force fields such as Amber, CHARMM, GROMOS, and OPLS-AA have been recognized as the most widely used for biomolecular simulations, specifically for protein dynamics. There are two types of interactions that need to be accounted for when dealing with proteins; bonded and non-bonded, and the ways their corresponding energies are calculated differ from force field to force field [37].

Let us first take a look at bonded interactions. Homochirality at the biochemical level defines the function and substrate-enzyme recognition, so using a force field that successfully reproduces the proper chirality of proteins is desired. When speaking of chirality, the force fields mentioned above perform different calculations to account for the so-called improper dihedral energy. For this particular case, CHARMM has the capability to reproduce a more accurate result [38].

Moving to non-bonded interactions, these are made up of both electrostatic and dispersion interactions and these compose a key issue when performing molecular simulations. Non-bonded interactions are long-ranged, which means that when setting up distance cutoff values for the calculations, we can encounter artifacts derived from this decision, especially in regard to water. However, the takeaway is that there is no such thing as a unique solution for the optimal set of functions and parameters that will lead without fail to the closest-to-reality result.

Another interesting topic to briefly touch on is the development of polarizable force fields, such as AMOEBA and Drude and their relatively novel applications for protein simulations. Drude is available for CHARMM, NAMD, and OpenMM. AMOEBA, on the other hand, was initially developed for water, but it has been extended to organic molecules ranging from simple hydrocarbon chains to more complex aromatic systems. Both of these force fields have aided the simulation of many conformational changes in ion channels and phenomena regarding membrane permeation as well as interesting observations in substrate-enzyme interactions [39].

Available Software

Due to the nature of numerical simulations, several options are available. Computationally speaking, the first choice is the computing language of choice; these include Fortran, C, C++, Python, or Java. However, in most cases, the simulation software is mostly based on Fortran and C. Recent packages now include wrappers to facilitate implementation or to provide flexibility. This trait is often desired as object-oriented languages, which opens the possibility to framework development.

From the wide ecosystem of simulation packages, there is no ultimate pick as each option has its limitations and strengths. The maturity of a given tool is often reached with the aid of a strong community. Sometimes software is designed to solve or provide a particular implementation of molecular dynamics. In practice, this often leaves newcomers with a choice: learn whichever package currently in use within their group; or learn whichever tool they may know or have heard of.

Nonetheless, a down road realization is made: Learning the generalities of at least three different engines is best for most cases.

Still, recurrent difficulties such as simulation postprocessing and special cases persist. For example, simulating protein-drug, protein-polymer of supramolecular complexes out of the box may not be possible for most packages. Moreover, different tools may use the same engine as a backend for simulation; while providing direct support for specific methods or applications.

Common packages used for protein simulation are presented in Table (**2**).

Table 2. Common packages for MDS.

Software	Force Fields	GPU Compatibility	License	Works Published to Date*	Implemented Protocols
CHARMM	AMBER, CHARMM, OPLS	No	Commercial	664	Replica exchange Umbrella sampling Adaptive sampling Metadynamics Accelerated
GROMACS	AMBER, CHARMM, GROMOS, OPLS	Yes	Free	619	Replica exchange Umbrella sampling Adaptive sampling Metadynamics Accelerated Steered QM/MM
NAMD	AMBER, CHARMM	No	Free	223	Replica exchange Umbrella sampling Accelerated
Desmond	OPLS	Yes	Free and commercial	105	Replica exchange Metadynamics
LAMMPS	AMBER, CHARMM, COMPASS. DREIDING	Yes	Free	85	Temperature accelerated
OpenMM	AMBER, AMOEBA. CHARMM	Yes	Free	55	Replica exchange Umbrella sampling Metadynamics Accelerated
YASARA	NOVA, AMBER, YASARA	Yes	Free and commercial	38	Adaptive sampling

Now the authors are going to share some of their experiences with simulation software.

Desmond

Part of the Schrödinger suite, Desmond is also available for academic use. In recent versions, it has dropped CPU compatibility. At the time of the writing of this chapter, version 2018.4 is still available and is the last to offer CPU support. Moreover, contrary to OpenMM or GROMACS, Desmond's implementation for GPGPU is only based on CUDA. Meaning that it is only compatible with NVIDIA GPUs. Advantages over other packages include the Maestro GUI, automatic parametrization (using the OPLS_2005 force field) and great performance.

The main difference between academic and commercial versions of the software is the parameters available for non-protein species. In addition, the commercial version allows for parameter fitting and selection if the OPLS3e parameters are insufficient. In academia, Viparr can be used for the reparametrization of systems with other well-known force fields (https://github.com/DEShawResearch/viparr-ffpublic). Another important note is that for proper preprocessing of protein or protein-ligand complexes, a commercial version is needed. Nonetheless, it is possible to use an independent preprocessing program and use the output with Desmond [40].

The software also implements different algorithms for integration and enhanced sampling methods, such as replica exchange and metadynamics. Computation of free energy is also possible, with alchemical transformations and well-tempered metadynamics.

As for flexibility, Desmond is quite limited, as in-depth knowledge of the software is required to implement plugins. Another notable disadvantage is the documentation, as it is outdated, and the user base is really small compared to other packages. Nonetheless, early versions of the software's manual are a great resource to grasp the basics of molecular dynamics. Summing up, Desmond may be a niche program, but it offers a balanced starting point if the user still finds command-line or programming daunting.

GROMACS

A staple of the field, currently in its eleventh version (2022.4). Its documentation is comprehensive and exhaustive. GROMACS manual is a great place to gain good notions for molecular simulation. Moreover, there are several tutorials on basic usage and common cases. Hardware wise, it offers great compatibility making it a go to option for installations ranging from laptops to HPC nodes.

GROMACS can be used to implement a wide array of simulation protocols such as umbrella sampling, replica exchange, steered molecular dynamics, metadynamics, and, more recently, QM/MM protocols. It also comes with several built-in tools to facilitate protein preparation and simulation post-processing [41].

Another positive trait is the number of included force fields out of the box, such as the AMBER family of force fields, GROMOS, OPLS, and CHARMM. The addition of external or user-built force fields is also possible.

On the downside, GROMACS may have a steep learning curve as the number of options and settings may overwhelm new users. Another drawback is the need for compilation by the user if any additional functionality is to be implemented. Hence, we recommend GROMACS to users who already have a good grasp of the concepts and implementation of molecular dynamics.

NAMD

Potentially, one of the better-known software along with GROMACS, NAMD offers a user friendly experience to allow beginners the implementation of an array of different systems ranging from a very simple equilibration to a more complex system. Although catching errors during the system set-up may not be the easiest of tasks, the documentation provided is sufficiently resourceful for troubleshooting.

This software package has been extensively used for diverse protocols such as dynamics, accelerated dynamics, metadynamics, umbrella sampling, and adaptive sampling. It is currently running on the 3.0 version, and it is available for use on a free license.

NAMD also has great interoperability with VMD, extending the features and capabilities. Additionally, this package has the possibility to perform simulations in both CPU and GPU based architectures [42].

OpenMM

This package has taken the field of molecular simulation by storm. Currently, in its eighth version, OpenMM is an open-source alternative mostly intended for object-oriented use (Python). The main highlights are good compatibility with input files from other tools, flexibility, and the fact that it is built on top of a rich ecosystem of molecular modelling tools.

For newcomers, OpenMM offers intuitive usage and a rather low learning curve. Molecular simulations can be started with a few lines of code with data being obtained and analyzed on the fly. For the experienced user, OpenMM offers very flexible implementations with the aid of custom forces. Moreover, it comes with a wide array of settings and choices, making it easy to implement non-conventional simulations [43].

In terms of hardware, compatibility is wide, as it supports both OpenCL and CUDA for GPUs, while CPUs still have notable support and performance. Thus, OpenMM is a viable option for workstations or HPC clusters. On top of this, native output formats are hardware agnostic, making it easier to resume or extend simulations in different environments.

A noteworthy aspect of this software is the fact that it serves as the backend engine for different packages, such as ACEMD, Sire, Flare and Tinker, to name a few. Similarly, OpenMM has been used in efforts to make simulations more accessible or directly educational, such as Making it rain [44] and TeachOpenCADD [45]. Thanks to all of this, OpenMM has gained a strong user base in recent years, which in turn has extended the capabilities and implementation of the package. Examples include the implementation of machine learning potentials [46].

YASARA

A rather obscure alternative, at least when compared to other members of this listing. YASARA offers a complete suite for molecular modelling, including built-in tools for homology modelling, molecular docking, and of course, molecular dynamics. Out of the box, YASARA offers the AMBER family of force fields, but the inclusion of additional ones is possible [47].

A major advantage it possesses is an automatic parametrization protocol with an algorithm known as AutoSMILES (http://www.yasara.org/autosmiles.htm). This allows easy parameter assignment for ligands or exotic species *via* the GAFF force field. Moreover, its graphical interface is very intuitive and streamlined. System preparation and simulation takes only a couple of steps. If more control over settings is needed, YASARA uses its own macro language: Yanaconda; somewhat reminiscent of Python syntax. Thanks to this, YASARA has great capabilities for postprocessing and framework implementation. These features make it very appealing for beginner usage; however, care must be taken as it may result in a black box.

Building the System

After careful preparation of the protein structure for simulation and selection of the force field, comes the system building. This process is very similar, in most cases, the structure is parametrized based on atom typing, then the system is solvated by buffering in a water box which implies that the solvent must have its own parameters, *i.e.*, the need of a water model. For protein simulation, there are several families, which include: SPC, TIP and, more recently, OPC. Each one gives a different treatment to the electrostatics of water, resulting in differences in the physical properties of the liquid. Meaning that model selection needs to comply with the intended methodology or analysis. This point should not be taken lightly, as the results can be severely affected by the water model. However, a prevalent choice for most instances is the TIP3P model.

Solvent padding must ensure at least two solvation shells, thus, common values for boxes are between 10 and 12 Å. Of course, this is no rule of thumb, as bigger systems would need a bigger box. This takes us to the subject of boundaries and geometry.

Simulating a solute in a physical cell means that a surface and interface will be created. This is troublesome as the atoms on the surface would have higher energies than those in the bulk liquid. Shall we continue with this venture, the simulation would suffer from artifacts created by the effects of the surface. Moreover, our main interest is the modelling of bulk properties. Therefore, an accurate representation of the bulk liquid is not possible in a closed container.

To avoid this complication altogether, periodic boundary conditions are implemented in MD simulations. In general terms, this means that the simulation cell is mirrored as a means to create a lattice. Only the original system is allowed to move, the rest of the mirror images just follow its movement. There is a crucial aspect to this; however, the interaction between the mirror images and the original system is still possible. In practice, this means that a cut-off for non-bonded interactions must be implemented. As a rule of thumb, the cut-off must be implemented so that self-interaction does not occur.

This would suggest that additional artifacts are introduced *via* the cut-off scheme. It must be stated that the term does not necessarily imply an actual truncation of the electrostatics. As doing this would compromise the accuracy and fidelity of the simulation. Cut-off only refers to a shift done in the calculations based on short or long-range electrostatics [48].

The system may be delimited in any given shape. Common cell shapes include cubic, orthorhombic, triclinic, hexagonal, prism, and rhombic dodecahedron (either hexagonal or squared). At first, these exotic shapes were introduced gradually to minimize volume. Nevertheless, this notion was soon proven to be mistaken and unnecessary, as the shape may be interconverted by mathematical transformation. Still, cell geometry also has some influence on the results, as box shapes tend to increase the probability of self-interaction, whereas dodecahedron cells are a more robust choice for simulations [49]. Based on these observations, our suggestion is to choose a shape that closely resembles the protein under study.

Running a MD Simulation

Once the simulation cell has been built, one can proceed with simulations. Generally speaking, a common MDS protocol is composed of the following steps:

• Minimization

• Equilibration

• Production

• Analysis

Minimization is carried out to bring the system to a minimum in energy. This is done to have a reasonable starting point and begin the transition from potential to kinetic energy. This step is often implemented with numerical optimization, with two main approaches: step-descent and second-order methods. These utilize the gradient as a means to find the minimum of the potential function. The steep-descent algorithm is a multistep method that follows the gradient in a downhill manner:

$$x_{i+1} = x_i - \gamma_i \nabla U(x_i) \tag{23}$$

This means that the algorithm iterates for a number of steps following the negative value of the gradient, until one of the following criteria is satisfied:

$$||\nabla U(x_i)|| \approx 0; \ |U(x_i) - U(x_{i-1})| \approx 0; \ ||x_i - x_{i-1}|| \approx 0 \tag{24}$$

This algorithm is quite popular for optimization, as it is simple and easily implemented. However, it must be stated that its main flaw can be the overreliance on iteration. Meaning that convergence can be slow, so the tradeoff between accuracy and speed comes into play for large systems. In such instances, additional algorithms are often used after the steepest descent. A well-known example is the Broyden-Fletcher-Golfarb-Shannon (BFGS) algorithm. A quasi-newtonian gradient method which is better suited for optimization problems with many variables [50].

Once the system is at a minimum, equilibration starts. This process must ensure the following:

• Potential energy is mostly converted to kinetic energy

• The simulation is in the desired macroscopic values (temperature, pressure, *etc.*)

• The desired ensemble is being sampled

Considering that simulation protocols may use canonical or isobaric isothermal conditions, it follows that we provide some context on the operation and theory of thermostats and barostats. The main premise is similar for both, yet the precise nature of these is not to maintain pressure or temperature at a constant value. Rather, the aim is to ensure a target average value is reached. Moreover, the mechanism to ensure the target condition must be algorithmically compatible with the integration scheme.

For temperature control, several approaches exist, based on the implementation, these can be divided into three main categories: thermal constraints, weak coupling scheme and extended systems [51]. Now, we will provide an overview of commonly implemented thermostats in molecular dynamics software. First comes a general thermostat known as velocity rescaling [52]:

$$v_i \rightarrow \sqrt{\frac{T^\star}{T_i}} v_i \tag{25}$$

Where v_i is velocity at step i, T^* is the target temperature and T_i is the current temperature at step i. The general idea is quite simple; however, the absence of temperature fluctuation leads to inaccuracies. To overcome these limitations, Berendsen *et al.* [53] proposed the following scheme to apply rescaling:

$$\frac{dT(t)}{dt} = \frac{1}{\tau}[T^\star - T_i] \tag{26}$$

Where the tau teparameter determines the rate at which T_i decays to the target value. While both approaches are rather easy to implement in the context of simulations, these pose major problems. In principle, none of them sample the canonical ensemble. Moreover, it has been shown that rescaling algorithms present a phenomenon known as "the flying ice cube effect". Briefly, this makes reference to a violation of the equipartition theorem [54], which has different manifestations, ranging from kinetic energy accumulation to the development of temperature gradients [55]. For protein simulations, this is known as the "hot solvent-cold solute problem". Where water molecules tend to gain more thermal energy [56]. Thus, using either of these methods is generally unadvisable, as the recovered data can be severely compromised.

On the other hand, thermostats can be developed from the Nosé-Hoover equations of motion:

$$q_i = \frac{p_i}{m_i}, \dot{p_i} = -\frac{\partial U(q)}{\partial q_i} - p_i\frac{p_\eta}{Q}, \dot{p_\eta} = \sum_i^N \frac{p_i^2}{m_i} - Nk_BT, \eta = \frac{p_\eta}{Q} \qquad (27)$$

$$\mathcal{H}\ (p,q,\eta,p_\eta) = U(q) + \sum_i^N \frac{p_i^2}{2m_i} + \frac{p_\eta^2}{Q} + Nk_BT \qquad (28)$$

Which are expansions over the Hamiltonian and Lagrangian formulations of mechanics. These can be understood as a Maxwell demon, as the added terms p_η and Q act as an "operative", which scales the velocities based on the kinetic energy and the intended temperature T^*. To visualize this, consider that the original system is coupled to a heat reservoir, such as the "bath" proposed by Berendsen. It is pertinent to mention that the presented equations are not the original formulation made by Hoover. The addition of the ç term was proposed by Martyna *et al.* [57] to better analyze phase space distribution.

Nosé-Hoover thermostat is widely regarded as an optimal choice for simulation, as it possesses notable advantages: time-reversibility, sound thermodynamic description and is easy to implement in code [51]. Unfortunately, it is non-ergodic, and it experiences thermal discrepancies often referred to as Toda demon [58]. A proposal to overcome this is to include additional variables to achieve the correct distribution of phase space [14]. This modification is known as the Nosé-Hoover chain and has been proposed as a putative solution to the "hot solvent - cold solute problem" in protein simulation [59]. However, Nosé-Hoover chains still show nonergodic behavior in several cases [60 - 62].

Now we introduce a final "thermostat", in this case, derived from Langevin's mechanics. This formulation has very interesting mathematical properties and physical implications. As the description is sensible enough to be applied in varying scales, from micro to macro-sized models [63]. In a classical mechanics context, the Generalized Langevin Equation (GLE) is:

$$\vec{F} = -\nabla U(p,q) - \gamma m_i q_i + \vec{R}(t-t'), \vec{R}(t-t') = 2m_i\gamma k_BT\delta(t-t') \qquad (29)$$

It is important to note that Langevin's equation was proposed to describe Brownian motion as a consequence of thermal fluctuations. At first glance, this seems rather obvious, as the equation can be split into viscous and noise terms. Thus, as the viscous term approaches 0, we recover Newton's mechanics. On the other hand, overdamped values yield a diffuse regime or Brownian motion [64]. A more general perspective would be Langevin's equation is constructed from a reasonable equation of motion with interaction terms described as a random force with statistical properties [65].

Therefore, the temperature "control" comes from the said random force, acting as a friction term (γ). This addition is often described as a means to simulate the random collisions between particles of the system and those in a heat bath [66]. The equation also has a memory kernel (R $\rightarrow$ (t-t')), as the friction acting on a given particle depends on velocities at an earlier time [67]. The nature of this "noise" is usually Markovian [68], but non-Markovian derivations are also possible and expand the applications of the equation [69].

Hence, Langevin's dynamics is a powerful tool to describe stochastic processes in thermal environments [70], yielding a canonical sampling of phase space [71]. It can be successfully applied to non-equilibrium systems [72]. For protein simulation, Langevin's dynamics has been proven as a robust choice, due to their overall stability [56, 73].

Pressure control, on the other hand, follows a similar reasoning, in this case, the volume of the system is allowed to fluctuate in order to yield a desired $<P>$ value [74]. A general implementation of molecular dynamics to achieve this comes from the following equation [75]:

$$P = \frac{Nk_BT}{V} + \frac{1}{3V} \sum_i r_i \vec{F}_i \tag{30}$$

which is basically the ideal gas equation for pressure, with the addition of the virial term to account for the internal forces in the system. Andersen was the first to introduce this extended formalism to sample the isobaric-isoenthalpic ensemble [76]. Thus, just like thermostats, we can classify barostats into two main classes: extended Lagrangian and weak coupling methods [77]. Each barostat implementation attempts to balance the difference between instantaneous pressure and the external one. In practice, barostats tend to have a notable influence on the system, either by affecting its dynamics or the compensation to new conditions [78]. Available barostats include the proposals made by:

• Andersen

• Parrinello-Rahman

• Nosé-Hoover

• Martyna-Tobyas-Tuckerman-Klein (MTTK)

• Langevin

• Chow-Ferguson (Monte Carlo Barostat)

In summary, temperature and pressure control in molecular dynamics cannot be executed in a fail-safe manner, each implementation has its advantages and weaknesses. For protein simulation, best practices include the implementation of grouping schemes and slow equilibration with different implementations to ensure the target conditions are met. Careful and thoughtful selection is a direct responsibility of the user and must ensure robust and accurate descriptions of the phenomenon under study [79].

Simulation Analysis

When simulations are done, there are a myriad of possible analysis and metrics to compute. Of course, this is limited to the study at hand and related methodologies. However, there are a couple of "sanity checks" that are carried out on a regular basis. A good practice is to visually inspect the trajectory using tools such as VMD, NGLView, *etc.* Any major artifact or critical mistake would be apparent and should not take more than 5-10 minutes to spot.

With this first round out of the way, you may proceed with more systematic checks, such as verifying the average conditions like temperature, pressure, energy, *etc.* These plots should have a clear average value with little fluctuation. Then comes the analysis for deviations, including position (Root Mean Squared Deviation; RMSD) and fluctuation (Root Mean Squared Fluctuation; RMSF).

RMSD is used to measure the overall movement of the system; usually, it spikes for a couple of nanoseconds. Then it should fluctuate but with no abrupt changes. Should any additional increments or decrease in value arise, this may be indicative of instability or an artifact. RMSF, on the other hand, serves to identify the most mobile regions or atoms, based on the delimitation of the system. Common practices include measuring the fluctuation of the backbone and side chains.

Joint analysis of both can serve to identify structural changes or putative mechanisms in folding or ligand recognition. To contrast these, the radius of gyration may be used. This measure can be related to the protein size and aids in the identification of folding changes associated with the compactness of the structure [80].

To integrate this information, consider Fig. (**8**), showing the analysis of a simulation. Notice that the RMSD plot shows a deviation of around 70 nanoseconds. Such an occurrence may be insignificant by itself. Therefore, a contrasting metric is the radius of gyration. We see that around the same period, its value increases, suggesting a folding event. Perhaps the observed deviation is interesting after all; finally, the fluctuation of the protein shows that the region comprising residues 30 to 40 is significantly mobile. Summing up, the analysis suggests that a folding event may be taking place between residues 30 to 40.

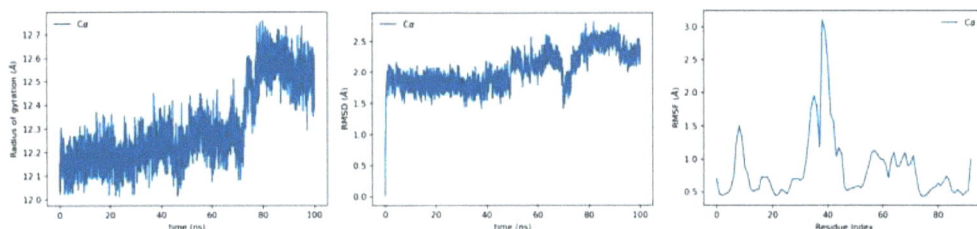

Fig. (8). Radius of gyration, RMSD, and RMSF plots as obtained from a MDS protocol using NPT ensemble with PFOR from *T. vaginalis*.

This trajectory belongs to the ferredoxin of *T. vaginalis*, this protein is the molecular target of metronidazole. As other ferredoxins, it has a sulfur-iron cluster to carry out redox processes. For this particular case, residues between 30 and 50 belong to a loop near the catalytic site. The transition of this loop has been identified as the main mechanism of the exposal of the [2Fe-2S] cluster [81]. Fig. (9) shows a comparison between the two frames of the trajectory and the crystal structure of PDBID 1L5P. Notice how the catalytic core is clearly exposed in the right panel.

RALTEGRAVIR: A CASE STUDY

Molecular dynamics simulations can have several applications in the development of drugs of different origins: synthetic, natural (small molecules), or biotechnological (macromolecules). The processes that can be evaluated using this technique range from target selection, structural knowledge of the target, drug-target interactions, molecular design, metabolism, and biopharmaceutical properties, as well as pharmaceutical development.

Over the past few years, computational approaches in combination with rational drug design have proven to be of great aid in drug discovery and in this sense, we compiled Table (**3**) as a short list of drugs whose discovery stories are intimately related to molecular dynamics.

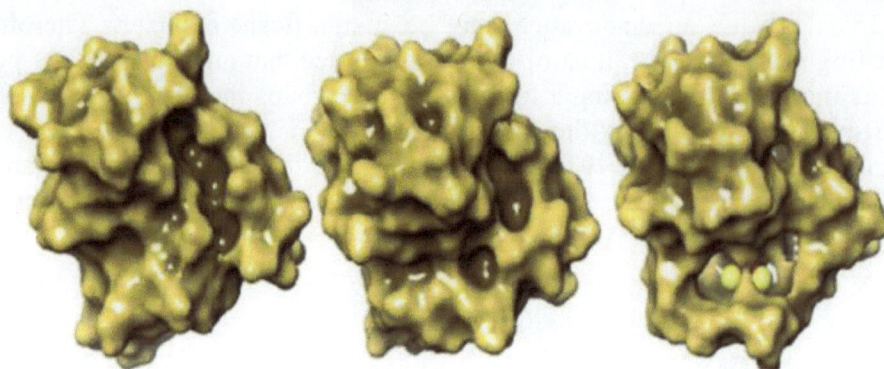

Fig. (9). Opening of the metallic cluster of PFOR from *T. vaginalis* as recovered from MDS. Crystal structure belonging to PFOR from *T. vaginalis* (PDBID 1L5P; left). Structure of PFOR from *T. vaginalis* prior to simulation (center). Trajectory snapshot obtained after the 70 ns mark (right).

A Look into the HIV-1 Integrase

The HIV-1 integrase (IN) plays a dual role in HIV-1 infection. In early stages, it is responsible for the insertion of viral DNA into the chromosome of a host cell, this process is essential in viral replication as it will be discussed in further detail in the following section. In recent years, IN has been linked to a second non-catalytic function in the regulation of viral replication; binding to viral RNA in virions, which is necessary for virion maturation. Consequently, this target has been studied extensively as it is attractive for the design of antiviral drugs. Additionally, this target has the advantage of not sharing a homologous enzyme in humans, thus increasing the odds of developing selective ligands for it [101].

Table 3. Drugs developed by MDS and the mechanisms of action.

Drug Name	Results of MDS Study	Ref.
Amprenavir (1999) HIV protease inhibitor – Antiretroviral HIV treatment	Amprenavir was designed based on the knowledge of his target structure (SBDD). On the other hand, molecular dynamics simulations have been used to elucidate the mechanisms of resistance and low effectiveness of amprenavir.	[82 - 86]
Imatinib (2001) Kinase inhibitor – Chronic myelogenous leukemia and gastrointestinal stromal tumors	By steered molecular dynamics it was stablished that the most favored pathway of dissociation between imatinib and its target c-Kit and the Abl channel is through an ATP channel over the allosteric pocket.	[87]

Drug Name	Results of MDS Study	Ref.
Sofarenib (2005) and Sunitinib (2006) VEGF inhibitors – Renal cell and hepatocellular carcinoma	The unbinding mechanisms between sorafenib and sunitinib and their binding sites were identified by steered molecular dynamics studies. The results showed that under tension, sunitinib dissociates from the ATP binding site without a breakpoint, while sorafenib moves in the opposite direction to the ATP binding pocket, causing a change in the orientation of the αC helix of VEGF. This could explain the longer half-life of sorafenib.	[88]
Raltegravir (2007) Integrase inhibitor – HIV treatment	Through a study of MDS and flexible molecular docking of the integrase inhibitor 5CITEP, an unknown trench adjacent to an active site that intermittently opens was discovered. This finding provided the basis for the development (structure-based drug design) of raltegravir.	[89]
Anti-lysozyme antibody (2007)	By a 1 ns molecular dynamics simulation of anti-hen egg white antibody, HyHEL63 (HH63), complexed with HEL, could find salt bridges and electrostatic forces, which are critical for the formation complex; it should be noted that these interactions could not be identified by X-ray crystallography because it is a static medium. According to the results, MDS can help to identify the so-called "hot-spot" regions at the antibody-antigen interface. Importantly, there is experimental evidence supporting the electrostatic contribution of salt bridges to the free energy of folding and binding.	[90 - 93]
Cytochrome p450 2C9 (2012)	The research group of Cojocaru *et al.* evaluated by MDS and RAMD simulations the different exit tunnels of CYP2C9 in complex with flurbiprofen and warfarin and found four main routes 2a, 2ac, 2c, and 2e. Moreover, we found that the occurrence of each depends on the nature (lipophilicity) of the ligand and the conformation of the F-G loop. This serves as a basis to understand how CYP mutations affect the metabolism of drugs and for the design of inhibitors that act through a specific pathway and interfere only in the metabolism of compounds using that pathway	[94]
c-Myc (2016) Oncoprotein – Anticancer target	c-Myc is a disordered protein (IDP), that, under physiological conditions, lacks a stable tertiary structure and exists only as conformational assemblies, depending on the ligand to which it binds. Through a replica exchange molecular dynamics (REMD) studied the conformational ensemble of the unbound disordered dimerization domain of the transcription factor c-Myc. In addition, by a conventional MDS study of the inhibitor-binding domain of c-Myc, binding interactions and constantly changing inhibitor binding sites in c-Myc were discovered. These results support the selected representative conformations of these different c-Myc binding sites to conduct a successful structure-based virtual screening study, which finds four active compounds that block c-Myc function in the cell.	[95, 96]
Naproxen sodium crystal (2017)	The hydration and dehydration processes of an active pharmaceutical ingredient (API) can directly influence the behavior of a drug during formulation, dissolution and even storage. Using density functional theory and two molecular dynamics methods, crystalline naproxen sodium and its hydrates were studied to determine the face-specific dehydration mechanisms. The results show a highly complex diffusion and nucleation behavior. Based on the above, this method provides a tool to explore dehydration pathways and predict new dehydrated crystal structures.	[97]

(Table 3) cont.....

Drug Name	Results of MDS Study	Ref.
Donepezil and Rivastigmine (2019)	MDS was used to assess the loading capacity of chitosan nanoparticles (NPs) as a nanocarrier for donepezil and rivastigmine drugs. The results of the simulation showed that in the presence of ions (Na$^+$), the overall drug loading capacity of rivastigmine is lower compared to donepezil in chitosan NPs.	[98]
Zaubrutinib (2019) and Ibrutunib (2022) Tyrosine kinase (BTK) inhibitor – Treatment of B-cell malignancies	BTK inhibitors ibrutinib and zanubrutinib were assayed against SARS-CoV-2 by Born surface area (MM/GBSA) calculations and molecular dynamics (MD). The results revealed that ibrutinib and zanubrutinib could act through different mechanisms in the viral entry and replication stage and could be repurposed as a possible treatment against SARS-CoV-2.	[99, 100]

Out of these examples, we will look into the case of raltegravir, a drug that acts as an inhibitor of the HIV-1 integrase.

IN is composed of three domains: the N-terminal domain (NTD), the catalytic core domain (CCD), and the C-terminal domain (CTD). Briefly, both the NTD and CTD domains are each responsible for mediating the DNA binding process which happens at either end of linear viral DNA to form something called the intasome complex. Importantly, the NTD features His and Cys residues that are conserved in all retroviral and retrotransposon INs (HHCC motifs), which coordinate with zinc and are essential for IN oligomerization and 3' processing and DNA strand transfer activities. On the other hand, the CCD contains a highly conserved D,D-35-E motif (DDE) in the active site of the enzyme that is necessary for catalytic activity, of which allows for the reverse transcribed viral DNA to be inserted into the host DNA [102]. The active site of the DDE is formed by a catalytic triad of electronegative residues coordinating the positions of two magnesium ions whose function is to activate the nucleophilic attack of water for processing 3′ and hydroxyl groups of viral DNA capsid-3′ for strand transfer and destabilize the cleaved phosphodiester [103].

The Drug Discovery Process

To this date, several inhibitors have been reported, and one of the most promising is 5CITEP [1-(5-chloroindole-3-yl)-3-hydroxy-3-(2*H*-tetrazol-5-yl)-propenone]; a diketo-acid derivative that belongs to the group of S-1360 Shionogi-GlaxoSmithKline integrase inhibitors. In further experiments, the X-ray crystal structure of the catalytic domain co-crystallized with 5CITEP inhibitor was obtained (PDB code: 1QS4). According to these experiments, 5CITEP binds centrally to the integrase active site and forms key interactions that appear to mimic the substrate-DNA-integrase interaction.

Unfortunately, the X-ray crystallography did not provide enough information, observing an anomaly in the flex loop (Ile141-Asn144) believed to be near the substrate during integration. Moreover, the structure of the complex integrase-DNA is not clear; finally, only the key residues were characterized [104]. To clarify the above, find other possible binding sites, and correctly elucidate the mode of binding of this inhibitor (5CITEP), the research group of Schames *et al.* [89] conducted a study of flexible molecular docking and MDS. The *in silico* study revealed two relevant conformations of 5CITEP (Fig. **10**) as the active site of the integrase, that an open conformation of the protein is necessary, and a new binding region (trench) adjacent to the active site was identified [105, 106]. Additionally, it was clarified that the trench is surrounded by the residues Ile141-Asn144 from the flex loop [89, 107].

Later, this was experimentally confirmed by X-ray crystallography. These findings prove the utility of MDS even with proteins that have ambiguous loops or are reconstructed and provide the basis for the development (structure-based drug design) of Raltegravir.

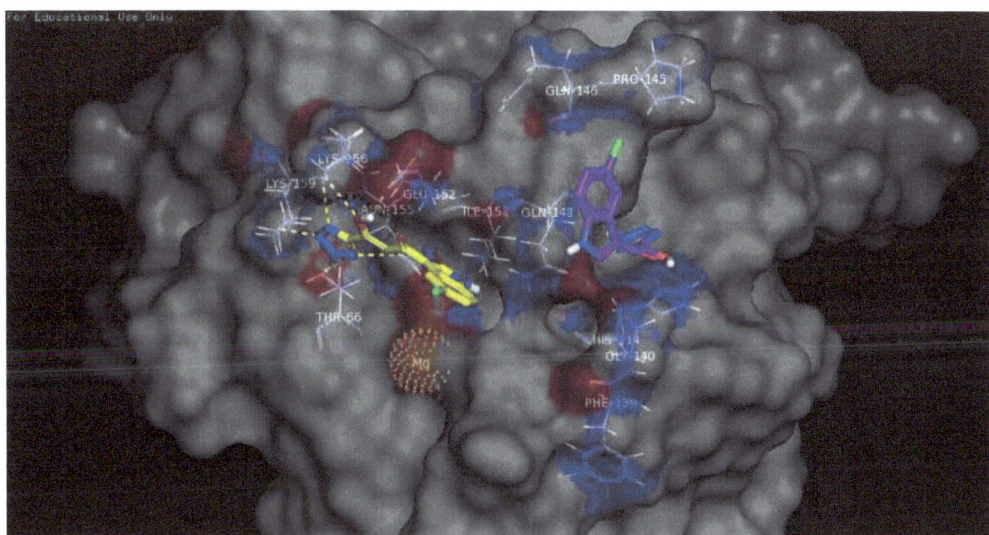

Fig. (10). 3D binding model of 5CITEP, with the active site (yellow) and at the new binding site of HIV-1 IN (PDB: 1QS4) (purple) discovered by MDS study.

Raltegravir (MK-0518) was the first antiretroviral drug to be marketed in the integrase strand transfer inhibitor class and was developed by Merck & Co. This drug was approved by the Food and Drug Administration (FDA) in 2007 for the treatment of HIV-1-infected ARV-experienced patients, patients without previous treatment and adolescents [108].

The Development of Raltegravir

The design of raltegravir (Fig. **11**) started with a series of 4-aryl-2-4-diketobutanoic acids with different β-diketo acid substituents, which are in coordination with the metal ions present in the active site. This class of compounds show selective and potent inhibitory activity against the DNA strand transfer integration step.

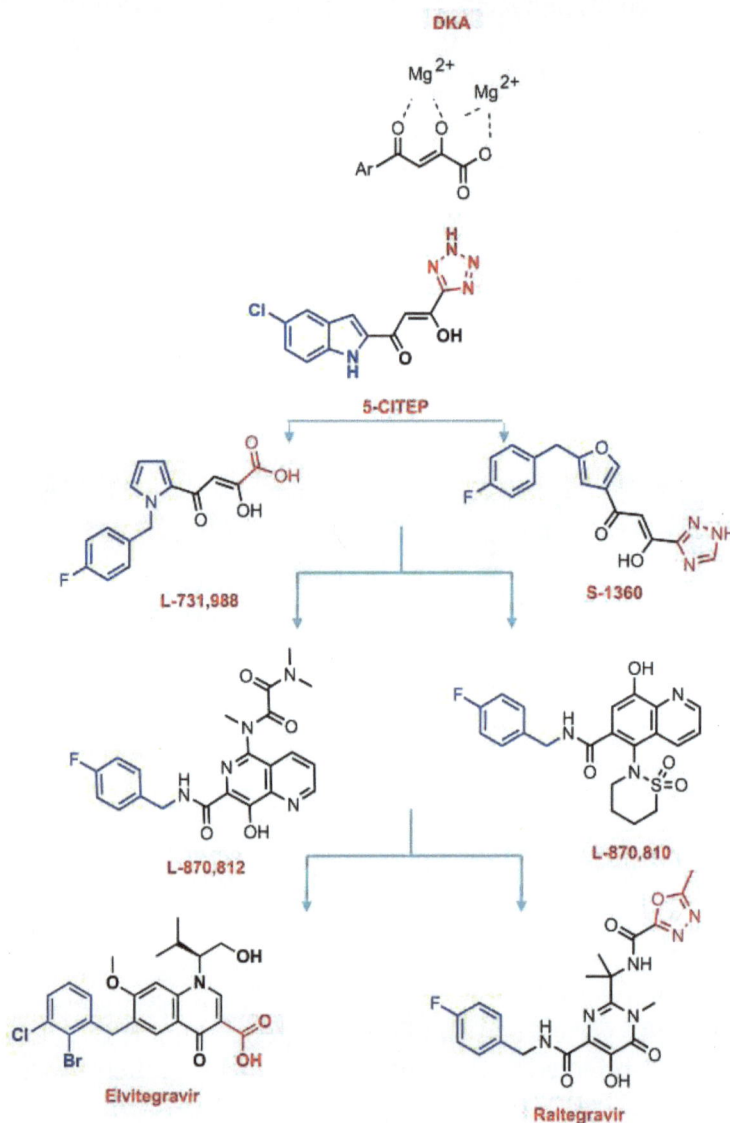

Fig. (11). Scheme showing the design process that eventually led to raltegravir.

Compound L-731,988, one of the most potent inhibitors, displayed a 70-fold higher IC50 value of 6 μM for 3'-processing compared to its 80 nM IC50 value for strand transfer inhibition and was the first structure co-crystallized with this enzyme. Later, some heterocycles were introduced to the molecular design replacing the indole in 5CITEP, and the tetrazole rings were replaced by its corresponding carboxylic acid-type bioisostere, which gave rise to heterocyclic analogues, among them was the first clinically tested HIV-1 IN inhibitor the compound S-1360. This compound showed inhibition of the IN enzyme and HIV replication at the low nanomolar and micromolar range. On the other hand, it revealed a good toxicology profile in Phase I trials. However, in the following clinical phases, the efficacy of S-1360 was not adequate, because it undergoes a metabolic reduction reaction at the carbon adjacent to the triazole, generating an inactive metabolite that is rapidly eliminated through its conjugation with glucuronic acid, and for this reason its study was not continued.

Subsequently, a substitution of the 1,3-diketoacid moiety (DKA) for 8-hydroxy-(1,6) naphthyridine carboxamide was carried out, this class of compounds retained their antiviral activity and strand transfer selectivity, which were also metabolically more stable. It should be noted that L870,810 was the most active inhibitor in the series, with IC50 values of 4 nM against multidrug-resistant viruses. Nevertheless, it manifested hepatotoxicity and nephrotoxicity after chronic administration in dogs, and because of this, its study and development were not continued.

Due to the above, new carboxylic acid analogues of *N*-alkyl hydroxypyrimidinone were designed, which displayed in vitro nanomolar activity against HIV-1 IN and an acceptable pharmacokinetic profile in rats [109]. Further modifications of these compounds led to the synthesis of Raltegravir, commercially known as Isentress™, a pyrimidinone carboxamide derivative [110 - 114].

To determine the antiviral activity, raltegravir was screened against a panel of HIV isolates (primary isolates from a variety of subtypes, isolates resistant to protease inhibitors, nucleoside reverse transcriptase inhibitors, non-nucleoside reverse transcriptase inhibitors, and simian immunodeficiency virus.) The IC_{50} values obtained reveal that the compounds are potent inhibitors of the strand transfer activity of purified HIV-1 integrase with values in the low nanomolar range IC_{50} of 2–7 nM, being 1000 times more selective against the HIV-1 integrase vs other phosphoryltransferases tested, including the polymerase and RNase H activities of HIV-1 reverse transcriptase and the human polymerases α, β, and γ.

Based on the above, the mechanism of action of this molecule is described as an inhibitory effect of the integrase of human immunodeficiency virus type 1), whose function is to catalyze the insertion of viral DNA [115] into the chromosomes of infected CD4+ lymphocytes.

However, one of the more common culprits of antiviral drugs becoming ineffective is drug resistance. Although Raltegravir at the time was still considered a relatively novel compound, by 2010, it had already been found HIV-1 integrase mutations that led to resistance to this drug. Considering this, newer attempts with molecular dynamics were developed.

In 2010, Perryman *et al.* described a restrained molecular dynamics protocol to produce a more accurate model of the active site of the HIV-1 integrase using the wild-type sequence as well as the reported mutations [116]. In short, they introduced NMR-type restraints during the equilibration and production phases, used the SHAKE algorithm to constrain any bonds with a hydrogen atom with snapshots recorded every 500 steps, each of 2 femtoseconds. Each simulation was 20 nanoseconds long. The results concluded that raltegravir displayed both the "primary" and "flipped" modes against only the wild-type protein. Among these, the better-known mutation responsible for raltegravir's drug resistance is N155H, however, during clinical trials, this was further explored by identifying Q148H/K/R and Y143C/R mutations as well as the importance of the combination of N155H with E92Q mutations in the same enzyme [117, 118]. The results observed from this MD protocol fall in accordance to what was observed experimentally.

Furthermore, in 2012, Xue *et al.* [119] found that using molecular dynamics followed by MM-GBSA calculations allowed them to explore how raltegravir interacted with the mutant and wild-type HIV-1 integrase, thus concluding that the Tyr143 residue plays an anchoring role for raltegravir at the binding site. Additionally, they also found that the movement of 3'-adenosine takes on different conformations in the wild type and the mutant protein and how this could potentially explain the lack of efficacy for these variants.

Moreover, molecular dynamics simulations have been used in the repurposing of approved drugs, allowing clinical trials to be carried out immediately in accordance with the guidelines and standards of the new pharmacological activity being studied. This strategy has been particularly successful in the discovery and use of rapidly spreading emerging disease drugs such as SARS-CoV-2. In 2020, the research group of Y. Kumar *et al.* performed molecular dynamic simulations of the complex coupled raltegravir with the X-ray crystal structure of SARS-CoV- 2 Mpro (PDB ID: 6Y2F) to study the conformational stability of the complex. The MD simulations were carried out (after molecular docking) through 10 nanoseconds with 101 snapshots with the AMBER14 force field, additionally being also used 1-femtosecond time steps and periodic boundaries in one simulation box. The results of molecular docking revealed raltegravir displayed interaction with the residues His164, Arg188, Gln192, Glu166, Met49, Met165, Phe140, Pro168, and Leu167 and MD simulations support the stability and conformational flexibility of this molecule in the enzyme active site [120].

WORKING EXAMPLE: MOLECULAR DYNAMICS TUTORIAL USING DESMOND

As previously mentioned, Desmond may not be the top choice when compared to current alternatives. Nonetheless, we consider that other resources have a stock of descriptions and instructions for their basic usage. Hence, in the spirit of diversity of choice, we present a brief tutorial on the basics of Desmond.

Generalities

DESMOND is composed of several tools, of which multisim can be considered the main backend. In practice, it can be used from the command line or *via* the Maestro GUI. For the purposes of this tutorial, we will be using Maestro. However, if the reader wishes to learn about command line usage, we refer them to the user's manual, as this subject falls beyond the scope of this brief tutorial. For context, in general, whenever quotation marks ("") or the dollar sign ($) are used, it signals the command prompt syntax. As mentioned before, Desmond is intended for GPU use only from its 2019 version and onwards, however, the CPU version of the software has minor differences, and these will be briefly discussed.

The installation procedure is described in the README file, from which we only need the absolute path to the created directory. Having that said, in order to gain access to the software tools, it is necessary to state an environment variable such as "SCHRODINGER" for which the next command is used:

"$export SCHRODINGER= /path/to/Desmond/directory"

From this point forward and until we close the terminal, we can use the different tools by typing "$SCHRODINGER". For local use, it is more convenient to define this variable in the file ".bashrc".

Making an Atomistic Simulation of Crambin

Crambin is a very small protein (only 46 residues), a member of the thionin family. Its sequence and structure bears notable homology to plant toxins, such as viscotoxin [121]. Due to its size and secondary structure distribution, it makes for a good test case. In the next sections, we present a basic simulation protocol accounting for tens of nanoseconds.

Setting Up the System

1. Open a terminal (Ctrl + Alt + T is an usual shortcut) and browse the desktop, then create a folder named "MD":

"$ cd Desktop &&mkdir MD && cd MD"

Now we will use the environment variable to execute Maestro:

"$ $SCHRODINGER/maestro -SGL"

Note: The use of the SGL flag is recommended as there are cases where a GPU is unavailable or GPU is installed, but maestro crashes due to driver issues.

2. In the main window, we browse to File -> Get PDB

When clicked, a new window will appear in which we will type the code "1CRN" and choose the option *Auto*:

3. With the file loaded, click on *Protein Preparation Wizard* in the shortcut bar.

4. In the dialog box, make sure the following options are checked: *Assign Bond orders*, *Use CCDC database*, *Add Hydrogens*, *Create zero-order bonds to metals*, *Create disulfide bonds* and *Cap Termini*. Next, choose *Preprocess*.

5. For this example, the protein structure is complete. In other cases, there may be missing atoms, residues, or even loops. For such systems prior preprocessing is due, unfortunately a commercial license of Prime and Glide is needed for these calculations. In such scenarios, available options for this include Ambertools, GROMACS, or PDBfixer.

6. Note that Maestro creates new "versions" of the system after each preparation step. Next, we will move to the Refine tab, verify that *Use PROPKA* is selected, and click to *Optimize*.

7. Finally, we can corroborate that the system is ready by choosing once again *Preprocess*. A confirmation pop-up box will appear stating that the protein doesn't have any issues to be fixed.

Building the Simulation Box

1. Now with the prepped protein selected, browse to TASKS -> *Desmond* (in the upper right-side corner).

2. In the following menu, select *System Builder*. Here is where we set up the system, check that the cell type is set to *Orthorhombic*. Other available options are cubic, truncated octahedron, dodecahedron, and triclinic cells.

3. Now proceed to the *Ions* tab to *Neutralize* the cell and add salt if needed.

4. All that is left is to assign a filename, which is done at the bottom bar. At the end of the process, the folder MD will have a subfolder with the given name. Additionally, in the Maestro window, the cell system will be added.

Reparametrization Using the AMBER99SB-ILDN Force Field

Now you should have a system ready for MD simulation. However, the default parametrization relies on the OPLS_2005 force field. Up next, we will comment on the use of Viparr, which allows reparametrization to other force fields. Of note, the Viparr utility included in most versions of Academic Desmond is outdated. Nonetheless, it is still possible to use it for parameter assignment and is the one described here. A newer version is available from Github: https://github.com/DEShawResearch/viparr. General usage of this newer version is described elsewhere [122, 123].

1. Open a new terminal and browse to the directory of the system created with Maestro

2. Execute the "ls" command and you will see several files. The one we need has the "-out.cms" extension.

3. The next step is to execute Viparr over this file. This script counts with many options, which are displayed by running:

"$ $SCHRODINGER/run -FROM desmond viparr.py -h"

4. Viparr uses preloaded templates located in the installation directory. When the user assigns a force field with -f flag, Viparr searches for residues that are similar and reassigns the parameters by homology. If Viparr is unable to find similarities, it will return an error message. In such cases, it is necessary to create files including the topology, parameters and templates compatible with the script.

5. For this example, AMBER99SBD-ILDN and TIP3P parameters will be used. To do this issue, the following command:

"$ $SCHRODINGER/maestro run -FROM desmond viparr.py -f amber99SB-ILDN -f tip3p system_name-out.cmssystem_amber.cms"

6. Of note, Viparr deletes any geometric constraints in the box. To recover those, run:

"$ $SCHRODINDER/run -FROM desmond build_constraints.py system_amber.cmssystem_amber-out.cms"

Initializing the Simulation Protocol: Minimization

1. In the Maestro window, import the reparametrized system (Ctrl + I) (system_amber-out.cms).

2. Again, browse to TASKS ->*Desmond-> Minimization.*

3. Make sure that the reparametrized system is selected (with both the circle and name in blue).

4. In the *Minimization* window, press *Load* and the name of the file will appear next to the button.

5. A notable difference between GPU and CPU versions is the approach for minimization. In the former, the system is simulated under Brownian dynamics with a slow timestep and low temperature. For the latter, a mixture of the steepest descent and L-BFGS algorithm is used.

6. In the advanced options tab, additional constraints can be set up using ASL selection. For this example, add a constraint for C-alpha atoms with 25 kcal/mol: Force constant 25.0; selection: protein AND atom type "CA". Select apply, choose a proper filename, and run minimization.

7. Wait for the system incorporation and select it.

Equilibration and Dynamics

With the system minimized, we can now move to the equilibration stage and lastly begin the simulation.

1. Once again, browse for TASKS ->*Desmond->Molecular Dynamics.* Make sure that the already minimized system is selected.

2. Select *Load* and make sure the name of the system is loaded. Now you may proceed to stipulate the simulation characteristics.

3. For the particular case of Desmond, the equilibration is specified in a file with the ".msj" extension. The syntax and options of this file go beyond the scope of this tutorial. However, Desmond uses a default protocol whenever *Relax System* is checked. This default procedure is designed to be robust enough for simple simulations. Even so, its use is not to be assumed as universal nor failsafe.

4. Once the parameters are defined, all that is left is to choose a file name and specify how many CPUs are to be used. By default, Desmond only uses one CPU, this can be modified within the menu at the bottom (the one with a gear icon). When pressed, a new dialog box will pop up and you will be able to increase the number of CPUs. After this, you may hit the *Run* button.

5. Simulation will commence shortly, and it should not take long, even on a laptop. If you desire to monitor the process, just click the green icon next to TASKS.

Trajectory Analysis

Analysis of molecular dynamics simulation is not a straightforward task, as very diverse measures can be obtained to monitor a particular event of interest. In general, a first step usually consists of determining if the system remains "stable" during the simulation. This does not refer to thermodynamics but rather physical stability. In this sense, there is no better sanity check than trajectory playback. It does not take long and should evidence any serious artifacts, if any. Of course, a proper approach is to examine if the desired conditions for the simulation were achieved. Another common set of tests includes root mean squared deviation and fluctuation of alpha carbons or side chains and gyration radius of the protein.

In Maestro, several tools are available, including command line scripts; in the GUI, available tools are *Simulation Quality Analysis*, *Simulation Event Analysis* and *Simulation Interaction Diagram*.

Note: Before we begin, note that whenever a file belonging to a simulation is imported, it will display a blue button with a "T". Clicking on it allows trajectory playback with a plethora of options to improve visualization.

The trajectory icon is also useful to visualize the simulation.

1. Move to TASKS-> *Desmond->Simulation Quality Analysis*.

2. Press *Browse*, go to the dynamics folder select the file with extension ".ene". Press Analyze and then Plot. The newly created Plot will appear with the tendencies for different system variables. Try changing the values and replot.

3. Now, onto *Simulation Event Analysis*.

4. This allows us to load directly any displayed trajectory.

5. Let's take a look at an example in the RMSD tab, it is possible to calculate this metric using predefined selection and pressing Add. Note how a new list appears in the keywords category.

6. Other analyses are possible by browsing other tabs. In particular, *Measurements* allow the evaluation of distance, angles, and dihedrals.

7. Once we select RMSD and RMSF of the alpha carbons and gyration radius, you may select *Analyze All* below.

8. When the script is finished, the right-hand options will be activated and will allow plot visualization by selecting the data types. Additionally, the button *Export Results* saves a CSV file with all raw data.

Even with the provided tools, there may be instances where a very specific set of measures would be needed. For cases like this, flexibility is crucial. Now, for the sake of completeness, we will comment on the analysis of Desmond simulations with open-source tools. In this regard, VMD is fully compatible with Desmond's trajectory format (dtr). Another (open source) alternative is MDTraj, which is our next example.

Analysis with MDTraj

Prior to any analysis, MDTraj will need data on the system's topology. This can be accomplished with VMD. Launch the program and select *New molecule*. Then select the file with the '-out.cms' extension. You will see the simulation cell, now select *Save coordinates*. Ensure that pdb is the output format and that all atoms are selected. Give the output a proper name and exit VMD. We now continue with the analysis.

MDTraj is a Python library which can be easily obtained with anaconda:

"$ conda install -c conda-forge mdtraj"

For the full extent of MDTraj capabilities, we recommend checking its documentation: https://mdtraj.org/1.9.4/index.html. We now present a simple function which replicates the analyses carried out with Simulation Event Analysis:

import mdtraj as md

import matplotlib.pyplot as plt

def md_basic_analysis(trajectory, topology, output):

"""""

Perform RMSD, RMSF, and gyration radius analysis of an MD trajectory,

calculations are made for CA atoms.

Parameters

-----

traj: The trajectory to load, for Desmond this refers to the clickme.dtr file in the _trj directory

topology: The file contains structural information such as PDB

output: Base name for the output plots

"""""

traj = md.load(trajectory, top= topology) #Loading the system and topology

traj = md.Trajectory.superpose(traj,traj,frame=0) #Aligning the trajectory

ca = traj.atom_slice(traj.topology.select('name CA')) # Extracting a subsystem

rmsd = md.rmsd(ca,ca)

rmsf = md.rmsf(ca,ca)

rg = md.compute_rg(ca)

rmsd_plot = plt.figure(dpi = 300)

plt.plot(traj.time/1000,rmsd*10) #MDTraj uses ps and nm as default units, thus the conversion

plt.xlabel("time (ns)")

```
plt.ylabel("RMSD ($\\AA$)")

plt.legend(labels = ['C$\\alpha$'],loc='best')

plt.savefig('rmsd_'+str(output)+'.png')

rmsf_plot = plt.figure(dpi = 300)

plt.plot(rmsf*10)

plt.xlabel("Residue Index")

plt.ylabel("RMSF ($\\AA$)")

plt.legend(labels = ['C$\\alpha$'],loc='best')

plt.savefig('rmsf_'+str(output)+'.png')

rg_plot = plt.figure(dpi = 300)

plt.plot(traj.time/1000,rg*10)

plt.xlabel("time (ns)")

plt.ylabel("Radius of gyration ($\\AA$)")

plt.legend(labels = ['C$\\alpha$'],loc='best')

plt.savefig('rg_'+str(output)+'.png')
```

CONCLUSION

Molecular dynamics simulations are in a coming-of-age period. With many software options and powerful hardware currently available. For the last three decades, molecular dynamics has been successfully applied in structure-based drug design. Notable examples include the study of binding mechanisms, protein folding, and transient phenomena such as cryptic pockets.

This has led to wide adoption of the methodology; however, this poses the risk of trivializing the technique. As it has happened with molecular docking and the overarching perception of it as the ultimate predictive tool for drug design. Furthermore, naïve application of molecular dynamics could yield artifacts that may be mistaken for valid results. Therefore, we encourage the use of good practices, with careful selection of simulation parameters and always ensuring the compatibility of selected algorithms.

In addition, proper sampling of the free energy landscape remains a major challenge for MD. Attempts to bias simulations in meaningful ways have paved the way for more robust studies of proteins. Examples include the study of kinetics and the development of statistical models such as Markov state modelling.

There is no denying that the field will see notable improvements in the years to come. As such, we tried to present a comprehensive, yet concise view with a focus on critical aspects to account for drug discovery campaigns.

ACKNOWLEDGEMENTS

Fernando D. Prieto-Martínez is thankful to CONACyT for a postdoctoral fellowship No. 31146; granted to the project FORDECYT-PRONACES, No. 1561802.

REFERENCES

[1]   Salo-Ahen OMH, *et al.* Molecular dynamics simulations in drug discovery and pharmaceutical development. Processes 2021; 9(1): 71.
[http://dx.doi.org/10.3390/pr9010071]

[2]   Schlander M, Hernandez-Villafuerte K, Cheng C Y, Mestre-Ferrandiz J, Baumann M. How much does it cost to research and develop a new drug? A systematic review and assessment. Pharmacoeconomics 2021; 39(11): 1243-69.
[http://dx.doi.org/10.1007/s40273-021-01065-y]

[3]   Sun D, Gao W, Hu H, Zhou S. Why 90% of clinical drug development fails and how to improve it? Acta Pharm Sin B 2022; 12(7): 3049-62.
[http://dx.doi.org/10.1016/j.apsb.2022.02.002] [PMID: 35865092]

[4]   Gangrade D, Sawant G, Mehta A. Re-thinking drug discovery: *In silico* method. J Chem Pharmac Res 2016; 8(8): 1092-9. Available from: www.jocpr.com

[5]   Nyberg AM, Schlick T. Increasing the time step in molecular dynamics. Chem Phys Lett 1992; 198(6): 538-46.
[http://dx.doi.org/10.1016/0009-2614(92)85028-9]

[6]   Stocker U, Juchli D, van Gunsteren WF. Increasing the time step and efficiency of molecular dynamics simulations: Optimal solutions for equilibrium simulations or structure refinement of large biomolecules. Mol Simul 2003; 29(2): 123-38.
[http://dx.doi.org/10.1080/0892702031000065791]

[7]   Streett WB, Tildesley DJ, Saville G. Multiple time-step methods in molecular dynamics. Mol Phys 1978; 35(3): 639-48.
[http://dx.doi.org/10.1080/00268977800100471]

[8]   Predescu C, Lippert RA, Eastwood MP, *et al.* Computationally efficient molecular dynamics integrators with improved sampling accuracy. Mol Phys 2012; 110(9-10): 967-83.
[http://dx.doi.org/10.1080/00268976.2012.681311]

[9]   Lippert RA, Predescu C, Ierardi DJ, *et al.* Accurate and efficient integration for molecular dynamics simulations at constant temperature and pressure. J Chem Phys 2013; 139(16): 164106.
[http://dx.doi.org/10.1063/1.4825247] [PMID: 24182003]

[10]   Verlet L. Computer "experiments" on classical fluids. i. thermodynamical properties of lennard-jones

molecules. Phys Rev 1967; 159(1): 98-103.
[http://dx.doi.org/10.1103/PhysRev.159.98]

[11]   Zanette C, Bannan CC, Bayly CI, *et al.* Toward learned chemical perception of force field typing rules. J Chem Theory Comput 2019; 15(1): 402-23.
[http://dx.doi.org/10.1021/acs.jctc.8b00821] [PMID: 30512951]

[12]   Wereszczynski J, McCammon JA. Statistical mechanics and molecular dynamics in evaluating thermodynamic properties of biomolecular recognition. Q Rev Biophys 2012; 45(1): 1-25.
[http://dx.doi.org/10.1017/S0033583511000096] [PMID: 22082669]

[13]   Frenkel D, Smit B. Understanding Molecular Simulation. From Algorithms to Applications. Elsevier: Academic Press 2002.
[http://dx.doi.org/10.1016/B978-0-12-267351-1.X5000-7]

[14]   Tuckerman ME. Statistical mechanics: Theory and molecular simulation. 2nd., Oxford University Press 2010.

[15]   Frenkel D, Smit B. Understanding molecular simulation: From algorithms to applications. 2nd., Academic Press 2001.

[16]   Rapaport DC. The Art of Molecular Dynamics Simulation. 2nd., Cambridge University Press 2011.

[17]   Ramachandran KI, Gopakumar D, Namboori K. Computational Chemistry and Molecular Modeling. Berlin, Heidelberg: Springer Berlin Heidelberg 2008.
[http://dx.doi.org/10.1007/978-3-540-77304-7]

[18]   Murthy KPN. Metropolis and wang-landau algorithms. BRNS School on Computational Methodologies across Length Scales, August 28 - September 09, 2017, BARC, Mumbai.

[19]   Paquet E, Viktor HL. Molecular dynamics, monte carlo simulations, and langevin dynamics: A computational review. BioMed Res Int 2015; 2015: 1-18.
[http://dx.doi.org/10.1155/2015/183918] [PMID: 25785262]

[20]   Shaw DE, Maragakis P, Lindorff-Larsen K, *et al.* Atomic-level characterization of the structural dynamics of proteins. Science 2010; 330(6002): 341-6.
[http://dx.doi.org/10.1126/science.1187409]

[21]   Hansen HS, Daura X, Hünenberger PH. Enhanced conformational sampling in molecular dynamics simulations of solvated peptides: Fragment-based local elevation umbrella sampling. J Chem Theory Comput 2010; 6(9): 2598-621.
[http://dx.doi.org/10.1021/ct1003059] [PMID: 26616064]

[22]   Sugita Y, Okamoto Y. Replica-exchange molecular dynamics method for protein folding. Chem Phys Lett 1999; 314(1-2): 141-51.
[http://dx.doi.org/10.1016/S0009-2614(99)01123-9]

[23]   Bowman GR, Ensign DL, Pande VS. Enhanced modeling *via* network theory: Adaptive sampling of Markov state models. J Chem Theory Comput 2010; 6(3): 787-94.
[http://dx.doi.org/10.1021/ct900620b] [PMID: 23626502]

[24]   Tiwary P, Parrinello M. From metadynamics to dynamics. Phys Rev Lett 2013; 111(23): 230602.
[http://dx.doi.org/10.1103/PhysRevLett.111.230602] [PMID: 24476246]

[25]   Wereszczynski J, McCammon JA. Accelerated molecular dynamics in computational drug design. Methods Mol Biol 2012; 819: 515-24.
[http://dx.doi.org/10.1007/978-1-61779-465-0_30]

[26]   Kokh DB, Amaral M, Bomke J, *et al.* Estimation of drug-target residence times by τ-random acceleration molecular dynamics simulations. J Chem Theory Comput 2018; 14(7): 3859-69.
[http://dx.doi.org/10.1021/acs.jctc.8b00230] [PMID: 29768913]

[27]   Procacci P. Multiple Bennett acceptance ratio made easy for replica exchange simulations. J Chem Phys 2013; 139(12): 124105.

[http://dx.doi.org/10.1063/1.4821814] [PMID: 24089748]

[28]   Groenhof G. Introduction to QM/MM simulations. Methods Mol Biol 2013; 924: 43-66.
[http://dx.doi.org/10.1007/978-1-62703-017-5_3]

[29]   Panteva MT, Giambaşu GM, York DM. Comparison of structural, thermodynamic, kinetic and mass transport properties of Mg^{2+} ion models commonly used in biomolecular simulations. J Comput Chem 2015; 36(13): 970-82.
[http://dx.doi.org/10.1002/jcc.23881] [PMID: 25736394]

[30]   Li P, Merz KM Jr. Metal Ion modeling using classical mechanics. Chem Rev 2017; 117(3): 1564-686.
[http://dx.doi.org/10.1021/acs.chemrev.6b00440] [PMID: 28045509]

[31]   Liao Q, Pabis A, Strodel B, Kamerlin SCL. Extending the nonbonded cationic dummy model to account for ion-induced dipole interactions. J Phys Chem Lett 2017; 8(21): 5408-14.
[http://dx.doi.org/10.1021/acs.jpclett.7b02358] [PMID: 29022713]

[32]   Li P, Merz KM Jr. MCPB.py: A python based metal center parameter builder. J Chem Inf Model 2016; 56(4): 599-604.
[http://dx.doi.org/10.1021/acs.jcim.5b00674] [PMID: 26913476]

[33]   Friedrich NO, de Bruyn Kops C, Flachsenberg F, Sommer K, Rarey M, Kirchmair J. Benchmarking commercial conformer ensemble generators. J Chem Inf Model 2017; 57(11): 2719-28.
[http://dx.doi.org/10.1021/acs.jcim.7b00505] [PMID: 28967749]

[34]   Zhu S. Validation of the generalized force fields GAFF, CGenFF, OPLS-AA, and PRODRGFF by testing against experimental osmotic coefficient data for small drug-like molecules. J Chem Inf Model 2019; 59(10): 4239-47.
[http://dx.doi.org/10.1021/acs.jcim.9b00552] [PMID: 31557024]

[35]   Qiu Y, Smith DGA, Boothroyd S, *et al.* Development and benchmarking of open force field v1.0.0—the parsley small-molecule force field. J Chem Theory Comput 2021; 17(10): 6262-80.
[http://dx.doi.org/10.1021/acs.jctc.1c00571] [PMID: 34551262]

[36]   Shaw DE, Deneroff MM, Dror RO, *et al.* Anton, a special-purpose machine for molecular dynamics simulation. Commun ACM 2008; 51(7): 91-7.
[http://dx.doi.org/10.1145/1364782.1364802]

[37]   Lopes PEM, Guvench O, MacKerell AD Jr. Current status of protein force fields for molecular dynamics simulations. Methods Mol Biol 2015; 1215: 47-71.
[http://dx.doi.org/10.1007/978-1-4939-1465-4_3] [PMID: 25330958]

[38]   Guvench O, Mackerell A D. Comparison of protein force fields for molecular dynamics simulations. Methods Mol Biol 2008; 443: 63-88.
[http://dx.doi.org/10.1007/978-1-59745-177-2_4]

[39]   Fujii N, Saito T. Homochirality and life. Chem Rec 2004; 4(5): 267-78.
[http://dx.doi.org/10.1002/tcr.20020] [PMID: 15543607]

[40]   Bowers KJ, Chow DE, Xu H, *et al.* Scalable algorithms for molecular dynamics simulations on commodity clusters. ACM/IEEE SC 2006 Conference (SC'06), IEEE, Nov. 2006, pp. 43–43
[http://dx.doi.org/10.1109/SC.2006.54]

[41]   Abraham MJ, Murtola T, Schulz R, *et al.* GROMACS: High performance molecular simulations through multi-level parallelism from laptops to supercomputers. SoftwareX 2015; 1-2: 19-25.
[http://dx.doi.org/10.1016/j.softx.2015.06.001]

[42]   Phillips JC, Hardy DJ, Maia JDC, *et al.* Scalable molecular dynamics on CPU and GPU architectures with NAMD. J Chem Phys 2020; 153(4): 044130.
[http://dx.doi.org/10.1063/5.0014475] [PMID: 32752662]

[43]   Eastman P, Swails J, Chodera JD, *et al.* OpenMM 7: Rapid development of high performance algorithms for molecular dynamics. PLOS Comput Biol 2017; 13(7): e1005659.

[http://dx.doi.org/10.1371/journal.pcbi.1005659] [PMID: 28746339]

[44]   Arantes PR, Polêto MD, Pedebos C, Ligabue-Braun R. Making it rain: Cloud-based molecular simulations for everyone. J Chem Inf Model 2021; 61(10): 4852-6.
[http://dx.doi.org/10.1021/acs.jcim.1c00998] [PMID: 34595915]

[45]   Sydow D, Rodríguez-Guerra J, Kimber TB, *et al.* TeachOpenCADD 2022: Open source and FAIR Python pipelines to assist in structural bioinformatics and cheminformatics research. Nucleic Acids Res 2022; 50(W1): W753-60.
[http://dx.doi.org/10.1093/nar/gkac267] [PMID: 35524571]

[46]   Doerr S, Majewski M, Pérez A, *et al.* TorchMD: A deep learning framework for molecular simulations. J Chem Theory Comput 2021; 17(4): 2355-63.
[http://dx.doi.org/10.1021/acs.jctc.0c01343] [PMID: 33729795]

[47]   Krieger E, Vriend G. New ways to boost molecular dynamics simulations. J Comput Chem 2015; 36(13): 996-1007.
[http://dx.doi.org/10.1002/jcc.23899] [PMID: 25824339]

[48]   Braun E, Gilmer J, Mayes HB, *et al.* Best practices for foundations in molecular simulations [Article v1.0]. Living J Comput Mol Sci 2019; 1(1): 5957.
[http://dx.doi.org/10.33011/livecoms.1.1.5957] [PMID: 31788666]

[49]   Wassenaar TA, Mark AE. The effect of box shape on the dynamic properties of proteins simulated under periodic boundary conditions. J Comput Chem 2006; 27(3): 316-25.
[http://dx.doi.org/10.1002/jcc.20341] [PMID: 16358324]

[50]   Schlick T. Molecular Modeling and Simulation: An Interdisciplinary Guide. New York, NY: Springer New York 2010; 21.
[http://dx.doi.org/10.1007/978-1-4419-6351-2]

[51]   Sri Harish M, Patra PK. Temperature and its control in molecular dynamics simulations. Mol Simul 2021; 47(9): 701-29.
[http://dx.doi.org/10.1080/08927022.2021.1907382]

[52]   Woodcock LV. Isothermal molecular dynamics calculations for liquid salts. Chem Phys Lett 1971; 10(3): 257-61.
[http://dx.doi.org/10.1016/0009-2614(71)80281-6]

[53]   Berendsen HJC, Postma JPM, van Gunsteren WF, DiNola A, Haak JR. Molecular dynamics with coupling to an external bath. J Chem Phys 1984; 81(8): 3684-90.
[http://dx.doi.org/10.1063/1.448118]

[54]   Harvey SC, Tan RKZ, Cheatham TE. The flying ice cube: Velocity rescaling in molecular dynamics leads to violation of energy equipartition. J Comput Chem 1998; 19(7): 726-40.
[http://dx.doi.org/10.1002/(SICI)1096-987X(199805)19:7<726::AID-JCC4>3.0.CO;2-S]

[55]   Braun E, Moosavi SM, Smit B. Anomalous effects of velocity rescaling algorithms: The flying ice cube effect revisited. J Chem Theory Comput 2018; 14(10): 5262-72.
[http://dx.doi.org/10.1021/acs.jctc.8b00446] [PMID: 30075070]

[56]   Mor A, Ziv G, Levy Y. Simulations of proteins with inhomogeneous degrees of freedom: The effect of thermostats. J Comput Chem 2008; 29(12): 1992-8.
[http://dx.doi.org/10.1002/jcc.20951] [PMID: 18366022]

[57]   Martyna GJ, Klein ML, Tuckerman M. Nosé–Hoover chains: The canonical ensemble *via* continuous dynamics. J Chem Phys 1992; 97(4): 2635-43.
[http://dx.doi.org/10.1063/1.463940]

[58]   Holian BL, Voter AF, Ravelo R. Thermostatted molecular dynamics: How to avoid the Toda demon hidden in Nosé-Hoover dynamics. Phys Rev E Stat Phys Plasmas Fluids Relat Interdiscip Topics 1995; 52(3): 2338-47.
[http://dx.doi.org/10.1103/PhysRevE.52.2338] [PMID: 9963676]

[59]   Cheng A, Merz KM. Application of the nosé–hoover chain algorithm to the study of protein dynamics. J Phys Chem 1996; 100(5): 1927-37.
[http://dx.doi.org/10.1021/jp951968y]

[60]   Patra PK, Bhattacharya B. Nonergodicity of the nose-hoover chain thermostat in computationally achievable time. Phys Rev E Stat Nonlin Soft Matter Phys 2014; 90(4): 043304.
[http://dx.doi.org/10.1103/PhysRevE.90.043304] [PMID: 25375620]

[61]   Brańka AC. Nosé-Hoover chain method for nonequilibrium molecular dynamics simulation. Phys Rev E Stat Phys Plasmas Fluids Relat Interdiscip Topics 2000; 61(5): 4769-73.
[http://dx.doi.org/10.1103/PhysRevE.61.4769] [PMID: 11031517]

[62]   Watanabe H, Kobayashi H. Ergodicity of a thermostat family of the Nosé-Hoover type. Phys Rev E Stat Nonlin Soft Matter Phys 2007; 75(4): 040102.
[http://dx.doi.org/10.1103/PhysRevE.75.040102] [PMID: 17500844]

[63]   Pomeau Y, Piasecki J. The langevin equation. C R Phys 2017; 18(9-10): 570-82.
[http://dx.doi.org/10.1016/j.crhy.2017.10.001]

[64]   Schuss Z. Brownian simulation of langevin's.Brownian Dynamics at Boundaries and Interfaces Applied Mathematical Sciences. New York, NY: Springer 2013; 186: pp. 89-109.
[http://dx.doi.org/10.1007/978-1-4614-7687-0_3]

[65]   Balakrishnan V. The langevin equation.Elements of Nonequilibrium Statistical Mechanics. Cham: Springer International Publishing 2021; pp. 10-23.
[http://dx.doi.org/10.1007/978-3-030-62233-6_2]

[66]   Pastor RW. Techniques and applications of langevin dynamics simulations.The Molecular Dynamics of Liquid Crystals. Dordrecht: Springer Netherlands 1994; pp. 85-138.
[http://dx.doi.org/10.1007/978-94-011-1168-3_5]

[67]   Lü JT, Hu BZ, Hedegård P, Brandbyge M. Semi-classical generalized Langevin equation for equilibrium and nonequilibrium molecular dynamics simulation. Prog Surf Sci 2019; 94(1): 21-40.
[http://dx.doi.org/10.1016/j.progsurf.2018.07.002]

[68]   Ferrari L. Test particles in a gas: Markovian and non-Markovian Langevin dynamics. Chem Phys 2019; 523: 42-51.
[http://dx.doi.org/10.1016/j.chemphys.2019.03.011]

[69]   Loos SAM. The Langevin Equation.Stochastic Systems with Time Delay. Springer 2021; pp. 21-75.
[http://dx.doi.org/10.1007/978-3-030-80771-9_2]

[70]   Sekimoto K. Stochastic Energetics. Berlin, Heidelberg: Springer Berlin Heidelberg 2010; 799.
[http://dx.doi.org/10.1007/978-3-642-05411-2]

[71]   Leimkuhler B, Matthews C. Numerical methods for stochastic molecular dynamics.Molecular Dynamics. springer 2015; pp. 261-328.
[http://dx.doi.org/10.1007/978-3-319-16375-8_7]

[72]   Chen JC, Kim AS. Brownian dynamics, molecular dynamics, and monte carlo modeling of colloidal systems. Adv Colloid Interface Sci 2004; 112(1-3): 159-73.
[http://dx.doi.org/10.1016/j.cis.2004.10.001] [PMID: 15581559]

[73]   Ruiz-Franco J, Rovigatti L, Zaccarelli E. On the effect of the thermostat in non-equilibrium molecular dynamics simulations. Eur Phys J E 2018; 41(7): 80.
[http://dx.doi.org/10.1140/epje/i2018-11689-4] [PMID: 29955976]

[74]   Heyes DM. Molecular dynamics at constant pressure and temperature. Chem Phys 1983; 82(3): 285-301.
[http://dx.doi.org/10.1016/0301-0104(83)85235-5]

[75]   Kahk JM, Tan BH, Ohl CD, Loh ND. Viscous field-aligned water exhibits cubic-ice-like structural motifs. Phys Chem Chem Phys 2018; 20(30): 19877-84.

[http://dx.doi.org/10.1039/C8CP02697A] [PMID: 29968884]

[76]   Van Eijck BP. Pressure calculation in molecular dynamics simulations of molecular crystals. Mol Simul 1994; 13(3): 221-30.
[http://dx.doi.org/10.1080/08927029408021985]

[77]   Paci E, Marchi M. Constant-pressure molecular dynamics techniques applied to complex molecular systems and solvated proteins. J Phys Chem 1996; 100(10): 4314-22.
[http://dx.doi.org/10.1021/jp9529679]

[78]   Rogge SMJ, Vanduyfhuys L, Ghysels A, *et al.* A comparison of barostats for the mechanical characterization of metal–organic frameworks. J Chem Theory Comput 2015; 11(12): 5583-97.
[http://dx.doi.org/10.1021/acs.jctc.5b00748] [PMID: 26642981]

[79]   Cerutti DS, Duke R, Freddolino PL, Fan H, Lybrand TP. A vulnerability in popular molecular dynamics packages concerning langevin and andersen dynamics. J Chem Theory Comput 2008; 4(10): 1669-80.
[http://dx.doi.org/10.1021/ct8002173] [PMID: 19180249]

[80]   Lobanov MY, Bogatyreva NS, Galzitskaya OV. Radius of gyration as an indicator of protein structure compactness. Mol Biol 2008; 42(4): 623-8.
[http://dx.doi.org/10.1134/S0026893308040195] [PMID: 18856071]

[81]   Weksberg TE, Lynch GC, Krause KL, Pettitt BM. Molecular dynamics simulations of Trichomonas vaginalis ferredoxin show a loop-cap transition. Biophys J 2007; 92(10): 3337-45.
[http://dx.doi.org/10.1529/biophysj.106.088096] [PMID: 17325017]

[82]   Wang Y, Shaikh SA, Tajkhorshid E. Exploring transmembrane diffusion pathways with molecular dynamics. Physiology 2010; 25(3): 142-54.
[http://dx.doi.org/10.1152/physiol.00046.2009] [PMID: 20551228]

[83]   Meiselbach H, Horn AHC, Harrer T, Sticht H. Insights into amprenavir resistance in E35D HIV-1 protease mutation from molecular dynamics and binding free-energy calculations. J Mol Model 2007; 13(2): 297-304.
[http://dx.doi.org/10.1007/s00894-006-0121-3] [PMID: 16794810]

[84]   Chen J, Liang Z, Wang W, Yi C, Zhang S, Zhang Q. Revealing origin of decrease in potency of darunavir and amprenavir against HIV-2 relative to HIV-1 protease by molecular dynamics simulations. Sci Rep 2014; 4(1): 6872.
[http://dx.doi.org/10.1038/srep06872] [PMID: 25362963]

[85]   Wang RG, Zhang HX, Zheng QC. Revealing the binding and drug resistance mechanism of amprenavir, indinavir, ritonavir, and nelfinavir complexed with HIV-1 protease due to double mutations G48T/L89M by molecular dynamics simulations and free energy analyses. Phys Chem Chem Phys 2020; 22(8): 4464-80.
[http://dx.doi.org/10.1039/C9CP06657H] [PMID: 32057044]

[86]   Chen J, Wang X, Zhu T, Zhang Q, Zhang JZH. A comparative insight into amprenavir resistance of mutations V32I, G48V, I50V, I54V, and I84V in HIV-1 protease based on thermodynamic integration and MM-PBSA methods. J Chem Inf Model 2015; 55(9): 1903-13.
[http://dx.doi.org/10.1021/acs.jcim.5b00173] [PMID: 26317593]

[87]   Yang LJ, Zou J, Xie HZ, Li LL, Wei YQ, Yang SY. Steered molecular dynamics simulations reveal the likelier dissociation pathway of imatinib from its targeting kinases c-Kit and Abl. PLoS One 2009; 4(12): e8470.
[http://dx.doi.org/10.1371/journal.pone.0008470] [PMID: 20041122]

[88]   Capelli AM, Costantino G. Unbinding pathways of VEGFR2 inhibitors revealed by steered molecular dynamics. J Chem Inf Model 2014; 54(11): 3124-36.
[http://dx.doi.org/10.1021/ci500527j] [PMID: 25299731]

[89]   Schames JR, Henchman RH, Siegel JS, Sotriffer CA, Ni H, McCammon JA. Discovery of a novel

binding trench in HIV integrase. J Med Chem 2004; 47(8): 1879-81.
[http://dx.doi.org/10.1021/jm0341913] [PMID: 15055986]

[90]   Sinha N, Li Y, Lipschultz CA, Smith-Gill SJ. Understanding antibody–antigen associations by molecular dynamics simulations: Detection of important intra- and inter-molecular salt bridges. Cell Biochem Biophys 2007; 47(3): 361-75.
[http://dx.doi.org/10.1007/s12013-007-0031-8] [PMID: 17652781]

[91]   Shimba N, Kamiya N, Nakamura H. Model building of antibody–antigen complex structures using GBSA scores. J Chem Inf Model 2016; 56(10): 2005-12.
[http://dx.doi.org/10.1021/acs.jcim.6b00066] [PMID: 27618247]

[92]   Xu D, Lin SL, Nussinov R. Protein binding versus protein folding: The role of hydrophilic bridges in protein associations.J Mol Biol 1997; 265(1): 68-84.
[http://dx.doi.org/10.1006/jmbi.1996.0712]

[93]   Waldburger CD, Schildbach JF, Sauer RT. Are buried salt bridges important for protein stability and conformational specificity? Nat Struct Mol Biol 1995; 2(2): 122-8.
[http://dx.doi.org/10.1038/nsb0295-122] [PMID: 7749916]

[94]   Cojocaru V, Winn PJ, Wade RC. Multiple, ligand-dependent routes from the active site of cytochrome P450 2C9. Curr Drug Metab 2012; 13(2): 143-54.
[http://dx.doi.org/10.2174/138920012798918462] [PMID: 22208529]

[95]   Jin F, Yu C, Lai L, Liu Z. Ligand clouds around protein clouds: A scenario of ligand binding with intrinsically disordered proteins. PLOS Comput Biol 2013; 9(10): e1003249.
[http://dx.doi.org/10.1371/journal.pcbi.1003249] [PMID: 24098099]

[96]   Yu C, Niu X, Jin F, Liu Z, Jin C, Lai L. Structure-based inhibitor design for the intrinsically disordered protein c-Myc. Sci Rep 2016; 6(1): 22298.
[http://dx.doi.org/10.1038/srep22298] [PMID: 26931396]

[97]   Larsen AS, Ruggiero MT, Johansson KE, Zeitler JA, Rantanen J. Tracking dehydration mechanisms in crystalline hydrates with molecular dynamics simulations. Cryst Growth Des 2017; 17(10): 5017-22.
[http://dx.doi.org/10.1021/acs.cgd.7b00889]

[98]   Mousavi SV, Hashemianzadeh SM. Molecular dynamics approach for behavior assessment of chitosan nanoparticles in carrying of donepezil and rivastigmine drug molecules. Mater Res Express 2019; 6(4): 045069.
[http://dx.doi.org/10.1088/2053-1591/aafec6]

[99]   Roschewski M, Lionakis MS, Sharman JP, *et al.* Inhibition of bruton tyrosine kinase in patients with severe COVID-19. Sci Immunol 2020; 5(48): eabd0110.
[http://dx.doi.org/10.1126/sciimmunol.abd0110] [PMID: 32503877]

[100] Kaliamurthi S, Selvaraj G, Selvaraj C, Singh SK, Wei DQ, Peslherbe GH. Structure-based virtual screening reveals ibrutinib and zanubrutinib as potential repurposed drugs against COVID-19. Int J Mol Sci 2021; 22(13): 7071.
[http://dx.doi.org/10.3390/ijms22137071] [PMID: 34209188]

[101] Pommier Y, Johnson AA, Marchand C. Integrase inhibitors to treat HIV/Aids. Nat Rev Drug Discov 2005; 4(3): 236-48.
[http://dx.doi.org/10.1038/nrd1660] [PMID: 15729361]

[102] Elliott JL, Kutluay SB. Going beyond integration: The emerging role of HIV-1 integrase in virion morphogenesis. Viruses 2020; 12(9): 1005.
[http://dx.doi.org/10.3390/v12091005] [PMID: 32916894]

[103] Engelman AN, Kvaratskhelia M. Multimodal functionalities of HIV-1 integrase. Viruses 2022; 14(5): 926.
[http://dx.doi.org/10.3390/v14050926] [PMID: 35632668]

[104] Goldgur Y, Craigie R, Cohen GH, *et al.* Structure of the HIV-1 integrase catalytic domain complexed

with an inhibitor: A platform for antiviral drug design. Proc Natl Acad Sci 1999; 96(23): 13040-3.
[http://dx.doi.org/10.1073/pnas.96.23.13040] [PMID: 10557269]

[105] Singh SK, Ed. Innovations and Implementations of Computer Aided Drug Discovery Strategies in Rational Drug Design. Singapore: Springer 2021.
[http://dx.doi.org/10.1007/978-981-15-8936-2]

[106] Durrant JD, McCammon JA. Molecular dynamics simulations and drug discovery. BMC Biol 2011; 9(1): 71.
[http://dx.doi.org/10.1186/1741-7007-9-71] [PMID: 22035460]

[107] Lins RD, Briggs JM, Straatsma TP, *et al.* Molecular dynamics studies on the HIV-1 integrase catalytic domain. Biophys J 1999; 76(6): 2999-3011.
[http://dx.doi.org/10.1016/S0006-3495(99)77453-9] [PMID: 10354426]

[108] Shamroe CL, Bookstaver PB, Rokas KEE, Weissman SB. Update on raltegravir and the development of new integrase strand transfer inhibitors. South Med J 2012; 105(7): 370-8.
[http://dx.doi.org/10.1097/SMJ.0b013e318258c847] [PMID: 22766666]

[109] Serrao E, Odde S, Ramkumar K, Neamati N. Raltegravir, elvitegravir, and metoogravir: The birth of "me-too" HIV-1 integrase inhibitors. Retrovirology 2009; 6(1): 25.
[http://dx.doi.org/10.1186/1742-4690-6-25] [PMID: 19265512]

[110] Hazuda D, Blau CU, Felock P, *et al.* Isolation and characterization of novel human immunodeficiency virus integrase inhibitors from fungal metabolites. Antivir Chem Chemother 1999; 10(2): 63-70.
[http://dx.doi.org/10.1177/095632029901000202] [PMID: 10335400]

[111] Espeseth AS, Felock P, Wolfe A, *et al.* HIV-1 integrase inhibitors that compete with the target DNA substrate define a unique strand transfer conformation for integrase. Proc Natl Acad Sci 2000; 97(21): 11244-9.
[http://dx.doi.org/10.1073/pnas.200139397] [PMID: 11016953]

[112] Pais GCG, Zhang X, Marchand C, *et al.* Structure activity of 3-aryl-1,3-diketo-containing compounds as HIV-1 integrase inhibitors. J Med Chem 2002; 45(15): 3184-94.
[http://dx.doi.org/10.1021/jm020037p] [PMID: 12109903]

[113] Embrey MW, Wai JS, Funk TW, *et al.* A series of 5-(5,6)-dihydrouracil substituted 8-hydrox--[1,6]naphthyridine-7-carboxylic acid 4-fluorobenzylamide inhibitors of HIV-1 integrase and viral replication in cells. Bioorg Med Chem Lett 2005; 15(20): 4550-4.
[http://dx.doi.org/10.1016/j.bmcl.2005.06.105] [PMID: 16102965]

[114] Summa V, Petrocchi A, Bonelli F, *et al.* Discovery of raltegravir, a potent, selective orally bioavailable HIV-integrase inhibitor for the treatment of HIV-AIDS infection. J Med Chem 2008; 51(18): 5843-55.
[http://dx.doi.org/10.1021/jm800245z] [PMID: 18763751]

[115] Min S, Song I, Borland J, *et al.* Pharmacokinetics and safety of S/GSK1349572, a next-generation HIV integrase inhibitor, in healthy volunteers. Antimicrob Agents Chemother 2010; 54(1): 254-8.
[http://dx.doi.org/10.1128/AAC.00842-09] [PMID: 19884365]

[116] Perryman AL, Forli S, Morris GM, *et al.* A dynamic model of HIV integrase inhibition and drug resistance. J Mol Biol 2010; 397(2): 600-15.
[http://dx.doi.org/10.1016/j.jmb.2010.01.033] [PMID: 20096702]

[117] Delelis O, Malet I, Na L, *et al.* The G140S mutation in HIV integrases from raltegravir-resistant patients rescues catalytic defect due to the resistance Q148H mutation. Nucleic Acids Res 2008; 37(4): 1193-201.
[http://dx.doi.org/10.1093/nar/gkn1050] [PMID: 19129221]

[118] Nakahara K, Wakasa-Morimoto C, Kobayashi M, *et al.* Secondary mutations in viruses resistant to HIV-1 integrase inhibitors that restore viral infectivity and replication kinetics. Antiviral Res 2009; 81(2): 141-6.

[http://dx.doi.org/10.1016/j.antiviral.2008.10.007] [PMID: 19027039]

[119] Xue W, Qi J, Yang Y, Jin X, Liu H, Yao X. Understanding the effect of drug-resistant mutations of HIV-1 intasome on raltegravir action through molecular modeling study. Mol Biosyst 2012; 8(8): 2135-44.
       [http://dx.doi.org/10.1039/c2mb25114k] [PMID: 22648037]

[120] Kumar Y, Singh H, Patel CN. *In silico* prediction of potential inhibitors for the main protease of SARS-CoV-2 using molecular docking and dynamics simulation based drug-repurposing. J Infect Public Health 2020; 13(9): 1210-23.
       [http://dx.doi.org/10.1016/j.jiph.2020.06.016] [PMID: 32561274]

[121] Schmidt A, Teeter M, Weckert E, Lamzin VS. Crystal structure of small protein crambin at 0.48 Å resolution. Acta Crystallogr Sect F Struct Biol Cryst Commun 2011; 67(4): 424-8.
       [http://dx.doi.org/10.1107/S1744309110052607] [PMID: 21505232]

[122] Lupyan D. How to assign CHARMM parameters to Desmond-generaed system with viparr4. Protocols 2020.

[123] Mast T, Lupyan D. How to assign AMBER parameters to Desmond-generated system with viparr4. Protocols 2020.

CHAPTER 8

Quantum Chemistry in Drug Design: Density Function Theory (DFT) and Other Quantum Mechanics (QM)-related Approaches

Samuel Baraque de Freitas Rodrigues[1], Rodrigo Santos Aquino de Araújo[2], Thayane Regine Dantas de Mendonça[3], Francisco Jaime Bezerra Mendonça-Júnior[2], Peng Zhan[4] and Edeildo Ferreira da Silva-Júnior[1,3,*]

[1] *Institute of Chemistry and Biotechnology, Federal University of Alagoas, Lourival Melo Mota Avenue, AC. Simões campus, 5587072-970, Alagoas, Maceió, Brazil*

[2] *Laboratory of Synthesis and Drug Delivery, Department of Biological Sciences, State University of Paraiba, João Pessoa 58429-500, PB, Brazil*

[3] *Laboratory of Medicinal Chemistry, Federal University of Alagoas, Lourival Melo Mota Avenue, AC. Simões campus, 5587072-970, Alagoas, Maceió, Brazil*

[4] *Department of Medicinal Chemistry, Key Laboratory of Chemical Biology (Ministry of Education), School of Pharmaceutical Sciences, Cheeloo College of Medicine, Shandong University, 44 West Culture Road, 250012 Jinan, Shandong, PR China*

Abstract: Drug design and development are expensive and time-consuming processes, which in many cases result in failures during the clinical investigation steps. In order to increase the chances to obtain potential drug candidates, several *in silico* approaches have emerged in the last years, most of them based on molecular or quantum mechanics theories. These computational strategies have been developed to treat a large dataset of chemical information associated with drug candidates. In this context, quantum chemistry is highlighted since it is based on the Schrödinger equation with mathematic solutions, especially the Born-Oppenheimer approximation. Among the Hartree-Fock-based methods, the Density Functional Theory (DFT) of Hohenberg-Kohn represents an interesting and powerful tool to obtain accurate results for electronic properties of molecules or even solids, which in many cases are corroborated by experimental data. Additionally, DFT-related methods exhibit a moderate time-consuming cost when compared to other *ab initio* methods. In this chapter, we provide a deep overview focused on the formalism behind DFT, including historical aspects of its development and improvements. Moreover, different examples of the application of DFT in studies involving GABA inhibitors, or catalytic mechanisms of enzymes, such as RNA-dependent RNA polymerase (RdRp) of SARS-CoV-2, and different proteases

* **Corresponding author Edeildo Ferreira da Silva-Júnior:** Institute of Chemistry and Biotechnology, Federal University of Alagoas, Lourival Melo Mota Avenue, AC. Simões campus, 5587072-970, Alagoas, Maceió, Brazil; & Laboratory of Medicinal Chemistry, Federal University of Alagoas, Lourival Melo Mota Avenue, AC. Simões campus, 5587072-970, Alagoas, Maceió, Brazil; Tel.: +55-82-9-9610-8311; E-mail: edeildo.junior@iqb.ufal.br

associated impacting diseases, such as malaria, Chagas disease, human African trypanosomiasis, and others. Moreover, the role of metal ions in catalytic enzymatic mechanisms is also covered, discussing iron-, copper-, and nickel-catalyzed processes. Finally, this chapter comprises several aspects associated with the elucidation of catalytic mechanisms of inhibition, which could be used to develop new potential pharmacological agents.

Keywords: Catalytic Mechanism, Copper, Hydrolase, Nickel, Protease, Quantum Mechanics.

INTRODUCTION TO THE HISTORY OF QUANTUM CHEMISTRY (QC)

Currently, there is a constant rise in the need for growing efficiency in a drug design and discovery campaign or even during the lead optimization since the central idea is reducing the time costs, yielding more effective drugs in the pipeline. Thus, this increased necessity requests accurate software for processing a large amount of information in a limited time, overstimulating software upgrades, and the development of novel protocols using well-known programs. In this context, different methods have been developed to treat a large dataset of chemical information, aiming to fill the lack that emerged during the development of a new drug. Well, before discussing how some computer-aided methods can help to elucidate essential information for designing drugs, we need a better understanding of what formalism these methods are based on, as well as their possible applicability. In the next pages, this chapter will lead the reader on a journey from the emergence of the most important computer methods and their current utilization focused on drug design and development, starting from the Schrödinger equation.

The main point of *quantum chemistry* (QC) is the obtainment of solutions for the Schrödinger equation to accurately determine the chemical properties of atoms and even more complex molecular systems. Then, we typically are searching solutions for stationary states that could involve different approximation methods. Thus, QC methods depend not only on computer advances but also on the development of new theories or methodologies. Currently, there are several methods involving QC for solving chemical problems associated with molecules, among them, the *ab initio* Hartree-Fock (HF) has been used to provide great approximated solutions for *many-electron problems*. Its theory treats the electrons individually, moving in an average field for all other electrons and nuclei, which allows the generation of a set of electron-coupled equations. Years later, *semi-empirical methods* emerged to reduce computational time-consuming. Otherwise, chemical problems that previously were treated with HF approximation are currently frequently treated by using the *Density Functional Theory* (DFT) calculations, resulting in values even closer to the experimental data. It has been used to study the electronic properties of molecules and solids. Furthermore, the development of more precise exchange correlation functionals and efficient algorithms of numerical integration has contributed to the development of the DFT method.

In 1927, Max Born and J. Robert Oppenheimer formulated the *Born-Oppenheimer approximation*, which assumes that the nuclei are much heavier than electrons and, as a consequence, they move more slowly, making this theory considered the heart of QC [1]. Considering this approximation, its main problem still remains in solving the non-relativistic time-independent Schrödinger equation:

$$\hat{H} \, | \, \Phi \rangle = \varepsilon \, | \, \Phi \, \rangle$$

in which,

$$\hat{H} = - \sum_{i=1}^{N} \frac{\hbar^2}{2m} \nabla_i^2 - \sum_{i=1}^{N} \sum_{A=1}^{M} \frac{Z_A e^2}{4\pi\epsilon_0 r_i A} + \sum_{i=1}^{N} \sum_{j>i}^{N} \frac{e^2}{4\pi\epsilon_0 r_{ij}}$$

Where m represents the electron mass, Z_A means the atomic number of the nucleus A, r_{ij} is the distance between i and j electrons, whereas r_{iA} means the distance between electron i and nucleus A. Finally, N and M represent the number of electrons and nuclei in the system, respectively. The above equation expresses the electronic term for the molecular *Hamiltonian operator* $\hat{H}$. Since the electrons in a molecule are considered moving faster than nuclei, the second term of this equation (kinetic energy of the nuclei) can be neglected. Moreover, the repulsion between the nuclei (the last term) is taken to be constant. Thusly, the remaining terms are called the *electronic Hamiltonian* [2]:

$$\hat{H}_{elec} = - \sum_{i=1}^{N} \frac{1}{2} \nabla_i^2 - \sum_{i=1}^{N} \sum_{A=1}^{M} \frac{Z_A}{r_{iA}} + \sum_{i=1}^{N} \sum_{j>i}^{N} \frac{1}{r_{ij}}$$

Then, HF approximation has an essential role in the development of modern QC concepts [2]. Douglas Hartree's methods were guided by some earlier semi-empirical methods of the early 1920s set in the old quantum theory of Bohr [3]. In general, HF approximation substitutes the many-electron problem with a one-electron problem, considering the electron-electron repulsion term as an average way [2]. In this context, the *Self-Consistent-Field* (SCF) method is used as a procedure for solving the HF equation. In essence, this approach creates an initial guess for the spin orbitals, from which it can calculate the average field seen by each electron, solving the eigenvalue equation for a new set of spin orbitals.

Thusly, SCF uses these new orbitals, generating new fields, in which this procedure will be repeated until self-consistency is reached when these fields no longer change and the spin orbitals used to generate the Fock operator are the same as its eigenfunctions [2]. In 1931, Henry Eyring and Michael Polanyi used, for the first time in theoretical chemistry, the semiempirical term when they were trying to combine quantum mechanics (QM), valence *bond theory*, and thermodynamics and kinetic parameters [4, 5]. A study performed by Eyring and Polanyi by mixing theory and experimental data demonstrated interesting insights into the mechanisms of adiabatic reactions, resulting in important concepts involving transition state and activated complex [5]. *Semi-empirical* methods emerged as an alternative to reduce the computational costs for calculations of large molecular systems (hundreds of atoms), being able to predict physicochemical parameters [6, 7]. These methods were initially developed to be capable of treating chemical problems, mainly those related to electronic properties, which normally were only estimated by using *ab initio* approaches. In general, semiempirical methods are faster than *ab initio* but much lower than molecular mechanics (MM) methods, even still having great accuracy [8].

In recent decades, *Density Functional Theory* (DFT) has been demonstrated as an essential tool in studies involving the investigation of electronic properties of molecules or even solids. Problems that traditionally were treated with HF and post-HF methods. The emergence of DFT calculations allowed the improvement of the results of the calculations, corroborating with the experimental data. The main advantage of DFT is associated with the fact of from moderate to big systems can be treated with this level of theory, providing chemically accurate results within a moderate time-consuming cost, especially when compared to other methods (*e.g. perturbation theory* and *coupled-cluster*) [9].

HOHENBERG-KOHN-SHAM THEOREM – DENSITY FUNCTIONAL THEORY (DFT)

Early in the 20th century, the utilization of the electronic density *p(r)* as a fundamental variable for describing an electronic system was initially introduced when Drude, Sommerfeld, Thomas, Fermi, and Dirac applied the theory of gases to metals. Thus, these were considered as a *homogenous gas of electrons*, allowing the formulation of their thermic and electric conduction theories. Posteriorly, the Thomas-Fermi model approximated the distribution of an electron gas, leading to the *energy functional*. Furthermore, this model was improved to include an electron gas developed by Thomas-Dirac [10, 11]. Finally, the energy functional (*E*) of Thomas-Fermi-Dirac (TFD) is given by:

$$E_{TFD}[\rho] = C_F \int \rho(r)^{5/3} dr + \int \rho(r)v(r)dr + \frac{1}{2}\int\int \frac{\rho(r_1)\rho(r_2)}{|r_1 - r_2|}dr_1 dr_2 - C_x \int \rho(r)^{4/3} dr$$

in which,

$$C_F = \frac{3}{10}(3\pi^2)^{\frac{2}{3}} \quad and \quad C_X = \frac{3}{4}\left(\frac{3}{\pi}\right)^{\frac{1}{2}}$$

In the energy potential equation, the second term represents the kinetic energy, whereas the external potential (typically formed by nuclei, positions and charges of atoms in the molecule), Coulomb interaction, and exchange energy, respectively. Moreover, ρ and r represent the electronic density and its coordinates, respectively. Unfortunately, this equation is too simple to correctly describe the quantum structure of shells of atoms [10]. Then, the use of the electronic density $\rho(r)$ term was legitimated in the publication of two theorems developed by Hohenberg and Kohn in 1964, which will be further discussed.

To better understand the DFT formalism, it will be necessary to know that the electronic density $\rho(r)$ is defined by the following equation:

$$\rho(r) = \int \cdots \int \psi(r_1 r_2 \cdots r_N)^* \, \psi(r_1 r_2 \cdots r_N)dr_2 dr_3 \cdots dr_N$$

where $\psi(r_1, r_2, ..., r_N)$ is a solution for the ground state of Hamiltonian. However, the total energy (E_0) of the system is still obtained by:

$$E_0 = \int \psi(r_1 r_2 \cdots r_N)^* \, \hat{H}_{BO} \psi(r_1 r_2 \cdots r_N)dr_1 dr_2 \cdots dr_N = \langle \psi | \hat{H}_{BO} | \psi \rangle.$$

Then, the external potential can be separated into a trivial functional for $\rho(r)$ and, in this case, E_0 equation will be written as:

$$E_0 = \langle \psi | \hat{T} + \hat{V}_e | \psi \rangle + \int \rho(r)v(r)dr.$$

Regarding these both equations, it is evidenced that the number of electrons, N, and the external potential completely describe the many-electron system, named the Hamiltonian system [9].

Furthermore, the DFT was proved by Hohenberg-Kohn-Sham, showing that the density of electrons ρ is proportional to the energy in the ground state and the number of electrons (N) in the system. This proof is composed of three different theorems, the Hohenberg-Kohn Existence Theorem, Hohenberg-Kohn Variational Theorem, and Kohn-Sham Self-Consistent Field Methodology.

Hohenberg-Kohn Existence Theorem

Hohenberg and Kohn proved that the ground-state density must determine the Hamiltonian, the number of electrons, and the external potential by a *reductio ad absurdum* (Latin for *"reduction to absurdity"*). The hypothesis is that two different potentials v_a and v_b have the same *nondegenerate* ground-state density ρ_o. Each potential with different Hamiltonian operators, which were associated with a ground-state wave function and its associated eigenvalue E_o.

$$E_{0,a} < \langle \psi_{0,b}|H_a|\psi_{0,b}\rangle$$

$$E_{0,a} < \langle \psi_{0,b}|H_a - H_b + H_b|\psi_{0,b}\rangle$$

$$E_{0,a} < \langle \psi_{0,b}|H_a - H_b|\psi_{0,b}\rangle + \langle \psi_{0,b}|H_b|\psi_{0,b}\rangle$$

$$E_{0,a} < \int [v_a(r) - v_b(r)]\rho_0(r)dr + E_{0,b}$$

It's the same as:

$$E_{0,b} < \int [v_b(r) - v_a(r)]\rho_0(r)dr + E_{0,a}$$

Adding the inequalities:

$$E_{0,a} + E_{0,b} < \int [v_b(r) - v_a(r)]\rho_0(r)dr + \int [v_a(r) - v_b(r)]\rho_0(r)dr + E_{0,a} + E_{0,b}$$

$$E_{0,a} + E_{0,b} < \int [v_b(r) - v_a(r) + v_a(r) - v_b(r)]\rho_0(r)dr + E_{0,a} + E_{0,b}$$

$$E_{0,a} + E_{0,b} < E_{0,a} + E_{0,b}$$

By absurd it is concluded that the ground-state density must determine the external potential, Hamiltonian, and the number of electrons since we are left with an impossible result that the sum of two energies is less than itself.

Therefore, the energy of an electronic system is a one-to-one proportion of the electronic density, $\rho(r)$, because the energy of the system is calculated by solving the Schrödinger equation, HBOψ = Eψ.

$$E = E_v[\rho]$$

Then, the v term is introduced to explain the dependence on the external potential $v(r)$.

Hohenberg-Kohn Variational Theorem

Hohenberg and Kohn assumed that a regular candidate density integrates the proper number of electrons and, by the first theorem, this candidate determines the Hamiltonian, the number of electrons, and the energy. This energy can be evaluated from an energy expectation value that must be greater than or equal to the ground-state energy because obeys the variational principle of Molecular Orbital Theory.

$$\langle \psi_{cand} | H_{cand} | \psi_{cand} \rangle = E_{cand} \geq E_o$$

Because of that, we can define a universal potential, in which T and V_e can be universally applied to all electronic systems.

$$E_o = E_v[\rho] = F[\rho] + \int \rho(r)v(r)dr \leq E_v[\overline{\rho}] = F[\overline{\rho}] + \int \overline{\rho}(r)v(r)dr$$

Kohn-Sham Self-Consistent Field Methodology

Kohn-Sham in a breakthrough thought considered the Hamiltonian as a non-interacting system of electrons. Considering a fictitious system of non-interacting electrons that has the same overall density as a real system that the electrons do interact. The Hamiltonian can be expressed as a sum of one-electron operators, having eigenfunctions that are Slater determinants of the individual one-electron eigenfunctions, and eigenvalues that are simply the sum of the one-electron eigenvalues. Moreover, we divide the energy functional in components that make it easy to analyze:

$$E[\rho(r)] = \sum_i^N \left(\left\langle \chi_i \left| \frac{1}{2}\nabla_i^2 \right| \chi_i \right\rangle - \left\langle \chi_i \left| \sum_k^{nuclei} \frac{Z_k}{|r_i - r_k|} \right| \right\rangle \right) + \sum_i^N \left\langle \chi_i \left| \frac{1}{2} \int \frac{\rho(r')}{|r_i - r'|} dr' \right| \chi_i \right\rangle + E_{xc}[\rho(r)]$$

Where N is the number of electrons and the ΔT and ΔVee have been united in the exchange-correlation energy, E_{xc}. The exchange correlation energy includes the effects of quantum mechanical exchange and correlation and the correction for the classical self-interaction energy. It also uses the density for Slater-determinant wave function:

$$\rho = \sum_{i=1}^N \langle \chi_i | \chi_i | \rangle$$

When we use the usual way to find the orbital χ that minimizes the energy, we find that it satisfies the *pseudoeigenvalue* equation.

$$h_i^{KS} \chi_i = \varepsilon_i \chi_i$$

$$h_i^{KS} = -\frac{1}{2}\nabla_i^2 - \sum_k^{nuclei} \frac{Z_k}{|r_i - r_k|} + \int \frac{\rho(r')}{|r_i - r'|} dr' + V_{xc}$$

and

$$V_{xc} = \frac{\delta E_{xc}}{\delta \rho}$$

The theory of Kohn-Sham (KS) for spin-polarized consists of the generalization of *KS* theory for a compensated spin [12]. The electronic density $\rho(r)$ is substituted by the α and β, $\rho^\alpha(r)$ and $\rho^\beta(r)$ electron density, as fundamental variables. Then,

$$\rho(r) = \rho^\alpha(r) + \rho^\beta(r).$$

By following the same thoughts, it is possible to obtain the Schrödinger equation shown below:

$$\left(-\frac{1}{2}\nabla^2 + v_{ef}^\sigma(r)\right)\psi_i^\sigma = \varepsilon_i^\sigma \psi_i^\sigma,$$

in which σ represents the α or β spin. Lastly, the effective potential vcfσ(r) is written as:

$$v_{ef}^\sigma(r) = v(r) + \int \frac{\rho(r_1)}{|r - r_1|} dr_1 + \frac{\delta E_{xc}[\rho^\alpha \rho^\beta]}{\delta \rho^\sigma(r)}.$$

CHEMICAL REACTIVITY INDEXES BY DENSITY FUNCTIONAL THEORY (DFT)

Chemical Potential and Electronegativity

Parr and Yang established an association involving the computational and DFT concepts when interpreting the Lagrange multiplier [13]. Additionally, they demonstrated that the partial derivative of the system's energy can be written as μ, which is the number of electrons submitted to an external potential, $v(r)$, fixed:

$$\mu = \left(\frac{\partial E}{\partial N}\right)_{v(r)}$$

Approximating this equation with the definition of *chemical potential*, and taking the finite difference in consideration, then we can write it as:

$$\mu = -\chi_M = -\frac{I + A}{2}$$

in which I and A represent the *ionization potential* and *electronic affinity*, respectively. Then, the external potential is maintained as a constant term.

Besides, it is known that the chemical potential is obtained by the inverse of electronegativity (χM) [14, 15], which posteriorly this concept was modified by the *principle of equalization of electronegativity*. It states that *"when two or more atoms, having initially different electronegativities, will connect to form compounds, then their electronegativities become an adjusted intermediate value into the compound"* [16, 17]. In other words, the chemical potential in an equilibrated system is equal in all its parts. Furthermore, the electronegativity is a property associated with the status of a molecular system, which can be determined by DFT calculations, by using the equation aforementioned.

The *hard* and *soft* acids and bases *(HSAB) principle of Pearson* affirms that *"soft acids prefer soft bases, whereas hard acids prefer hard bases"* [18, 19]. This definition was incorporated in DFT fundamentals [20]. Lastly, these important properties can be determined according the following equations below:

Hardness

$$\eta = \left(\frac{\partial \mu}{\partial N}\right)_{v(r)} = \left(\frac{\partial^2 \mu}{\partial N^2}\right)_{v(r)} \approx I - A$$

Softness

$$S = \left(\frac{\partial \mu}{\partial N}\right)_{v(r)} = \frac{1}{I - A}$$

Fukui Functions

The *Frontier function*, which is also named as *Fukui function, f(r)*, considers the electronic density of an atom (or even molecules) in its valence shell to measure how an external perturbation can affect the chemical potential of a determined molecular system [21], having the form:

$$f(r) = \left(\frac{\partial \mu}{\partial v(r)}\right)_N = \left(\frac{\partial \rho(r)}{\partial N}\right)_{v(r)}$$

Typically, this function is used to determine the chemical reactivity of a species towards a *nucleophilic* or *electrophilic* attack or even a *radical* reactant. Since there is a discontinuity for N equal to an entire number, then there are three different numerical definitions for Fukui functions [22], as written below:

$$f^+(r) = \left(\frac{\partial p(r)}{\partial N}\right)_{v(r)}^+ \approx \rho_{N=N_o+1}(r) - \rho_{N=N_o}(r) = \rho_{LUMO}(r)$$

$$f^-(r) = \left(\frac{\partial p(r)}{\partial N}\right)_{v(r)}^- \approx \rho_{N=N_o}(r) - \rho_{N=N_o-1}(r) = \rho_{HOMO}(r)$$

$$f^0(r) = \left(\frac{\partial p(r)}{\partial N}\right)_{v(r)}^0 \approx \frac{1}{2}(f^+(r) + f^-(r))$$

These shown above equations describe a nucleophilic (f+r), electrophilic (f-r), and radicalar (f0r) attack, respectively. In this context, we should know that an electrophilic attack is related to the *Highest-Occupied Molecular Orbital* (HOMO), ρHOMO*(r)*; whereas a nucleophilic attack is associated with the *Lowest-Occupied Molecular Orbital* (LUMO), ρLUMO*(r)*. Finally, a radical attack is a result of the average involving f+r and f-r functions [22]. Then, other energy derivatives and functions emerged from these concepts, taking into account that E is distinct from N and $v(r)$ for superior orders [23, 24]. Thusly, the total energy of an electronic system can be written as functions of these terms:

$$E[N, v] = N\mu - \frac{1}{2}N^2\eta + \int v(r)[\rho(r) - Nf(r)]dr + \cdots$$

As a result, the total energy of an electronic system is connected to the chemical potential (μ), hardness (η), and to the Fukui function ($f(r)$).

DENSITY FUNCTIONAL THEORY (DFT)-RELATED APPROACHES

Hybrid Method: Quantum Mechanics / Molecular Mechanics (QM/MM)

QM/MM method is a mix of *Quantum Mechanics* (QM) and MM methods, in which a minimum area of interest is described using a reliable electronic structure method that allows the investigation of all participants of the chemical reaction and is treated with QM, semiempirical, or Empirical Valence Bond levels, and the rest of the molecule is described using MM, with the purpose of catching the contribution that does not need such accurate description. It is a very promising tool but still is not as usual as *Molecular Dynamics* (MD) and QM calculations because of its complex approximation base. An important link between experimental data and theoretical calculations is the *Transition State Theory* (TST), in which the *Transition State* (TS) is an ensemble of the high-energy structure that dictates whether the reaction, with an equal probability, proceeds to either the reactant or product valleys. TS is identified as a stationary point of a potential energy hypersurface:

$$k \; = \; \gamma \left(\frac{k_b T}{h}\right) \; \exp\left(-\frac{\Delta G^{\top}}{k_b T}\right)$$

Where γ is the transmission coefficient, k_b and h are Boltzmann and Planck constants, respectively; while T represents the thermodynamic temperature.

Usually, QM/MM methods divide the system into two regions, allowing two approaches to be used, the additive approach or the subtractive approach. In the additive approach, the energy of QM core (E_{QM}), the MM energy (E_{MM}), which does not interact with the QM core, and the interaction between the energy of QM core and the MM energy interaction is then summed as total energy (E_{tot}):

$$E_{tot} \; = \; E_{QM} \; + \; E_{MM} \; + \; E_{QM/MM}$$

The subtractive approach, usually called *Our own N-layered Integrated molecular Orbital + Molecular Mechanics* (ONIOM), divides the system into layers and subtracts double-counted energies of the smaller layer, where E_{model}^{MM} represents the energy of the model containing both QM and MM regions, which is calculated at a lower level, such as MM; E_{model}^{MM} the QM core was calculated with a QM method, while E_{model}^{MM} is the energy of the QM core calculated at MM level:

$$E_{tot} \; = \; E_{real}^{MM} \; + \; E_{model}^{QM} \; + \; E_{model}^{MM}$$

An essential step in QM/MM method is to clearly define the borders and coupling of QM and MM areas, choosing a poorly the boundary or/and coupling will affect the accuracy of the calculation, and this decision is arbitrary.

DFT CALCULATIONS IN DRUG DESIGN & DEVELOPMENT

Drug-Target Interactions

Some molecular chemical properties are used as a base for structural prediction by using computational tools. The knowledge of the atomic constitution and its connections in a molecule allows an understanding of the behavior of its molecular electronic density and how it influences the intra- and intermolecular interactions, as well as its distribution in a given medium. QC equations allow the prediction of molecular electronic density by DFT studies and the generation of models able to explain how this density can govern some parameters, such as interactions, bonds, geometry (*e.g.* length, strength, angles), dipole moments, and molecular polarizabilities [25 - 27].

The understanding of molecular electronic density allows us to identify zones of electronic repulsion and attraction. Therefore, one of the main aims of their use is known as these electrostatic interactions assist in the spatial distribution of the chemical groups in a molecule, in the determination of its geometry. DFT studies are utilized in the energetic evaluation of a variety of possible conformers, providing those more energetically stable. For this evaluation, a possible model is given by the determination of HOMO-LUMO energy gap, by analyzing the frontier molecular orbitals, in which the greater this gap, more stable the molecule will be [28 - 30]. For molecular polarizability, electrostatic potential maps are generated, differentiating regions that concentrate a greater electronic density, usually due to the presence of more electronegative atoms and/or groups, from those regions with a greater electronic deficit [27, 29, 31]. Somehow, the study of this parameter also contributes to the understanding of molecular geometry from electrostatic repulsion/attraction models, however, additionally, it has a central role in the generation of dipole moments and identification of intra- and intermolecular interactions, which also include the modes of recognition of a chemical compound by its molecular target. In DFT, analyzes such as the *Natural Bond Orbital* (NBO) method helps to determine the possible points for these interactions *via* different mechanisms, such as electronic conjugation, charge and/or proton transfer, the nature of chemical bond formation (covalent and noncovalent) or even metal-chelating interactions [32, 33]. Here, the frontier orbitals analyze, previously mentioned, helps to comprise the relationship between bonding and anti-bonding orbitals [34, 35].

Determination of noncovalent interactions between two components can occur by using the *Reduced Density Gradient* (RDG) function. Ganesan *et al.* [34] used this function to comprehend the interaction between a ligand and farnesyl pyrophosphate synthetase (FPPS) in *E. coli*, which leads to the generation of a graph of RDG function *versus* electronic density, allowing the demonstration of regions where hydrogen bonds form, van der Waals interactions and electrostatic repulsions [36]. Therefore, the use of RDG function makes it possible to understand the geometry of ligands when interacting with the protein target, which can be utilized in conjunction with molecular docking approaches to comprise the types of interaction between the compound and the residues of the target. Still, in the study of Ganesan *et al.*, an important donor role in hydrogen-bonding interaction involving Lys[320] residue and the ligand was revealed, in addition to an acceptor role in hydrogen bonding with Pro[25] residue and hydropho

-bic interactions with Leu[28] and Val[32] residues. These interactions suggest some stability in the complex formation [34].

The knowledge of these ligand-biological-target interactions, through the elucidation of their stable three-dimensional structures, with the support of DFT studies, can also help in the identification of structural changes that trigger the resistance of some drugs, mainly for viral targets, which exhibit higher mutation rates. The evaluation of structural changes that may occur in the binding site of a target can explain the reduction of affinity for drugs, whose spatial geometry is maintained by the decrease in the interactions between them, reflected in the reduction in affinity energy during the formation of the complex [37]. To exemplify the use of computational tools based on DFT studies for this purpose, Saen-oon *et al.* analyzed the energetic, structural, and dynamics parameters in the formation of complexes involving saquinavir (an antiviral drug) and mutant and non-mutant HIV-1 proteases [38]. Then, they obtained the spatial arrangement of saquinavir, understanding how it approximates the protein binding residues at the B3LYP/6-31G** level of theory, which is widely reported in the literature for these purposes [38, 39]. With the obtaining of the three-dimensional structure of the ligand, its modes of interaction with the biological target are evaluated, based on their interaction energies, obtained by the equation below:

$$\Delta E = E_{Complex} - E_{Protease} - E_{Saquinavir}$$

In this sense, it was possible to observe that a reduction in energetic affinity in the formation of the drug-target complex was consistent with the level of mutation in the protein, explaining the increased resistance of mutant targets to the analyzed drug. The mutations caused a different three-dimensional arrangement for the enzyme, modifying the mode of interaction with the drug and, consequently, increasing the distances between important interaction zones. Furthermore, Saen-oon *et al.* standardized a distance among a donor-acceptor pair (≤ 3.5 Å) for effecting a hydrogen bond and a donor$\cdots$H$\cdots$acceptor angle ($\geq 120°$), in which the structural change of the complex generated with the mutant targets demonstrated a decrease in the strength of hydrogen bonds [38].

The evaluation of the three-dimensional structure of a ligand before and after binding to the molecular target can be done by comparing DFT studies and molecular docking. A low or no variation in the results of this ligand indicates that the optimized ligand structure is maintained in its binding with the target, being additionally stabilized by interactions. Frequently, these tools are used together to compare the types of intermolecular interactions and the distances maintained by the chemical groups of a ligand with a target, in order to validate the proposed model. In the comparison between these models, each isolated energy for each contribution during the formation of a ligand-target complex is obtained, allowing us to conclude about the greater contributions for the energetic stability of a complex [40]. The presence of electronegative heteroatoms in a molecule works as a direct conditional in determining its electronic distribution and interference in parameters, such as its dipole moment and polarizability. The existence of atoms, such as oxygen and nitrogen, contributes not only to these parameters but represents direct points of interaction due to charge transfer, ionization, and generation of conformational energies and of the border orbitals [39]. Thus, parameters such as electronegativity, electrophilicity, hardness, softness, and chemical potential of ligands in the formation of its complex with targets can be comprised by DFT studies [41]. Of these parameters, the softness is associated with the polarizability of the molecular electron density, providing points of dipole interaction with the target due to ease of charge transfer [39], while the existence of electronegative atoms determines regions of greater and lesser electronic density, such as points of interaction with complementary regions of other molecules [39, 42]. In this sense, softness (S) is inversely proportional to hardness (η).

Hyperpolarizability parameters and dipole moment proved to be essential, for example, in the study of the interaction of a $GABA_A$ antagonist with this receptor and its binding affinity similarities with the γ-aminobutyric acid (GABA) receptor. DFT studies allowed the obtaining of optimized stable conformers of minimum energy of bicuculline (Fig. **1**), and the observation that both the antagonist and the natural ligand have oxygen and nitrogen atoms with similar behaviors in the interaction with the receptor. In this sense, Fukui's quantum function could be used to determine these similarities and demonstrate an isosteric behavior of these groups in both ligands, through the formation of hydrogen bonds between them and Tyr[97], Arg[156], and Tyr[157] residues [39]. Although Fukui's quantum function equations are more related to the chemical reactivity of molecules, the nucleophilicity and electrophilicity characteristics described by them can be utilized to interpret intermolecular interactions, as well. More nucleophilic regions (higher relative nucleophilicity values), for example, relate to points more prone to receiving hydrogen bonds [39].

On a more significant number of occasions, when dealing with the recognition and affinity of a drug substrate for its target, we consider the many noncovalent interactions aforementioned. However, for a class of inhibitors (targeted covalent inhibitors), their covalent interactions are the most important parameter in their binding with a target [43 - 45]. As these are stronger interactions involving the formation and breaking of chemical bonds, studies are addressed to the energy analysis of the reaction in progress, and the magnitude of the energy barriers of rate-limiting steps becomes the central focus of DFT-related approaches [46, 47].

For these cases, the prediction of the evolution of the reactional transition states is necessary, based on the description of the noncovalent intermolecular interactions, which help in the approximation and maintenance of the target-ligand complex until the formation of the covalent bond [45]. The utilization of DFT studies allows, thus, the comparison of the inhibition reaction rate constant (noncovalent step) with the inactivation reaction rate constant (covalent step):

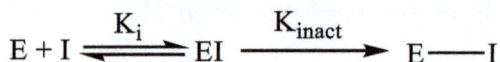

$$E + I \underset{}{\overset{K_i}{\rightleftharpoons}} EI \xrightarrow{K_{inact}} E—I$$

Where E means the enzyme energy term, whereas I represents the inhibitor (or ligand) energy term. So, as performed by Mihalovits *et al.*, the binding free energy calculations between a ligand and a target are accomplished by considering two steps, the use of thermodynamic integration to calculate the binding free energy in the formation of the noncovalent complex and the energy barrier for the formation of the covalent ligand-target complex [45]. Therefore, molecular physical-chemical properties are used as a basis for DFT studies for structural prediction based on the electronic parameters of a molecule and how it may behave when interacting with other molecules or acting as a ligand on a molecular target of interest.

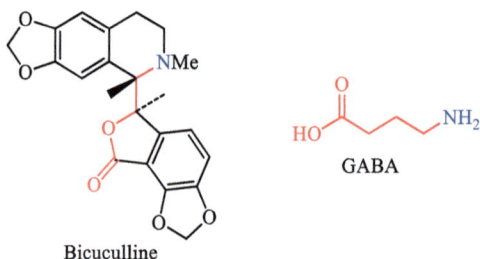

Fig. (1). Bicuculline and GABA structures, with the highlight of the isosteric atoms of oxygen and nitrogen.

GABA$_A$ Receptor Inhibition

Arylpiperazine derivatives are known as agonists of the GABA receptor, inducing synergistic antianxiety and antistress effects. In this context, the GABA$_A$ receptor is an ionotropic receptor modulated by Na^+ and K^+ ions, being a *Ligand-gated Ion Channel* (LIC) [48 - 50]. Since most drugs modulate endogenous ligands [51], DFT calculations and docking studies can provide a better understanding of the binding mode of drugs to this target. Then, Onawole *et al.* reported a computational investigation of electronic properties for 1-(4-(3-metho-y-4-nitrophenyl)piperazin-1-yl)ethanone (MNPE), a synthetic compound [52]. Initially, MNPE was characterized by spectrophotometric methods, exhibiting an infrared band of 1363 cm^{-1} (attributed to the CH$_2$ oscillation mode of the piperazine ring, similar to the value predicted (1361 cm^{-1}) by DFT using 6-311++G** level of theory), 1308 cm^{-1} in Raman line (associated with the phenyl ring breathing), 1242 cm^{-1} in Raman line, and 1092 cm^{-1} infrared band (both associated with C−N stretching), as key modes in their vibrational spectra. Still, its reactivity properties were also evaluated using the concepts of *Average Local Ionization Energy* (ALIE) and Fukui function. Thus, the results obtained for the MNPE revealed that the benzene ring and NO$_2$ group are prone to possible electrophilic attacks. This implies that the electrons in the regions close to C^8 and C^{12} carbon atoms of the benzene ring, N^1 nitrogen atom of the piperazine ring, and oxygen atoms of the NO$_2$ group are less strongly bound. In all these regions, ALIE values were slightly less than 200 kcal/mol, suggesting possible reactive sites. Furthermore, according to the results shown in Fig. (**2**), the increased electron density due to the charged oxygen atoms is located over the NO$_2$ group, according to the Fukui f^+ function. In contrast, the methyl group close to the benzene ring and close to the O^{13} oxygen atom was identified as a potential reactive center. So, the observed low HOMO-LUMO energy gap is associated with high chemical reactivity and the existence of charge transfer throughout the molecule. The probability of degradation by autooxidation and hydrolysis of MNPE was estimated from the calculated bond dissociation energies and radical distribution functions, predicting that MNPE should be readily biodegradable in aqueous solutions. In this sense, all values of *bond dissociation energy* (BDE) related to hydrogen abstraction are greater than 92 kcal/mol, indicating that MNPE is highly stable outdoors in the presence of an oxygen atmosphere. On the other hand, it was predicted that two BDE values for the remainder of the single acyclic bonds would have values of ~66 and ~69 kcal/mol, indicating that the acetyl-phenyl bond and the O CII$_3$ bond are dissociable and, therefore, can be a site of potential degradation. Since this compound can act as a potential agonist of the human GABA$_A$ receptor, the authors investigated this compound by using MD simulations and then docking. As a result, they verified that MNPE has a great affinity for the GABA$_A$ C binding site, performing π-π stacking interactions with Gln^{64} and Thr^{202} residues, π-alkyl interactions with Tyr^{62}, Phe^{200}, and Tyr^{205} residues, and finally, carbon-hydrogen bonding interactions with Asp^{43} and Tyr^{157} residues (Fig. **2**).

Understand Enzymatic Mechanisms of Catalysis

In general, an uncatalyzed chemical reaction in an aqueous medium is often related to the sizable negative activation entropy values, which may add a meaningful penalty ($-T\Delta S^{\ddagger}$) to the overall activation free energy, described as the term $\Delta G^{\ddagger} = \Delta H^{\ddagger} - T\Delta S^{\ddagger}$ [53]. This entropy penalty is frequently interpreted in terms of a loss of rotational and translational motions of the reactants as they pass through their TS [54]. In this context, it is known that in the reactant state, the molecules rotate freely, whereas the TS may need a specific angular requirement to overcome the energetic barrier and lead to the product [55]. Normally, biomolecular reactions are often intrinsically associated with more negative activation entropies than unimolecular ones, disregarding any possible solvent contributions [55]. Since enzymes can bind their substrates, it is expected that there are some rotational and translational entropy penalties. In this context, Jencks' "Circe effect" hypothesis postulates that the enzyme spends part of the binding free energy on destabilizing the substrate [56]. Nevertheless, it is assumed that this destabilization could involve enthalpic terms, such as electrostatic, desolvation, or conformation issues) [56]. Additionally, Wolfenden *et al.* demonstrated that enzymes typically, rather reduce $\Delta H^{\ddagger}$ than the $-T\Delta S^{\ddagger}$ penalty [57]. Besides, Bruice and Lightstone stated that enthalpy and not entropy is the driving force for the high fraction of *"near-attack conformation"* in enzymes [58].

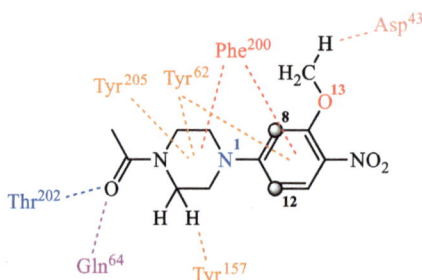

Fig. (2). The molecular interactions of MNPE and amino acids in the binding site C of GABA$_A$.

Cytidine Deaminase

Åqvist *et al.* calculated the overall activation entropies both in enzymes (cytidine deaminase) and solution (spontaneous cytidine deamination in water (Fig. **3**), comprising energetic contributions from both solvent and protein, rather than just the substrate entropies [59]. Then, the authors used an approach similar to the experimental construction of Arrhenius plots of the logarithm *versus* inverse temperature. Thusly, the activation parameters were then extracted by a linear regression equation, represented by $\Delta G^{\ddagger} / T = \Delta H^{\ddagger} / T - \Delta S^{\ddagger}$, in which the temperature dependence was obtained by performing extensive MD simulations. For their purposes, the authors applied DFT in a continuum solvent with different numbers of explicitly treated water molecules. Additionally, *empirical valence bond* (EVB) models were parametrized against DFT results and used for free energy simulations to provide thermodynamics activation parameters [59]. Considering the cytidine in the aqueous medium, the authors found values of 21.4 and −9.1 kcal/mol for $\Delta H^{\ddagger}$ and T$\Delta S^{\ddagger}$, respectively; closer than the experimental data are 22.1 and −8.3 kcal/mol, respectively [60 - 62]. Concerning its 5,6-

dihydrocytidine in water, the authors observed 12.7 and -10.9 kcal/mol for $\Delta H^{\ddagger}$ and $T\Delta S^{\ddagger}$, respectively, whereas the experimental data showed 13.4 and -10.1 kcal/mol, respectively [60 - 62]. Finally, the authors confirmed that created an interesting alternative method to determine the temperature dependence of free energy profiles applying DFT and MM/EVB methods, in which $\Delta H^{\ddagger}$ and $T\Delta S^{\ddagger}$ components can be obtained from a regular Arrhenius plot [59]. Still, they concluded that it is clear how the entropic effects have an essential role in enzymatic catalysis, in which frequently these are somewhat misdirected.

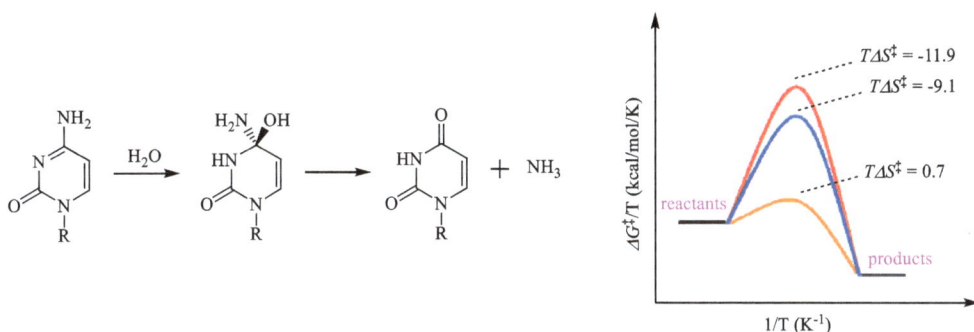

Fig. (3). Reaction for the spontaneous conversion of cytidine into uridine *via* deamination in an aqueous medium and the entropy of its transition states. *In orange, enzymatic deamination of cytidine; blue: concerted mechanism of cytidine deamination in water; red: stepwise cytidine deamination in water.*

RNA-dependent RNA Polymerase (RdRp) in Severe Acute Respiratory Syndrome Coronavirus 2 (SARS-CoV-2)

SARS-CoV-2 RNA-dependent RNA polymerase (RdRp) currently represents a promising target to be considered in drug design protocols targeting Coronavirus disease 2019 (COVID-19) [63]. In this context, molnupiravir (Fig. **4**), a broad-spectrum antiviral originally produced by Merck (coded by EIDD-2801) to treat influenza A infections, emerged as a potential drug candidate addressed to this pandemic virus. According to phase III clinical trials, this compound was capable of reducing hospitalizations by 50% [64]. Still, no deaths were registered within the study, making it be considered as a *"COVID-19 pill"* [64]. Molnupiravir is a prodrug administered via oral, which is a nucleoside analog metabolized and converted into the analog *β-D-N*4-hydroxycytidine (NHC, (Fig. **4**)). It promotes an increase in G to A and C to U transition mutations during the SARS-CoV-2 replication cycle [65]. Then, our research decided to investigate how temperature variations would affect the interactions of NHC with SARS-CoV-2 RdRp, using MD simulations (in 300, 310, and 313 K) and DFT B3LYP/6-311G* [66]. Then, it was verified that NHC exhibits a similar binding mode to remdesivir, another nucleoside analog that also acts by inhibiting RdRp of SARS-CoV-2 [67].

Additionally, it was seen that both NHC-RdRp complexes obtained in 300 and 310 K exhibit similar stability, verified by *Root-Mean Square Deviation* (RMSD) plots. In contrast, the NHC-RdRp complex at 313 K displayed slight variations in its RMSD plot, but it still is stable. These complexes were further investigated by DFT B3LYP/6-311G*, in which it was observed that the HNC-RdRp complex at 313 K is the most stable one, exhibiting a free energy of binding equal to −47.86 kcal/mol. Finally, the better stability of the NHC-RdRp complex at 313 K suggests that it could be effective in febrile patients, as observed in phase III studies.

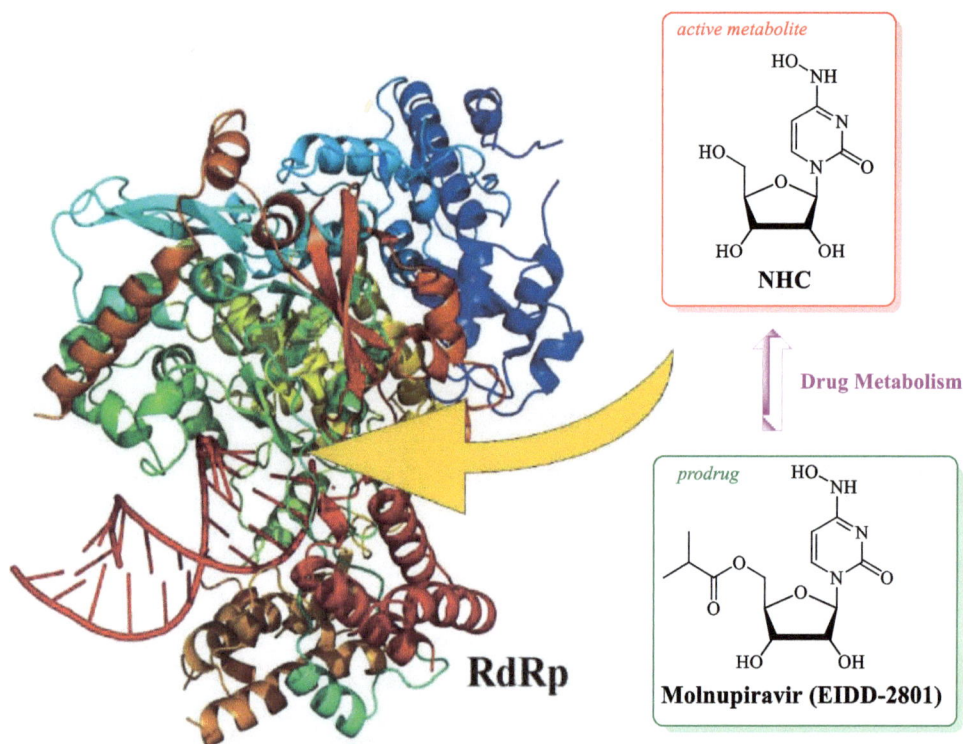

Fig. (4). SARS-CoV-2 RdRp, molnupiravir and its active metabolite.

Oxidized Polyvinyl Alcohol Hydrolase (OPH) from Pseudomonas O-3 Strain

Pseudomonas O-3 strain, discovered in 1973, is a bacterium capable of converting polyvinyl alcohol (PVA) into a sole source of carbon, *via the* decomposition of PVA [68]. Since then, different bacteria from the *Pseudomonas* genus have been used in this purpose [69]. PVA is a plastic-like polymer that has demonstrated properties in adhesion, film formation, tensile strength, flexibility, emulsifying, grease, and solvents, being broadly employed in textile and papermaking industries [70]. In some cases, pyrroloquinoline quinones act as a coenzyme to complete the degradation process [69]. It is known and accepted that the PVA degradation is performed by the action of dehydrogenase or even oxidase oxidizes the neighboring hydroxyl groups, and, posteriorly, a carbon-carbon bond cleaved reaction occurs with the participation of a hydrolase or an aldolase [71]. In accordance with experimental data, Ser172Ala mutation in the oxidized polyvinyl alcohol hydrolase (OPH) reduces the affinity of the enzyme for the substrate [72]. It is known that this oxidation process is divided into two steps, in which one enzyme is initially involved and then two other enzymes are involved as well. In this context, Sakai *et al.* isolated the secondary alcohol oxidase and the β-diketone hydrolase, revealing that the oxidase uses molecular oxygen as an oxidant agent to produce β-diketone in the PVA chain, whereas the covalent bond between β-diketone group undergoes a hydrolysis reaction by the action of β-diketone hydrolase enzyme [73]. Besides, Suzuki *et al.* purified another type of PVA-degrading enzyme from *Pseudomonas* sp. O-3 strain, which is able to oxidize the hydroxyl group to oxygen and catalyzes the bond between β-diketone function [74]. Considering the importance of PVA for the industry, Chen *et al.* decided to investigate the binding process associated with OPH and oxidized PVA by using *linear interaction energy* (LIE), molecular mechanics Poisson-Boltzmann surface area (MM/PBSA), MD simulations (20 ns), and then, the cleavage mechanism was investigated at the atomic level with QM/MM, using DFT focused on the catalytic triad (Ser-His-Asp) [75], while the rest of enzyme was treated as molecular mechanics (CHARMM22 and TIP3P model). Thus, M06-2X/-31G*//CHARMM22 level was used to optimize the geometry of the intermediates, while 6-311++G** was used for computing the single point energy. Furthermore, *Partitioned Rational Function Optimization* (P-RFO) method was employed to search for transition states. The QM region consists of residues Ser[172], Asp[253], His[298], Ser[173], Ser[66], Val[167], and acetylacetone (ACA), a dimer of oxidized PVA. Based on their calculations for four artificial mutations (Ser66Ala, Val67Ala, Trp255Ala, and Try270Ala), only Ser66Ala demonstrated that this residue has a great contribution in the binding with ligands. MM/PBSA calculations revealed that the van der Waals and electrostatic interactions promote the binding of ACA into OPH [75]. Finally, QM/MM calculations were capable of displaying the catalytic mechanism of OPH towards ACA (Fig. **5**).

Polyethylene Terephthalate Hydrolase (PETase) from Ideonella Sakaiensis

Jerves *et al.* described an atomistic and thermodynamic interpretation of the catalytic mechanism of Polyethylene terephthalate hydrolase (PETase) using umbrella sampling simulations at the robust PBE/MM MD level with a large QM region [76]. The catalytic mechanism of PET degradation by PETase was shown, using a PET dimer model as substrate. The reaction mechanism occurs in two stages, acylation and deacylation, which occur through a single, associative, concerted, and asynchronous step, as reported by Han *et al.* [77]. Additionally, transient tetrahedral conformations were observed as each reaction approaches the corresponding transition state. Then, it was verified that the acylation reaction occurred *via* a combined asynchronous pathway. The nucleophilic attack occurs after the deprotonation of Ser^{131} by His^{208} (an acid-base mechanism), and cleavage of the ester bond occurs after $Ser^{131}(O^-)$ nucleophilic attack. Protonation of the terephthalate leaving group, mono (2-hydroxyethyl) terephthalate (MHET), was the last event (Fig. **6**). In fact, conventional MD simulations of the first intermediate state were performed, where it was observed that the MHET molecule formed was diffused into the solvent, being replaced by molecules of bulk solvent water. Thus, MD simulations suggested that MHET diffusion leads to a second intermediate state, which should be an entropy-driven spontaneous process, and that, given the slightly endergonic acylation step and decreasing free energy after the first intermediate, the overall process must be exergonic. Unlike the acylation step, the deacylation step exhibits clear extremes, which were deduced to be related to the stronger binding of MHET at the PETase active site. In this step, at the end of the acylation step, a water molecule from the active site occupies the region that became available after the diffusion of the outgoing MHET to the bulk solvent, which carries out a nucleophilic attack on the enzyme substrate adduct, generating the product and restoring the resting state of the enzyme. The initial deacylation state (second intermediate) was similar to the first intermediate, but MHET had already left the active site. Water molecules were abundant in the active site, observed in MD simulations. The initial distance between $His^{208}N\varepsilon$ and the water molecule was favorable for the deprotonation of water (2.49 ± 0.92 Å), and the distance from the water molecule to C_4^1 also favored the nucleophilic attack (3.27 ± 0.12 Å). Furthermore, the hydrogen bond between Asp^{177} and His^{208} was stable (1.70 ± 0.14 Å). Unlike the acylation step, the nucleophilic attack of water occurs synchronized with its deprotonation by His^{208}, as the distance between the hydrogen of water and $His^{208}N\varepsilon$ was 1.33 ± 0.28 Å and the O−H bond of water was amplified to 1.46 ± 0.46 Å. In this sense, Ser^{131} $O_\gamma-C_4^1$ bond was definitely broken (3.04 ± 0.11 Å), Ser^{131} deprotonated His^{208} on the way from the second transition state to the product, and a neutral Ser^{131} was regenerated. The bond between the oxygen atom of the water and the carbon C_4^1 atom is near (1.34 ± 0.03 Å), confirming the formation of the MHET product. The authors suggested that Asp^{83}, Asp^{89}, and Asp^{157} residues could be mutated by neutral or positively charged residues to stabilize the first transition state. In contrast, Arg^{61}, Lys^{66}, Arg^{94}, Glu^{175}, Asp^{177}, and Glu^{202} should be kept unchanged in the structure, as they would help to stabilize the charges of the polarized zones in the transition state.

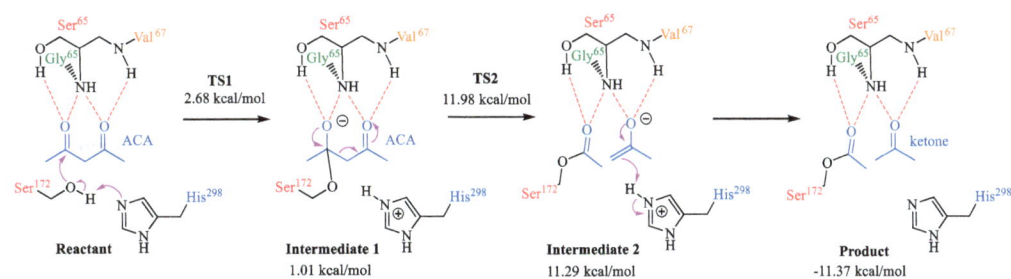

Fig. (5). The proposed mechanism of the biodegradation of acetylacetone by oxidized polyvinyl alcohol hydrolase (OPH), including transition states (TS), and energy.

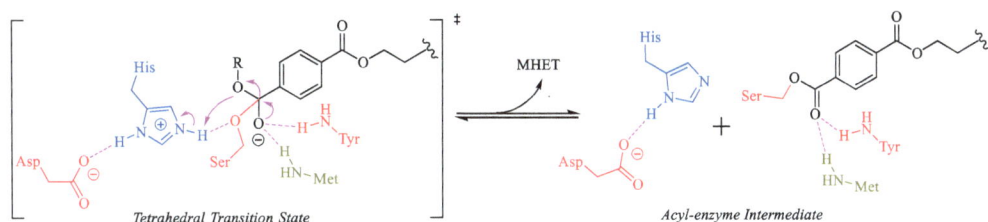

Fig. (6). Reaction Mechanism of PETase Proposed by Jerves *et al.*

Exploring Catalytic Reactions of Cysteine Protease (Papain-like Proteins)

Typically, cysteine proteases are enzymes that play essential roles in many life processes and most of them are targeted in several drug-design campaigns. This group of enzymes, along with serine proteases, represent the most abundant proteases found in the human body [78]. Cysteine proteases constitute a family of enzymes, named cathepsins, which are categorized into different families, considering their reaction mechanisms and structures, being the C1 family the most prevalent [79]. In this context, these enzymes are mainly found in lysosomes and are associated with protein degradation, apoptosis, and autophagy [79, 80]. In general, their enzymatic mechanisms of catalysis are complex processes, comprising many steps that are divided into two groups: acylation and deacylation reactions. Based on their activities, these enzymes represent druggable targets for many diseases, in which understanding their mechanisms is vital for drug development.

Papain Protease

Chemoenzymatic peptide synthesis has been applied for the production of peptides in chemical and bacterial synthesis, representing an effective and alternative for obtaining these compounds [81, 82]. Typically, this type of synthesis requires protease enzymes, acting as a biocatalyst [83, 84]. In general, this strategy provides high yields, atom-economy, stereo- and regioselectivity, and there is no need for (de)protection reactions, in contrast to the broadly used *solid-phase peptide synthesis* (SPPS) [84, 85]. It is known that proteases catalyze the cleavage of peptide bonds, but, surprisingly, chemoenzymatic peptide synthesis can make proteases able to form peptide bonds [85 - 87]. This fact is possible since enzymes cannot modify the thermodynamics of reactions, being capable of catalyzing both hydrolysis and aminolysis reversibly [88]. At physiological conditions, the hydrolysis reaction takes place instead aminolysis [88]. When a nucleophilic amino group of an amino acid monomer reacts with another monomer, then aminolysis occurs, leading to the fusion of the monomer to form a polypeptide, *via* polymerization [85, 89]. Nevertheless, when the concentration of this nucleophile decreases, then hydrolysis occurs instead of polymerization, cleaving the polypeptide previously formed [85, 89 - 91]. Still, the polymerization rate is associated with the specificity of the enzyme for the acyl donor group [92]. In this context, papain, a cysteine protease, has been broadly employed in chemoenzymatic peptide synthesis [81, 83, 93 - 95]. This enzyme has a catalytic dyad, Cys^{25} and His^{159} residues, which forms an active ion pair, where the protonated His^{159} is stabilized by the adjacent Asn^{175} residue, but it is not essential for the catalytic mechanism [96 - 98]. In general, papain polymerizes $_L$amino acids, converting these into polypeptides [99]. Thus, it is not able to polymerize $_D$amino acids [100]. Thus, Gimenez-Dejoz *et al.* investigated a reaction mechanism for papain-mediated chemoenzymatic peptide synthesis using $_L$Ala-OEt and $_D$Ala-OEt stereoisomer substrates to determine the specificity of this enzyme [101]. The authors used *adaptatively biased molecular dynamics* (ABMD) and DFT calculations, as well as experimental assays to prove their hypotheses (Fig. **7**). For the QM region, they used DFT B3LYP/6-31G*, while the MM region was treated with AMBER ff14SB force field. It was found transition states with values of 23 and 19 kcal/mol for $_L$Ala-OEt and $_D$Ala-OEt, respectively. Moreover, the authors verified that Gln^{19} residue acts by stabilizing the transition state formed. Still, they observed that the acylation step occurs *via* a concerted mechanism for both substrates. Finally, the orientation of the methyl group of the $_D$acyl intermediate causes a more hindered zone, explaining why it is less reactive than the $_L$acyl-intermediate [101].

Falcipain-2 from Plasmodium Falciparum

The *Plasmodium* genus comprises five protozoa species responsible for causing malaria in humans, P. *vivax*, *P. falciparum*, *P. malariae*, and *P. ovale*, in which *P. falciparum* represents the deadliest form of this disease, whereas *P. vivax* causes the most extend infection [102]. Typically, malaria protozoa are transmitted to humans by the bite of infected *Anopheles* spp. mosquitoes [103]. Different proteases from *P. falciparum* can catalyze the degradation of human hemoglobin, in which the resulting amino acid residues from this process are used as a source of energy for its metabolism or these also could be incorporated into its proteins [104]. In this context, its papain-like enzyme (from the CA clan), named falcipain-2, is involved in this metabolic process [105]. It is a cysteine protease expressed within the erythrocytic stage of the life cycle of this parasite [105], where the blockage of this target can be considered a promising alternative to prevent parasitic proliferation and growth [106]. Considering the inhibition of falcipain-2 by epoxysuccinate E64, a natural, potent and non-selective irreversible inhibitor of cysteine proteases, Arafet *et al.* decided to investigate its mechanism of catalysis by applying MD simulations and hybrid QM/MM approaches [107]. Then, the authors investigated the regioselectivity of the thiolate at the attack epoxide ring of E64, in which the nucleophilic attack can take place at either C^2 or C^3 (Fig. **8**), depending on the orientation of E64 at the binding site [108]. The QM region was carried out by using the AM1d method at M06-2X/6-31+G** level of theory, while the rest of the target and water molecules were described by OPLS-AA and TIP3P force fields, respectively. All these parameters were obtained from the fDYNAMO library. According to the authors, their method predicted that the more stable enzyme-E64 complex when Cys^{42} attacks C^3 than when it attacks C2, exhibiting −20.5 and −19.1 kcal/mol, respectively; even C^3 having a higher TS value than C^2, displaying 13.6 and 12.3 kcal/mol, respectively. This attack at C^3 takes place in a two-step mechanism, in which the first step involves the formation of the S−C^3 bond, which is the rate-limiting step. The second step involves the hydrogen transfer from the carboxylic acid function of E64 to the oxygen atom of the opened epoxide ring. Finally, the authors concluded that Gln^{36}, Trp^{43}, Asn^{81}, Asp^{170}, Gln^{171}, and His^{174} residues have an essential role in the orientation of E64 for the reaction to take place [107].

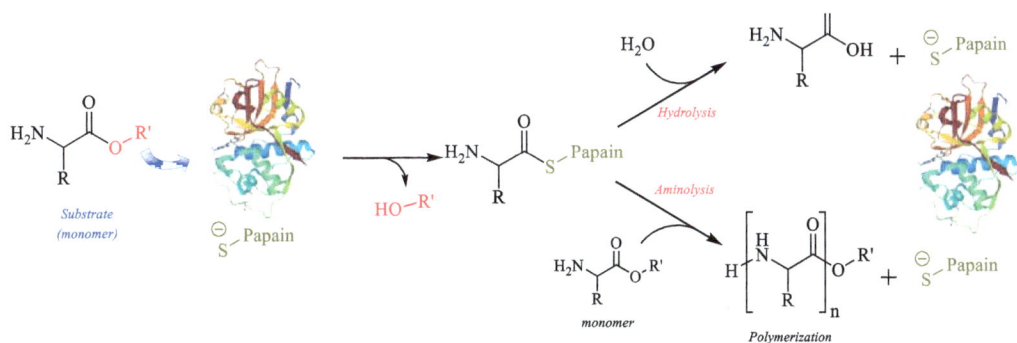

Fig. (7). Stereospecific pathway for $_D$Ala-OEt and $_L$Ala-OEt substrates with papain.

Fig. (8). Possible epoxide rings open at E64 towards the nucleophilic attack of the thiolate from falcipain-2 from *Plasmodium falciparum*.

Cruzain Protease from Trypanosoma cruzi, and Rhodesain from T. brucei

Chagas disease, or American trypanosomiasis, is an infectious disease caused by the protozoan parasite *Trypanosoma cruzi* and discovered in 1909 by the Brazilian doctor Carlos Chagas [109], which mainly endemic in Latin American countries [110]. There are no licensed vaccines, whereas there are only two *Food & Drug Administration* (FDA)-approved drugs, benznidazole and nifurtimox, which have several adverse effects and contra-indications (renal or hepatic failure, pregnancy, neuronal and psychiatric disorders). Additionally, both of these drugs are ineffective in the chronic stage of the disease [111, 112]. In this context, cruzain, which belongs to the papain-like family of cysteine proteases, is expressed in all stages during the life cycle of the parasite, being involved in the nutrition and evasion of the host immune system [113 - 115]. Cruzain promotes the hydrolysis of peptides and proteins, which is associated with its highly catalytic activity due to its nucleophilic character, enhanced by the formation of a Cys(S⁻)/His(H⁺) ion pair (Fig. **9**), *via* an acid-base reaction [116, 117]. However, there are many hypotheses and even questions remaining about such catalytic mechanisms. For example, Polgar *et al.* [118, 119] concluded that this ion pair is essential for the catalytic activity of papain cysteine protease, by performing spectrophotometric determinations of pH dependence of the reaction. Shafer *et al.* suggested that the ion pair is favored over the neutral residues [120, 121]. By using MD simulations, Engles *et al.* demonstrated that the ion pair is 7.2 kcal/mol lower in energy than the neutral one [122].

In 2015, Arafet and collaborators reported the first QM/MM studies to elucidate the inhibition mechanism of cruzain by two irreversible peptidyl halomethyl ketones, by investigating three possible mechanisms [123]. Thus, the inhibitor, Cys[25] and His[159] residues were described by using AM1d semiempirical Hamiltonian as QM region. Additionally, Gln[19] side-chain residue was also included in the QM area to investigate its involvement in a bond or charge transfer during the catalytic mechanism. For this purpose, M06-2X functional with the 6-31+G** basis set was employed in this step. In addition, during the rest of the system, protein residues and water molecules were treated as MM by means of OPLS-AA and TIP3P force fields, respectively. Additionally, *potential energy surfaces* (PESs) were used to describe the *inherent reaction coordinate* (IRC). The authors revealed that the Cys[25] residue promotes a nucleophilic attack on the halogenated methylene bond, *via* a concerted mechanism. Studying the free energy barriers and reaction free energies, determined at both levels of theory, revealed that the inhibition by Bz-Tyr-Ala-CH$_2$-Cl inhibitor (PClK) (Fig. **10**) is thermo and kinetically more favorable than Bz-Tyr-Ala-CH$_2$-F (PFK) (Fig. **10**), exhibiting energy barrier values of 10 and 29.5 kcal/mol, respectively. In contrast, large differences were obtained upon a comparison of the stabilized reaction energies, being −27.7 and −2.3 kcal/mol for PClK and PFK, respectively. However, the authors stated that their QM/MM overestimated the free energy barrier values when compared with those obtained by DFT method. Moreover, it was observed that their results suggested that PClK is an irreversible inhibitor of cruzain, based on its great chemical reactivity [123].

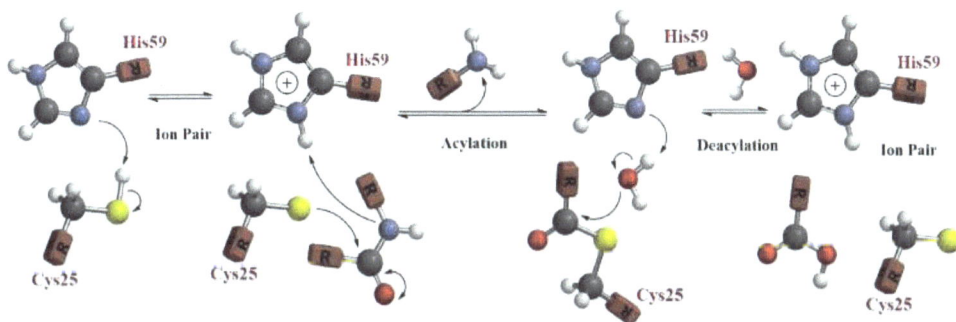

Fig. (9). General mechanism for peptide hydrolysis catalyzed by cruzain protease.

Posteriorly, Arafet *et al.,* 2018, reported a computational study focused on Michaelis complex formation towards cruzain cysteine protease from *Trypanosoma cruzi* [124], which is the etiological agent of Chagas disease. This target belongs to the papain family of proteins, which are essential for several other parasitic protozoa, making these macromolecules very attractive for designing new potential drugs [125]. In this context, the authors performed a QM/MM study to investigate three possible mechanisms of acylation catalyzed by cruzain Fig. (**11**). For this purpose, the authors treated the catalytic binding site with the AM1d semi-empirical Hamiltonian for the QM region, whereas we described the rest of the system (protein residues and water molecules) by using OPLS-AA and TIP3P force fields, respectively. Additionally, the M06-2X/- -31+G**:AM1d/MM was employed in their study. Thus, they proved that the acylation reaction is a two-step process, in which firstly, the proton for the cationic histidine $(His(H^+)^{159})$ residue is transferred to the N atom of the peptide (substrate), which posteriorly undergoes a nucleophilic attack by the anionic cysteine $(Cys(S^-)^{25})$ residue to the carbonyl carbon atom [124]. Subsequently, the diacylation reaction occurs *via* a concerted mechanism, which is the rate-limiting step of this catalytic process. It involves a water molecule to activate the cysteine residue, yielding the ion pair again. Still, the authors verified that this ion pair is more stable than the neutral form, with a low transition barrier of 2 kcal/mol for the free protease; whereas a value of 13.4 kcal/mol is observed in the presence of the substrate. Additionally, the analysis of the average of these two states revealed that the proton donor and acceptor are placed at an inter-atomic distance of 3.17 ± 0.15 Å, for free protease, whereas this distance increases to 3.97 ± 0.21 Å in the presence of the substrate. Furthermore, their findings improved of knowledge on this enzyme at the molecular level, being crucial for designing novel inhibitors based on the structures of the transition states or stable intermediates, revealing that pathway I is more plausible.

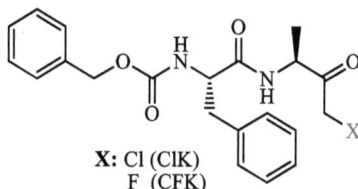

X: Cl (ClK)
    F (CFK)

Fig. (10). Potential irreversible ketone-halogenated inhibitors of cruzain from *Trypanosoma cruzi.*

Then, Oanca *et al.* performed a theoretical study focusing on the energy profile for acylation and part of the deacylation step for three cysteine proteases, cruzain, papain, Q19A-mutated papain toward the benzyloxycarbonyl-phenylalanylarginine-4-methylcoumaryl-7-amide (Fig. **12**) [126]. This compound was drawn and energetically minimized by using HF/6-31+G(d). The atomic charges were investigated from the *electrostatic surface potential* (ESP)-derived charges, which were generated at the HF/6-311+G** level, with the *Polarizable Continuum Model* (PCM). Posteriorly, the authors used the *restrained electrostatic surface potential* (RESP)-fitted charges, which were utilized for the substrate and reacting residues, whereas the Cys^{25} and His^{159} dyad were treated by the ENZYMIX force field. The reference system was solvated by a *Surface Constraint Atom Solvent* (SCAAS) sphere, with a radius of 20 Å. They observed that the ion pair formation *via* proton transfer from Cys^{25} to His^{159} with energy values of 4.5 ± 0.5 > 10, 2.3 ± 2.0, and 2.0 ± 0.4 kcal/mol for water (only both dyad), papain, Q19A-mutated papain, and cruzain in presence of substrate. In contrast, in the absence of substrate, values of 4.5 ± 0.4, 1.4 ± 0.9, and 2.1 ± 0.4, and −1.0 ± 0.5 kcal/mol were observed for water, papain, Q19A-mutated papain, and cruzain, respectively. By using the QM/MM method, the authors verified that all cysteine proteases present similar energy profiles, including their transition states ($\Delta G^{\ddagger}$), except for the water-catalyzed mechanism. In general, the authors verified that the complete mechanism involves a Cys^{25} to His^{159} proton transfer (−1.0 kcal/mol); substrate binding (−6.5 kcal/mol); nucleophilic attack and rearrangement ($\Delta G^{\ddagger}$: 6.7 and ΔG_0: 4.8 kcal/mol); His^{159} to *N*-amino proton transfer (−3.7 kcal/mol); bond breaking and rearrangement ($\Delta G^{\ddagger}$: 7.4 and ΔG_0: −9.7 kcal/mol); concerted proton transfer ($\Delta G^{\ddagger}$: 8.6 kcal/mol); and nucleophilic attack in the deacylation step (ΔG_0: 6.8 kcal/mol).

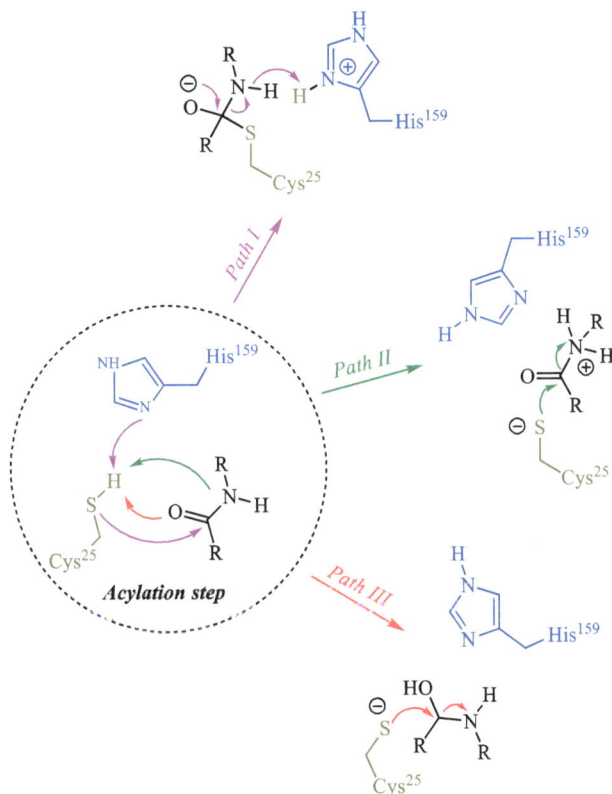

Fig. (11). Three different possible mechanisms of acylation by cruzain investigated by M06-2X/--31+G**:AM1d/MM.

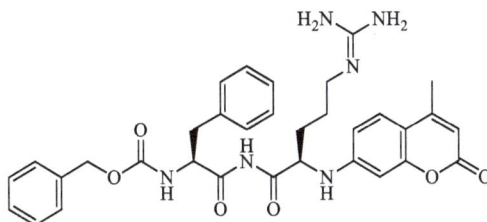

Fig. (12). Substrate of papain-like proteases investigated by Oanca *et al.,* 2020.

Recently, our research team developed inhibitors of cruzain and rhodesain (the *T. brucei* cathepsin-like protein) in a study using a *virtual fragment-based drug design* (vFBDD) approach focused on the development of naphthoquinone-based derivatives [127]. Then, we synthesized a promising naphthoquinone analog, JN-11 (Fig. **13**), towards both cruzain and rhodesian, exhibiting IC_{50} values of 6.3 and 1.8 µM, respectively. Moreover, both complexes were investigated by MD simulations and MM/PBSA calculations, revealing that JN11 presents the best stability with rhodesain and high affinity for this target, respectively; corroborating with its better activity upon this protease. Posteriorly, a mechanistic hypothesis (*via* C^3-Michael-addition reaction) involving a potential covalent binding mode of inhibition for JN-11 towards RhD was investigated by covalent molecular docking and then DFT B3LYP/6–31+G* calculations, suggesting that JN-11 could be attacked by the thiolate anion from the protease (at a distance of 3.95 Å). As a result, it was obtained a low activation energy barrier ($\Delta G^{\ddagger}$) and a stable product (ΔG), with values of 7.78 and −39.72 kcal/mol, respectively. In this context, suggested that this compound probably is not a covalent irreversible inhibitor. Then, to corroborate our findings, time-dependence inhibition and reversibility assays were performed, demonstrating that it is a reversible non-covalent inhibitor [127], and reinforcing the results of the DFT calculations.

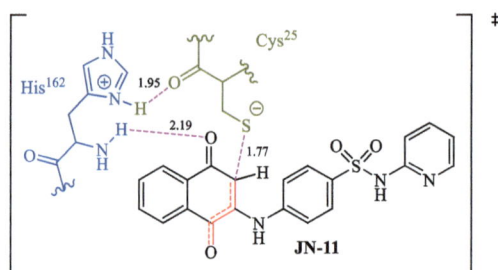

Fig. (13). The potential transition state for JN-11 with the rhodesain binding site investigated by Silva *et al.*, 2021, in which all distance values are shown in Ångström.

Main Protease (Mpro or 3CLpro) from Severe Acute Respiratory Syndrome Coronavirus 2 (SARS-CoV-2)

SARS-CoV-2, which was responsible for the 2019 global pandemic, has many structural and non-structural proteins, including its main protease, named Mpro (or 3CLpro) [128 - 131]. This protease cleaves the polyproteins pp1a and pp1ab at 16 different positions to generate structural (spike, envelope, membrane, and nucleocapsid) and non-structural proteins (several NSPs). Additionally, there are no similar proteases in human beings, making the drug design of mpro inhibitors an excellent strategy to fight against this virus [132 - 135]. The Mpro is a cysteine protease having a catalytic dyad, constituted by Cys^{145} and His^{41} residues. Based on this, there are some propositions stating that catalytic dyads of proteases are already in the thiolate-imidazolium ion pair state in the apo form [118]. In contrast, there is a proposal that states that Cys-His catalytic dyad remains as neutral residues in the apo form of the cysteine protease and then, after the binding of the substrate or inhibitor, the His activates the Cys residue by abstracting the proton from Sγ atom. Subsequently, this activated residue attacks the electrophilic group of the substrate or inhibitor [136, 137]. Still, there is another mechanism proposed, in which the proton transfer occurs concomitant with the nucleophilic attack, after the substrate or inhibitor binding [138]. All mechanisms are identical to those studied by Arafet *et al.* 2018 [124], as aforementioned (Fig. **11**). Then, Mondal & Warshel (2020) decided to investigate these proposed covalent mechanisms using an α-ketoamide inhibitor (Fig. **14**) [139]. They started from a covalently co-crystallized inhibitor of mpro, in which the covalent bond between sulfhydryl group of Cys^{145} and the inhibitor was previously removed. Posteriorly, the MOLARIS-XG was used as a solvation model to generate a water sphere based on SCAAS. In addition, B3LYP/6-31+G** level of theory using RESP was used to calculate the partial charges of the inhibitor. Posteriorly, ENZYMIX polarizable force field was used for all simulations within 300 ps. Initially, the authors verified two non-covalent mechanisms, one with all ionized residues (those that were ionizable) and another with neutral residues, obtaining ΔG_{noncov} values of −3.5 ± 0.5 and −3.6 ± 0.5 kcal/mol for before and after proton transfer, respectively. These values suggest that the ion pair formation is not essential for the non-covalent complex formation [139]. Additionally, DFT/MM (PDLD/S-LRA/β) calculations revealed that the transition state for proton transfer occurs with 2.9 ± 0.2 and 7.3 ± 1.7 kcal/mol for the protease in *apo* and *holo* forms, respectively. Moreover, it was observed that the transition state for the nucleophilic attack of Cys^{145} is most favorable in the protein than water environment. A-ketoamide is a reversible inhibitor that has an experimental free binding energy of −8.5 kcal/mol [132], corroborated by Mondal & Warshel, which found free binding energy values ranging from −8.1 to −9.2 kcal/mol for the covalent ligand-protein complex. The authors stated that the expected values of 3-4 kcal/mol were more negative, but they did consider their method as underestimated. Finally, they suggested that the ion pair is formed after binding with the inhibitor, and then the nucleophilic attack occurs to generate the final complex [139], as shown in Fig. (**14**).

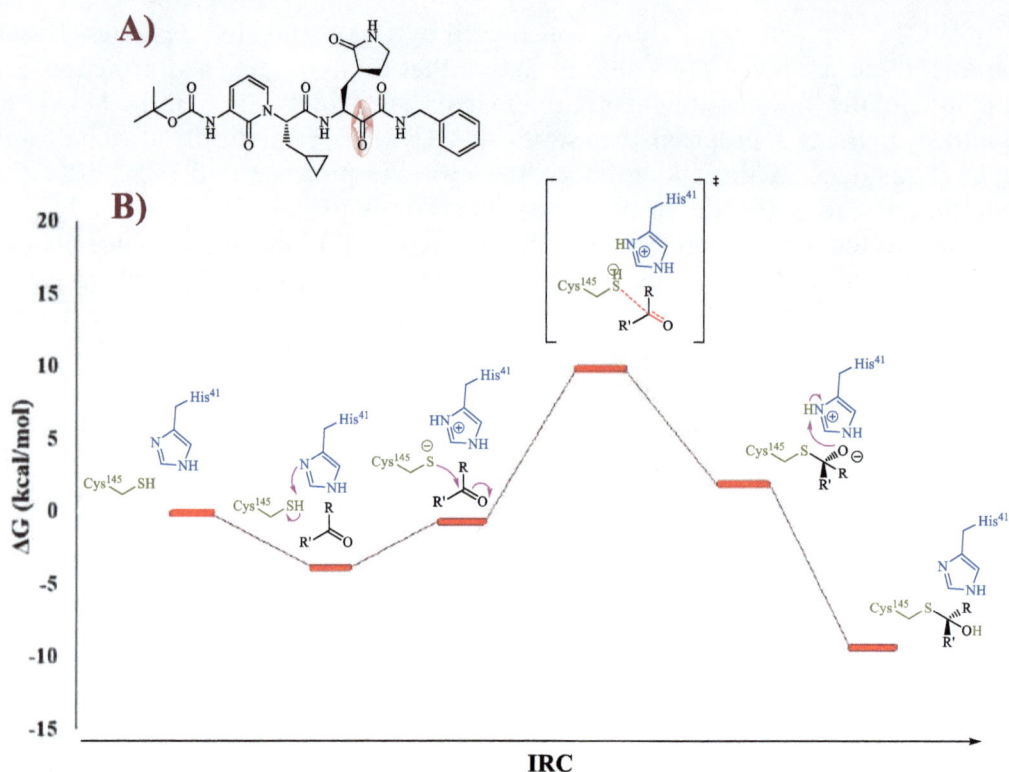

Fig. (14). Reversible α-ketoamide inhibitor of SARS-CoV-2 Mᵖʳᵒ (**A**) and the inherent reaction coordinate (IRC) and its inhibition mechanism (**B**). The susceptible electrophilic center is shown in red color.

In 2022, Eno *et al.* reported a detailed study involving DFT, molecular docking, and *multilinear regression analyze* (MLRA) of eight biologically active Rhenium(I)-tricarbonyl complexes [140] (Fig. **15**), they were selected and modeled based on the results of Karges *et al.* [141]. The authors performed the atomistic DFT modeling (6-31+G* for H, C, N, and O atoms, whereas the LanL2DZ basis set was used for Re atoms) to investigate reactivity, structural stability, and electronic properties based on *frontier molecular orbitals* (FMO), NBO, interaction energies, the *density of states* (DOS), charge distributions, and thermochemical parameters. Furthermore, molecular docking simulations were also performed to study the binding interactions between the selected biologically active complexes and SARS-CoV-2 3CLᵖʳᵒ. The *quantitative structure-activity relationship* (QSAR) was employed to demonstrate the correlations between the results calculated by DFT and *in vitro* biological activity (IC₅₀) of the Reᴵ-complexes [140]. In this sense, it was found that the carboxylic complexes are the most reactive, while the A₆ complex proved to be kinetically more stable. On the other hand, the A₃ complex presented the highest interaction energy, which was correlated with the type of functional group present in the molecule and, therefore, it was the one that had the highest inhibitory activity against the studied protein complex. The authors also mentioned that the dicarboxylic acid derivatives present in A_4 and A_7 complexes are electron-withdrawing groups, thus attracting the electrons of the bipyridine rings to themselves. This way, it reduces the energy gap between HOMO-LUMO, increasing the passage of electrons. This facilitates electronic mobility, which is responsible for the reactivity of carboxylic functionalized complexes. According to the NBO analysis, the A_4 complex presented a relatively high value of $E^{(2)}$, suggesting that it could be a compound for application as an enzyme inhibitor. Thus, carboxylic acid-functionalized complexes are destabilized as a result of the electron mobility within the structure, which subsequently increases the second-order energy correction of the perturbation, thereby increasing the reactivity of the complexes. Moreover, the enthalpy changes for A_7 and A_4 complexes revealed that these are relatively more exothermic, providing an additional explanation for their greater biological effects. In summary, all QM calculations investigated in this study showed excellent agreement with the data found experimentally and, thus, pointed to Re(I)-tricarbonyl complexes as promising candidates for the development of effective SARS-CoV-2 inhibitors.

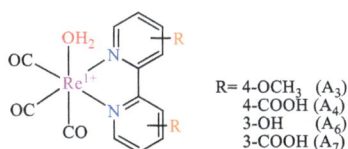

Fig. (15). Chemical structures of Re(I)-tricarbonyl complexes investigated by Eno *et al.*

METALLOPROTEASES

Metalloenzymes are catalytic proteins that have at least one metallic center as a cofactor in their active sites, which usually involve copper, zinc, cobalt, tin, iron, magnesium, and/or manganese ions. Because they are electron-deficient species, these ions participate in the formation of complexes with the enzyme's amino acids (usually catalytic amino acids) as electron acceptors. Therefore, the residues involved in this complex need to have electrons available for their donor effect to the metallic center, which can also be important in binding to a ligand, acting as an electronic acceptor. In this context, the formation of the complex is driven by electrostatic interactions, which can be simulated by using DFT calculations. In the dynamics of the electronic density of the entire complex, the residues involved in the charge transfer to the metallic center normally have side chains with atoms that have unshared pairs of electrons, which can participate in this conjugation effect, such as histidine, aspartate, and asparagine. The ligands that increase the formation of the complex with the enzyme, *via* interaction with the metallic center, also need to contain groups that have this electronic donor effect and, due to this characteristic, are increased by the metallic center, while water molecules often play an important role in stabilization of the three-dimensional structure of a metalloholoenzyme [142 - 145], or even the metal can work as a stabilizer of negative charges in regions close to the catalytic center [146]. The presence of metallic centers as coordinators of bonds with amino acid residues and with ligands demonstrates the importance of DFT calculations to determine their geometric optimization from an electron density acceptor center, thus, the interactions between HOMO-LUMO frontier orbitals are important for understanding the stabilization of the complex [143] and of its modifications along changes in the number of complexed ligands over the course of the catalyzed reaction [145]. Thus, the presence of metallic cofactors to intermediate binding of a ligand to their molecular target changes the dynamics of this interaction, in the sense that the metal works as a strong electronic acceptor and coordinates the oncoming between the ligand and amino acid residues based on charge and/or chemical group transferences.

For stabilizing role, Mg^{2+} ions are usually present at catalytic sites that have nucleic acids as substrate. DFT studies are used to demonstrate how this stabilization allows coordination with pyrophosphate groups of nucleotides to facilitate their release in polymerization by the action of HIV-1 reverse transcriptase, directing a mechanistic study of the reaction steps. In this context, *Root-Mean Square Fluctuation* (RMSF) plots are typically utilized for analysis of the positions of atoms in the catalytic center, with the explanation of interactions based on their distances. Compared to this, RMSD plots usually are applied for the determination of atomic positions of non-protein organic structures [147]. Metals that have more than one oxidation state often play an efficient catalytic role in the catalysis of oxidation-reduction reactions and can act by changing the electronic state of the atoms with which they interact, facilitating proton transfer steps between residues or between residues and ligands [142]. Thusly, DFT studies can be used to simulate the redox potential of catalytic centers in the presence of metallic cofactors by demonstrating their reactional electronic flow [148] or even for understanding their spin density transfer, which also aids in the uptake of their multiplicity of coordination [145]. By allowing us to understand the dynamics of the catalytic center by replacing interactions between the metal and water molecules with interactions involving a ligand, electron density simulations can support to knowing its changes in dielectric constants during catalytic structural modifications [142]. For many of these metal ions, their roles as electron density acceptors are made possible by the participation of their *d*-shell orbitals, where electron flow is observed when there are interactions between amino acid residues and ligands [145]. Regarding metal ions as enzymatic cofactors, histone deacetylases are important therapeutic targets in the development of antitumor and anti-inflammatory drugs, as well as for drugs targeting the prevention of neurodegenerative diseases that have a zinc-dependent catalytic center, which coordinates some important residues for the structural maintenance of the protein and mediates the complexation with a ligand or drug. Thus, the study of an inhibitor of this enzyme must contain zinc-binding heteroatoms as electron density donors, to be considered in the development of DFT studies [149, 150]. The existence of this metal in the enzymatic catalytic center provides a region of strong electrostatic polarization for the ligand being a direct interference in the binding free energy obtained by the formation of the ligand-target complex [151].

Iron(III)-Catalyzed Aerobic Degradation by Biphenyl 2,3-dioxygenase (BphA)

In order to elucidate the catalytic mechanism of biphenyl 2,3-dioxygenase (BphA), Zhu *et al.* implemented QM/MM method using a combination of triple-zeta LACV3P+* basis set on iron, 6-311++G** basis set on other atoms and CHARMM27 force field to investigate the chemical mechanism that leads to *cis*-diols. Since BphA is a Rieske-type enzyme, being the first enzyme in the aerobic degradation process, it plays a key role in the process of metabolizing aromatic biphenyl (BP)/polychlorinated biphenyl (BP-Cl) pollutants in the environment [152]. Since the hydroperoxo-iron (III) species are involved in the enzyme-catalyzed reaction, the authors explored the direct reaction mechanism of hydroperoxo-iron (III) species with biphenyl and 4-4'-dichlorobiphenyl. Thus, the dioxygenation process of the two substrates consisted of three elementary reactions. The rate-determining step was the formation of epoxide, in which the breakage of the O^1-O^2 bond in the hydroperoxide and the formation of new C^1-O^2 and C^2-O^2 bonds were considered a concerted step. This step was revealed to be the rate determination process, and the energy barriers of the BP and Cl-BP systems were 17.6 and 19.8 kcal/mol, respectively. Furthermore, it was observed that the distance of the C^2-O^2 bond gradually increased and then broke to form the intermediate carbocation in the second stage, which obtained energy barriers of 6.8 and 4.6 kcal/mol for BP and Cl-BP, respectively. In the final step, binding of the hydroxo ligand to the intermediate carbocation to produce diol occurred with low energy barriers of 23.1 and 20.7 kcal/mol for the BP and Cl-BP systems, respectively. For the Cl-BP system, the calculated barrier of each step was slightly higher than that of the BP system. The energy calculation results implied that, compared to the Cl-BP system, the BP system had more energy advantages and a higher probability of occurring. Fig. (**16**) shows the mechanism investigated by Zhu *et al.* The authors showed that during the process, the hydroperoxy-iron (III) species was found in the sextet spin (S = 5/2) state, according to Bassan *et al.* [153], it has more favorable energy than its low spin state (S = 3/2). Furthermore, the QM calculation highlighted the important role of the six residues in the system: BP (Asp[230], Gly[335], Asn[337], Thr[338], Ile[339], and Arg[340]) and Cl-BP system (Phe[227], Ile[336], Asn[337], Ile[339], Arg[340], and Phe[378]) in the process of dioxygenation. For the BP system, Arg[340] residue significantly suppressed the epoxide formation process, increasing the reaction barrier by about 4.1 kcal/mol. In addition, Asp[230], Gly[335], Asn[337], Thr[338], and Ile[339] residues facilitated the process by decreasing the reaction barrier by about 2.4, 2.3, 3.2, 1.6, and 3.9 kcal/mol, respectively. For the Cl-BP system, Phe[227], Ile[336], and Arg[340] suppressed the epoxide formation process by increasing the reaction barrier by about 1.3, 2.7, and 3.1 kcal/mol, respectively. As for residues Asn[337], Ile[339], and Phe[378], they promoted the process by decreasing the reaction barrier by about 2.3, 2.5, and 1.1 kcal/mol, respectively. This way, the results of the electrostatic influence analysis can provide candidates for future mutation studies [152].

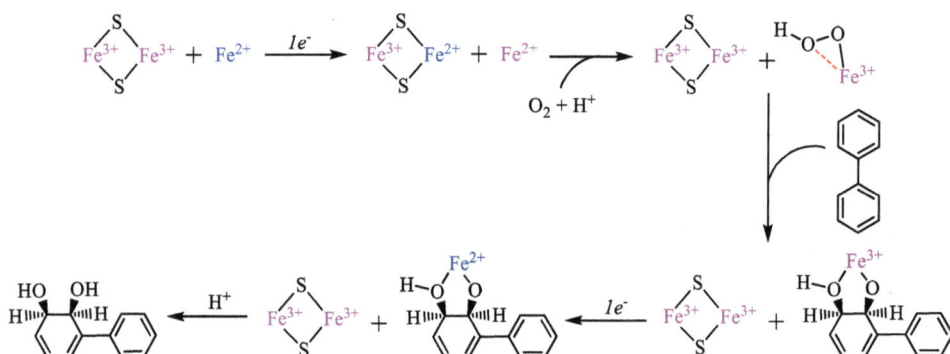

Fig. (16). Iron(III)-catalyzed mechanism for biphenyl 2,3-dioxygenase enzyme.

Mushroom Copper-Containing Tyrosinase

Polatoğlu and Karataş evaluated the molecular interactions between catechol and copper-containing tyrosinase in four different environments (vacuum, explicit water, ethanol/water, and acetone/water mixture) [154]. These interactions were modeled using the DFT faster functional B97-Dispersion (B97-D) and TZVP basis set. At first, the scan profiles of *o*-quinone formation start from catechol in PBS, PBS-ethanol, and PBS-acetone solutions at pH 6.5 and 25 °C. Furthermore, the maximum peak of the enzymatically produced *o*-quinone (λ: 390 nm) was monitored during the first minute to measure the enzyme activity in the UV-visible spectrophotometer. The authors also showed that the proximity of the catechol molecule to the center of the copper atoms and the consequent increase in the Cu−Cu distance led to facilitated coordination between catechol and *met-*

tyrosinase, which resulted in the formation of H−bonds and H−abstraction (catechol deprotonation) in the explicit water and ethanol/water mixtures, respectively. Ethanol significantly increased the molecular interaction between catechol and *met*-tyrosinase (−361.85 kJ/mol), probably because ethanol provided an ordered structure for water molecules without penetrating them, which allowed more favorable positions for the system. The distances between Cu−Cu and Cu−O in *met*-tyrosinase were also greater than those observed in a water environment, being 3.72 and 2.24 Å, respectively. On the other hand, acetone showed a tendency towards the copper center without any bond formation. In addition, they mentioned that the high interaction of acetone with water molecules that led to the irregular dispersion of water resulted in an unfavorable conformation for the catechol/*met*-tyrosinase interaction. It was also observed that the distances of H−bonds in the acetone environment, in most cases, were greater. Finally, the conformational change of histidine residues in the model containing acetone was another parameter that led to enzyme inhibition. The distances of the H−bonds are one of the most significant parameters that cause the interaction energy to be smaller in an acetone environment [154]. In Fig. (**17**), the investigated copper-catalyzed mechanism of tyrosinase is shown.

[NiFe] Hydrogenase from Desulfovibrio Gigas

In the study of Stein & Lubtz, spectroscopic data of paramagnetic intermediates of the enzymatic reaction cycle of metalloenzyme [NiFe] hydrogenase were determined using relativistic DFT within *Zero-Order Regular Approximation* (ZORA) [155]. Several models of structures were investigated for the paramagnetic states Ni−A, Ni−B, Ni−C, Ni−L, and Ni−CO. The authors observed that the nickel site activates H_2 and the bridged oxygenated ligand is released during the catalytic process. Protons leave the active site through a proton transfer channel or solvent molecule. They then suggested that water assists in the cleavage of molecular hydrogen in the mechanism of [NiFe] hydrogenase. In this sense, they mentioned that previous QM studies neglected the bridging ligand and/or its participation in the reaction mechanism (Fig. **18**). They also mentioned that its mechanism is similar to that of hydrogenase with only [Fe] suggested by Fan & Hall based on DFT calculations [156]. They found that the inclusion of a nitrogen-containing bridging ligand between the iron atoms in the hydrogenase active site presents a low-barrier route for H_2 cleavage and heterolytic formation. The authors also suggested that the CO molecule binds axially to the Ni atom. Since this inhibits the enzyme's catalytic ability, the observation provides support for the discovery that Ni is indeed the enzyme's active site. This is in agreement with the results obtained from the diffusion of Xe through the H_2 channels of [NiFe] hydrogenase from *Desulfovibrio sp* [157]. Furthermore, the study suggested that the H_2 inlet channel is directed to the Ni site. By a series of additional calculations, it was shown that the H_2 splitting barrier is strongly reduced by providing a base (amino acid residue), which helps in the coordination of the formed water molecule. Thus, the influence of this residue is mainly of a steric nature, and it does not alter the electronic states of the paramagnetic intermediates. It indirectly assists the heterolytic cleavage of the dihydrogen substrate by positioning the liberated bridging ligand (H_2O), which thus provides a barrier-free route for H_2 splitting and the uptake of the proton to be discarded. In addition, the terminal cysteine Cys^{530} acts as a temporary base. It is worth noting that until this work was carried out, not all details of the proposed reaction mechanism were supported by experimental data.

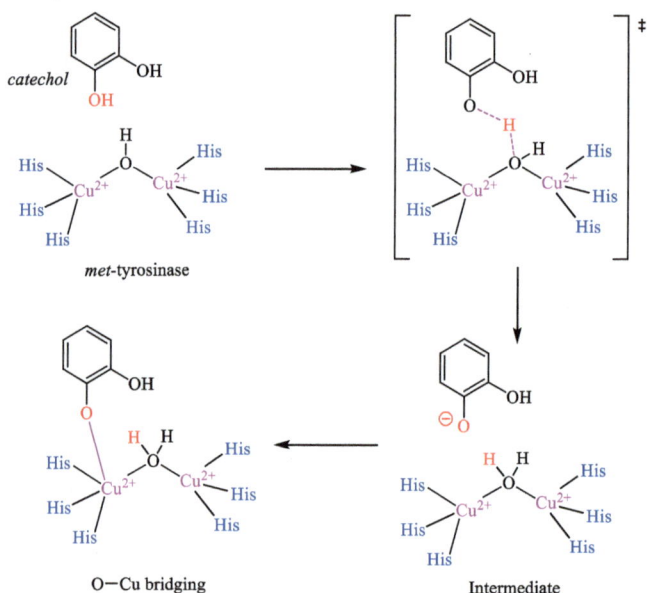

Fig. (17). Initial copper-catalyzed enzymatic mechanism of tyrosinase towards catechol, described by DFT studies.

Fig. (18). Structure of the active site of [NiFe] hydrogenase from *Desulfovibrio gigas* in the oxidized form. *The bridging ligand X is assumed to be an oxygenic species (O^{2-}, HO^-, or H_2O).*

QUANTUM CHEMICAL (QC) METHODS AND THEIR USES FOR DESIGNING DRUGS – VIEWPOINT AND COMPUTER REQUIREMENTS

QC calculations are not only challenging but also incredibly important for medicinal chemists engaged in the development of new potential drugs. These calculations offer a deep understanding of the underlying molecular mechanisms and interactions that govern drug behavior within the human body. By employing QM principles, researchers can investigate the intricate electronic structure, energy landscapes, and reactivity profiles of drug molecules. This knowledge allows medicinal chemists to make informed decisions when designing and optimizing drug candidates, leading to improved efficacy and reduced side effects. QC calculations enable the prediction and analysis of key drug properties. This knowledge helps in the fine-tuning of drug candidates, optimizing their pharmacokinetic and pharmacodynamic properties. Furthermore, quantum chemical calculations facilitate the exploration of chemical space, enabling the identification of novel scaffolds or motifs with desirable drug-like properties. These calculations can guide medicinal chemists in the selection of appropriate starting points for drug discovery efforts, saving time and resources by focusing on promising candidates. In summary, QC calculations serve as indispensable tools for medicinal chemists by providing deep insights into molecular properties and interactions. With their help, medicinal chemists can accelerate the drug discovery process, leading to the development of safer, more effective, and targeted therapeutic agents. Once fully aware of the immense potential and benefits offered by QM calculations, there arises an urgent need to utilize them in order to facilitate drug discovery processes. However, it is crucial to recognize that harnessing the full power of these tools requires significant configuration and computational time. However, time-consuming will depend on the number of atoms in a molecular system and also the basis set employed for calculations. In general, the minimum computer requirements for performing QM calculations can vary depending on the specific software and methods being used. However, here are some general guidelines:

1- Processor (CPU): QM calculations can be computationally intensive, so a modern multi-core processor is recommended. Higher clock speeds and more cores will generally result in faster calculations. Processors from Intel (*e.g.,* Core i7, Core i9) or AMD (*e.g.,* Ryzen) are commonly used.

2- Memory (RAM): The amount of RAM required depends on the size of the system being studied and the computational method employed. However, a minimum of 8 GB is typically recommended, with larger systems and more demanding calculations benefiting from 16 GB or more.

3- Hard Drive: A solid-state drive (SSD) is preferable over a traditional hard disk drive (HDD) due to its faster read/write speeds. QM calculations often involve reading and writing large amounts of data, so an SSD can significantly improve performance.

4- Graphics Processing Unit (GPU): While most QM calculations primarily rely on the CPU, certain calculations can be accelerated using GPUs. Not all software packages support GPU acceleration, but if a software does, a dedicated GPU from NVIDIA (*e.g.,* GeForce or Quadro series) can speed up calculations.

5- Software: It is needed a QM software package capable of performing the desired calculations. Some popular choices include Gaussian, NWChem, ORCA, and Q-Chem, among others. Each software package may have its own specific system requirements, so it's important to check the documentation provided by the software vendor.

6- Operating System: Most quantum chemistry software is compatible with Windows, macOS, and Linux operating systems. Ensure that the chosen software supports the operating system for the intended calculation.

It's worth noting that more complex systems, larger basis sets, and more accurate methods will require more computational resources. If it is planned to work with particularly challenging systems, it may be beneficial to have a high-performance computing (HPC) cluster or access to cloud-based computing resources to handle the computational demands. Therefore, it is imperative that the scientific community recognizes the urgent need to embrace and invest in these computational tools to drive innovation and advancement in the field of drug discovery.

CONCLUSION

In summary, Hohenberg-Kohn DFT is an HF-based technique that has been considered very promising in the field of drug design and development since it can increase the chances of success in developing a new drug candidate. Moreover, it has the main advantage of proving its results in a moderately time-consuming alternative when compared to other *ab initio* methods. Still, the DFT approach can provide excellent accuracy of electronic properties, which are directly associated with the level of theory employed in the calculations. Several historical aspects are associated with the development of the DFT method, including its formalism. In general, it is observed that the hybrid methods of QM/MM are frequently used to determine the behavior of ligands at the binding site of different enzymes having medicinal or industrial interests. In this context, searching for transition states represents the heart of catalytic mechanisms at an atomistic level of theory. Thusly, these have been applied to discuss or even elucidate the mechanisms of reactions involving inhibitors or specific substrates. Still, it was verified that B3LYP still is a broadly employed basis set, which has been often used with 6-311++G** level of theory, especially in studies involving the elucidation of electronic properties associated with the reactivity of intermediates or even transition states. Clearly, proteases (or proteinases) have been broadly investigated by these types of techniques, in which is possible to verify some correlations between proteases from different organisms, mainly those associated with the initial (or not) formation of an ion pair, *via* an acid-base mechanism. Since proteases are potential druggable targets for the development of novel bioactive compounds, these data about their catalytic mechanisms are crucial for designing electrophilic compounds that could undergo a nucleophilic attack of a catalytic residue. Furthermore, DFT-related methods have been employed to predict the diverse electronic properties of bioactive compounds, being able to explain their activity in biological assays. Moreover, these can be used to describe electronic properties that make some ligands favorable for metal coordination in metalloproteases. However, some of these studies exhibit a lack of experimental data to prove the hypotheses of the authors. Concerning the constant advances in computer sciences, mainly the increasing power of data processing, it is believed that this *in silico* approach will continue growing up in the medicinal chemistry field, especially focused on the development of novel drug candidates with distinct pharmacological applications targeting different macromolecules.

LIST OF ABBREVIATIONS

QC	Quantum Chemistry
HF	Hartree-Fock
DFT	Density Functional Theory
SCF	Self-Consistent-Field
MM	Molecular Mechanics
TFD	Thomas-Fermi-Dirac
KS	Kohn-Sham
HSAB	Hard And Soft Acids And Bases
HOMO	Highest-Occupied Molecular Orbital
LUMO	Lowest-Occupied Molecular Orbital
QM	Quantum Mechanics
QM/MM	Quantum Mechanics / Molecular Mechanics
MD	Molecular Dynamics
TST	Transition State Theory

TS	Transition State
ONIOM	Our Own N-Layered Integrated Molecular Orbital + Molecular Mechanics
NBO	Natural Bond Orbital
RDG	Reduced Density Gradient
FPS	Farnesyl Pyrophosphate Synthetase
GABA	Γ-Aminobutyric Acid
LIC	Ligand-Gated Ion Channel
MNPE	1-(4-(3-Methoxy-4-Nitrophenyl)Piperazin-1-Yl)Ethanone
ALIE	Average Local Ionization Energies
BDE	Bond Dissociation Energy
EVB	Empirical Valence Bond
RdRp	RNA-Dependent RNA Polymerase
SARS-CoV-2	Severe Acute Respiratory Syndrome Coronavirus 2
COVID-19	Coronavirus Disease 2019
NHC	B-D-N4-Hydroxycytidine
RMSD	Root-Mean Square Deviation
OPH	Oxidized Polyvinyl Alcohol Hydrolase
PVA	Polyvinyl Alcohol
LIE	Linear Interaction Energy
MM/PBSA	Molecular Mechanics Poisson-Boltzmann Surface Area
P-RFO	Partitioned-Rational Function Optimization
ACA	Acetylacetone
PETase	Polyethylene Terephthalate Hydrolase
MHET	Mono (2-Hydroxyethyl) Terephthalate
SPPS	Solid-Phase Peptide Synthesis
ABMD	Adaptively Biased Molecular Dynamics
FDA	Food & Drug Administration
PES	Potential Energy Surfaces
IRC	Inherent Reaction Coordinate
ESP	Electrostatic Surface Potential
PCM	Polarizable Continuum Model
RESP	Restrained Electrostatic Potential
SCAAS	Surface Constraint All Atom Solvent
vFBDD	Virtual Fragment-Based Drug Design
M^{pro}	Main Protease

MLRA	Multilinear Regression Analyzes
FMO	Frontier Molecular Orbitals
DOS	Density Of States
QSAR	Quantitative Structure-Activity Relationship
RMSF	Root-Mean Square Fluctuation
BphA	Biphenyl 2,3-Dioxygenase
BP	Biphenyl
BP-Cl	Polychlorinated Biphenyl
B97-Dispersion	B97-D
ZORA	Zero-Order Regular Approximation

ACKNOWLEDGMENTS

The authors thank Coordenação de Aperfeiçoamento de Pessoal de Nível Superior – CAPES, National Council for Scientific and Technological Development – CNPq, Financiadora de Estudos e Projetos – FINEP, and Fundação de Amparo à Pesquisa do Estado de Alagoas – FAPEAL for their scientific support to the Brazilian Post-Graduation Programs.

REFERENCES

[1] Born M, Oppenheimer R. On the quantum theory of molecules. Ann Phys 1927; 389(20): 457-84.
     [http://dx.doi.org/10.1002/andp.19273892002]

[2] Szabo A, Ostlund NS. Modern Quantum Chemistry : Introduction to Advanced Electronic Structure Theory. 1ª., New York, USA: Dover Publications 1989.

[3] Hartree DR. The wave mechanics of an atom with a non-coulomb central field. part ii. some results and discussion. Math Proc Camb Philos Soc 1928; 24(1): 111-32.
     [http://dx.doi.org/10.1017/S0305004100011920]

[4] Eyring H. The activated complex in chemical reactions. J Chem Phys 1935; 3(2): 107-15.
     [http://dx.doi.org/10.1063/1.1749604]

[5] Nye MJ. Laboratory practice and the physical chemistry of michael polanyi. In: Homes FL, Levere T, Eds. Instruments and Experimentation in the History of Chemistry. Cambridge: MIT Press 2000; pp. 367-400.

[6] Stewart JJP. Semiempirical molecular orbital methods. In: Lipkowitz KB, Boyd DB, Eds. Reviews in Computational Chemistry. New York, USA: VCH Publishing 1990; pp. 45-81.

[7] Zerner MC. Semiempirical molecular orbital methods. In: Lipkowitz KB, Boyd DB, Eds. Reviews in Computational Chemistry. New York, USA: VHC Publishing 1991; pp. 313-65.

[8] Reynolds CH, Semiempirical MO. Semiempirical MO methods: The middle ground in molecular modeling. J Mol Struct THEOCHEM 1997; 401(3): 267-77.
     [http://dx.doi.org/10.1016/S0166-1280(97)00028-6]

[9] Hohenberg P, Kohn W. Inhomogeneous electron gas. Phys Rev 1964; 136(3B): B864-71.
     [http://dx.doi.org/10.1103/PhysRev.136.B864]

[10] Thomas LH. The calculation of atomic fields. Math Proc Camb Philos Soc 1927; 23(5): 542-8.

[http://dx.doi.org/10.1017/S0305004100011683]

[11]   De S, Chakrabarty S. Thomas-fermi and thomas-fermi-dirac models in two-dimension : Effect of strong quantizing magnetic field. Eur Phys J D 2017; 71(1): 5.
[http://dx.doi.org/10.1140/epjd/e2016-70295-1]

[12]   Chen J, Krieger JB, Esquivel RO, Stott MJ, Iafrate GJ. Kohn-Sham effective potentials for spin-polarized atomic systems. Phys Rev A 1996; 54(3): 1910-21.
[http://dx.doi.org/10.1103/PhysRevA.54.1910] [PMID: 9913679]

[13]   Parr RG, Yang W. Density-Functional Theory of Atoms and Molecule. 1st., New York, USA: Oxford University Press 1989.

[14]   Mulliken RS. A new electroaffinity scale: Together with data on valence states and on valence ionization potentials and electron affinities. J Chem Phys 1934; 2(11): 782-93.
[http://dx.doi.org/10.1063/1.1749394]

[15]   Mulliken RS. Electronic structures of molecules xi. electroaffinity, molecular orbitals and dipole moments. J Chem Phys 1935; 3(9): 573-85.
[http://dx.doi.org/10.1063/1.1749731]

[16]   Sanderson RT. An interpretation of bond lengths in alkali halide gas molecules. J Am Chem Soc 1952; 74(1): 272-4.
[http://dx.doi.org/10.1021/ja01121a522]

[17]   Sanderson RT. Electronegativity and bond energy. J Am Chem Soc 1983; 105(8): 2259-61.
[http://dx.doi.org/10.1021/ja00346a026]

[18]   Pearson RG. Hard and soft acids and bases, HSAB, part II: Underlying theories. J Chem Educ 1968; 45(10): 643.
[http://dx.doi.org/10.1021/ed045p643]

[19]   Pearson RG. Hard and soft acids and bases, HSAB, part 1: Fundamental principles. J Chem Educ 1968; 45(9): 581.
[http://dx.doi.org/10.1021/ed045p581]

[20]   Pearson RG. Chemical hardness and density functional theory. J Chem Sci 2005; 117(5): 369-77.
[http://dx.doi.org/10.1007/BF02708340]

[21]   Fukui K, Yonezawa T, Shingu H. A molecular orbital theory of reactivity in aromatic hydrocarbons. J Chem Phys 1952; 20(4): 722-5.
[http://dx.doi.org/10.1063/1.1700523]

[22]   Yang W, Parr RG. Hardness, softness, and the fukui function in the electronic theory of metals and catalysis. Proc Natl Acad Sci 1985; 82(20): 6723-6.
[http://dx.doi.org/10.1073/pnas.82.20.6723] [PMID: 3863123]

[23]   Chermette H. Density functional theory. Coord Chem Rev 1998; 178-180: 699-721.
[http://dx.doi.org/10.1016/S0010-8545(98)00179-9]

[24]   Geerlings P, De Proft F, Langenaeker W. Conceptual density functional theory. Chem Rev 2003; 103(5): 1793-874.
[http://dx.doi.org/10.1021/cr990029p] [PMID: 12744694]

[25]   Gökce H, Öztürk N, Sert Y, El-Azab AS, AlSaif A, Abdel-Aziz AA-M. 4-[(1, 3-Dioxoisoindolin-2-Yl)Methyl]Benzenesulfonamide: Full structural and spectroscopic characterization and molecular docking with carbonic anhydrase II. ChemistrySelect 2018; 3: 10113-24.
[http://dx.doi.org/10.1002/slct.201802484]

[26]   Khan F, Syed F, Iqbal A, Khan ZUH, Ullah H, Khan S. Synthesis, spectral characterization and antibacterial study of schiff base metal complexes derived from 4-Bromo-N-[(E)-(5-chloro-2-hydroxyphenyl)methylidene]benzenesulfonamide. Asian J Chem 2016; 28(8): 1658-60.
[http://dx.doi.org/10.14233/ajchem.2016.19771]

[27] Elangovan N, Sowrirajan S. Synthesis, single crystal (XRD), Hirshfeld surface analysis, computational study (DFT) and molecular docking studies of (E)-4-((2-hydroxy-3,5-diiodobenzylidene)am-no)-N-(pyrimidine)-2-yl) benzenesulfonamide. Heliyon 2021; 7(8): e07724.
[http://dx.doi.org/10.1016/j.heliyon.2021.e07724] [PMID: 34458601]

[28] Kumar VS, Mary YS, Pradhan K, *et al.* Synthesis, spectral properties, chemical descriptors and light harvesting studies of a new bioactive azo imidazole compound. J Mol Struct 2020; 1199: 127035.
[http://dx.doi.org/10.1016/j.molstruc.2019.127035]

[29] Kumar VS, Mary YS, Mary YS, *et al.* Conformational analysis and DFT investigations of two triazole derivatives and its halogenated substitution by using spectroscopy, AIM and Molecular docking. Chemical Data Collections 2021; 31: 100625.
[http://dx.doi.org/10.1016/j.cdc.2020.100625]

[30] Mandal S, Pan A, Bhaduri R, Tarai SK, Kapoor BS, Moi SC. Theoretical investigation on hydrolysis mechanism of cis-platin analogous Pt(II)/Pd(II) complex by DFT calculation and molecular docking approach for their interaction with DNA & HSA. J Mol Graph Model 2022; 117: 108314.
[http://dx.doi.org/10.1016/j.jmgm.2022.108314] [PMID: 36041352]

[31] Feizi-Dehnayebi M, Dehghanian E, Mansouri-Torshizi H. Synthesis and characterization of Pd(II) antitumor complex, DFT calculation and DNA/BSA binding insight through the combined experimental and theoretical aspects. J Mol Struct 2021; 1240: 130535.
[http://dx.doi.org/10.1016/j.molstruc.2021.130535]

[32] Komeiji Y, Ishida T, Fedorov DG, Kitaura K. Change in a protein's electronic structure induced by an explicit solvent: Anab initio fragment molecular orbital study of ubiquitin. J Comput Chem 2007; 28(10): 1750-62.
[http://dx.doi.org/10.1002/jcc.20686] [PMID: 17340606]

[33] Sgrignani J, Magistrato A. First-principles modeling of biological systems and structure-based drug-design. Curr Computeraided Drug Des 2013; 9(1): 15-34.
[http://dx.doi.org/10.2174/1573409911309010003] [PMID: 23106775]

[34] Ganesan TS, Elangovan N, Vanmathi V, *et al.* Spectroscopic, Computational(DFT), Quantum mechanical studies and protein-ligand interaction of Schiff base 6,6-((1,--phenylenebis(azaneylylidene))bis(methaneylylidene))bis(2-methoxyphenol) from o-phenylenediamine and 3- methoxysalicylaldehyde. J Indian Chem Soc 2022; 99(10): 100713.
[http://dx.doi.org/10.1016/j.jics.2022.100713]

[35] Demircioğlu Z, Özdemir FA, Dayan O, Şerbetçi Z, Özdemir N. Synthesis, X-ray diffraction method, spectroscopic characterization (FT-IR, 1H and 13C NMR), antimicrobial activity, Hirshfeld surface analysis and DFT computations of novel sulfonamide derivatives. J Mol Struct 2018; 1161: 122-37.
[http://dx.doi.org/10.1016/j.molstruc.2018.02.063]

[36] Fathima AF, Jothi Mani R, Roshan MM, Sakthipandi K. Enhancing structural and optical properties of ZnO nanoparticles induced by the double co-doping of iron and cobalt. Mater Today Proc 2022; 49: 2598-601.
[http://dx.doi.org/10.1016/j.matpr.2021.06.433]

[37] Hong L, Zhang XC, Hartsuck JA, Tang J. Crystal structure of an *in vivo* HIV-1 protease mutant in complex with saquinavir: Insights into the mechanisms of drug resistance. Protein Sci 2000; 9(10): 1898-904.
[http://dx.doi.org/10.1110/ps.9.10.1898] [PMID: 11106162]

[38] Saen-oon S, Aruksakunwong O, Wittayanarakul K, Sompornpisut P, Hannongbua S. Insight into analysis of interactions of saquinavir with HIV-1 protease in comparison between the wild-type and G48V and G48V/L90M mutants based on QM and QM/MM calculations. J Mol Graph Model 2007; 26(4): 720-7.
[http://dx.doi.org/10.1016/j.jmgm.2007.04.009] [PMID: 17543558]

[39] Srivastava A, Rawat P, Tandon P, Singh RN. A computational study on conformational geometries,

chemical reactivity and inhibitor property of an alkaloid bicuculline with γ-aminobutyric acid (GABA) by DFT. Comput Theor Chem 2012; 993: 80-9.
[http://dx.doi.org/10.1016/j.comptc.2012.05.025]

[40]   Govindasamy H, Magudeeswaran S, Kandasamy S, Poomani K. Binding mechanism of naringenin with monoamine oxidase – B enzyme: QM/MM and molecular dynamics perspective. Heliyon 2021; 7(4): e06684.
[http://dx.doi.org/10.1016/j.heliyon.2021.e06684] [PMID: 33898820]

[41]   Vektariene A, Vektaris G, Svoboda J. A theoretical approach to the nucleophilic behavior of benzofused thieno[3,2-b]furans using DFT and HF based reactivity descriptors. ARKIVOC 2009; 2009(7): 311-29.
[http://dx.doi.org/10.3998/ark.5550190.0010.730]

[42]   Parr RG, Szentpály L, Liu S. Electrophilicity index. J Am Chem Soc 1999; 121(9): 1922-4.
[http://dx.doi.org/10.1021/ja983494x]

[43]   Mah R, Thomas JR, Shafer CM. Drug discovery considerations in the development of covalent inhibitors. Bioorg Med Chem Lett 2014; 24(1): 33-9.
[http://dx.doi.org/10.1016/j.bmcl.2013.10.003] [PMID: 24314671]

[44]   Backus KM, Correia BE, Lum KM, *et al.* Proteome-wide covalent ligand discovery in native biological systems. Nature 2016; 534(7608): 570-4.
[http://dx.doi.org/10.1038/nature18002] [PMID: 27309814]

[45]   Mihalovits LM, Ferenczy GG, Keserű GM. Affinity and selectivity assessment of covalent inhibitors by free energy calculations. J Chem Inf Model 2020; 60(12): 6579-94.
[http://dx.doi.org/10.1021/acs.jcim.0c00834] [PMID: 33295760]

[46]   Cee VJ, Volak LP, Chen Y, *et al.* Systematic study of the glutathione (GSH) Reactivity of *N* - Arylacrylamides: 1. effects of aryl substitution. J Med Chem 2015; 58(23): 9171-8.
[http://dx.doi.org/10.1021/acs.jmedchem.5b01018] [PMID: 26580091]

[47]   Lonsdale R, Burgess J, Colclough N, *et al.* Expanding the armory: Predicting and tuning covalent warhead reactivity. J Chem Inf Model 2017; 57(12): 3124-37.
[http://dx.doi.org/10.1021/acs.jcim.7b00553] [PMID: 29131621]

[48]   Miller PS, Aricescu AR. Crystal structure of a human GABAA receptor. Nature 2014; 512(7514): 270-5.
[http://dx.doi.org/10.1038/nature13293] [PMID: 24909990]

[49]   Schofield PR, Darlison MG, Fujita N, *et al.* Sequence and functional expression of the GABAA receptor shows a ligand-gated receptor super-family. Nature 1987; 328(6127): 221-7.
[http://dx.doi.org/10.1038/328221a0] [PMID: 3037384]

[50]   Sieghart W. GABAA receptors: Ligand-gated Cl− ion channels modulated by multiple drug-binding sites. Trends Pharmacol Sci 1992; 13(12): 446-50.
[http://dx.doi.org/10.1016/0165-6147(92)90142-S] [PMID: 1338138]

[51]   Pawson AJ, Sharman JL, Benson HE, *et al.* The IUPHAR/BPS guide to pharmacology: An expert-driven knowledgebase of drug targets and their ligands. Nucleic Acids Res 2014; 42(D1): D1098-106.
[http://dx.doi.org/10.1093/nar/gkt1143] [PMID: 24234439]

[52]   Onawole AT, Al-Ahmadi AF, Mary YS, *et al.* Conformational, vibrational and DFT studies of a newly synthesized arylpiperazine-based drug and evaluation of its reactivity towards the human GABA receptor. J Mol Struct 2017; 1147: 266-80.
[http://dx.doi.org/10.1016/j.molstruc.2017.06.107]

[53]   Schaleger LL, Long FA. Entropies of activation and mechanisms of reactions in solution. Adv Phys Org Chem 1963; 1: 1-33.
[http://dx.doi.org/10.1016/S0065-3160(08)60276-2]

[54]   Page MI, Jencks WP. Entropic contributions to rate accelerations in enzymic and intramolecular

reactions and the chelate effect. Proc Natl Acad Sci 1971; 68(8): 1678-83.
[http://dx.doi.org/10.1073/pnas.68.8.1678] [PMID: 5288752]

[55]   Kazemi M, Åqvist J. Chemical reaction mechanisms in solution from brute force computational Arrhenius plots. Nat Commun 2015; 6(1): 7293.
[http://dx.doi.org/10.1038/ncomms8293] [PMID: 26028237]

[56]   Jencks WP. Binding energy, specificity, and enzymic catalysis: The circe effect. Adv Enzymol Relat Areas Mol Biol 2006; 43: 219-410.
[http://dx.doi.org/10.1002/9780470122884.ch4] [PMID: 892]

[57]   Wolfenden R, Snider MJ. The depth of chemical time and the power of enzymes as catalysts. Acc Chem Res 2001; 34(12): 938-45.
[http://dx.doi.org/10.1021/ar000058i] [PMID: 11747411]

[58]   Bruice TC, Lightstone FC. Ground state and transition state contributions to the rates of intramolecular and enzymatic reactions. Acc Chem Res 1999; 32(2): 127-36.
[http://dx.doi.org/10.1021/ar960131y]

[59]   Åqvist J, Kazemi M, Isaksen GV, Brandsdal BO. Entropy and enzyme catalysis. Acc Chem Res 2017; 50(2): 199-207.
[http://dx.doi.org/10.1021/acs.accounts.6b00321] [PMID: 28169522]

[60]   Snider MJ, Gaunitz S, Ridgway C, Short SA, Wolfenden R. Temperature effects on the catalytic efficiency, rate enhancement, and transition state affinity of cytidine deaminase, and the thermodynamic consequences for catalysis of removing a substrate "anchor". Biochemistry 2000; 39(32): 9746-53.
[http://dx.doi.org/10.1021/bi000914y] [PMID: 10933791]

[61]   Kötting C, Gerwert K. Time-resolved FTIR studies provide activation free energy, activation enthalpy and activation entropy for GTPase reactions. Chem Phys 2004; 307(2-3): 227-32.
[http://dx.doi.org/10.1016/j.chemphys.2004.06.051]

[62]   Pinsent BRW, Pearson L, Roughton FJW. The kinetics of combination of carbon dioxide with hydroxide ions. Trans Faraday Soc 1956; 52: 1512.
[http://dx.doi.org/10.1039/tf9565201512]

[63]   Silva LR, da Silva Santos-Júnior PF, de Andrade Brandão J, *et al.* Druggable targets from coronaviruses for designing new antiviral drugs. Bioorg Med Chem 2020; 28(22): 115745.
[http://dx.doi.org/10.1016/j.bmc.2020.115745] [PMID: 33007557]

[64]   Merck. Merck and ridgeback's investigational oral antiviral molnupiravir reduced the risk of hospitalization or death by approximately 50 percent compared to placebo for patients with mild or moderate COVID-19 in positive interim analysis of phase 3 study. Available from: https://www.merck.com/news/merck-and-ridgebacks-investigational-oral-anti-iral-molnupiravir-reduced-the-risk-of-hospitalization-or-death-by-appro-imately-50-percent-compared-to-placebo-for-patients-with-mild-or-moderat/ (Accessed Dec 13, 2022).

[65]   Kabinger F, Stiller C, Schmitzová J, *et al.* Mechanism of molnupiravir-induced SARS-CoV-2 mutagenesis. Nat Struct Mol Biol 2021; 28(9): 740-6.
[http://dx.doi.org/10.1038/s41594-021-00651-0] [PMID: 34381216]

[66]   Silva L, Silva-Júnior E. Exploring molnupiravir (EIDD-2801) by molecular docking, temperature-dependent dynamics simulations, and DFT calculations on the RNA-dependent RNA polymerase (RdRp) from SARS-CoV-2. Proceedings of the Proceedings of 7th International Electronic Conference on Medicinal Chemistry,. MDPI: Basel, Switzerland, 2021; p. 11433.

[67]   Gottlieb RL, Vaca CE, Paredes R, *et al.* Early remdesivir to prevent progression to severe Covid-19 in outpatients. N Engl J Med 2022; 386(4): 305-15.
[http://dx.doi.org/10.1056/NEJMoa2116846] [PMID: 34937145]

[68]   Suzuki T, Ichihara Y, Yamada M, Tonomura K. Some characteristics of pseudomonas O-3 which

utilizes polyvinyl alcohol. Agric Biol Chem 1973; 37(4): 747-56.
[http://dx.doi.org/10.1271/bbb1961.37.747]

[69] Shimao M, Yamamoto H, Ninomiya K, *et al.* Pyrroloquinoline quinone as an essential growth factor for a polyvinyl alcohol)-degrading symbiont, pseudomonas Sp. VM15C. Agric Biol Chem 1984; 48: 2873-6.

[70] Teodorescu M, Bercea M, Morariu S. Biomaterials of poly(vinyl alcohol) and natural polymers. Polym Rev 2018; 58(2): 247-87.
[http://dx.doi.org/10.1080/15583724.2017.1403928]

[71] Shimao M. Biodegradation of plastics. Curr Opin Biotechnol 2001; 12(3): 242-7.
[http://dx.doi.org/10.1016/S0958-1669(00)00206-8] [PMID: 11404101]

[72] Yang Y, Ko TP, Liu L, *et al.* Structural insights into enzymatic degradation of oxidized polyvinyl alcohol. ChemBioChem 2014; 15(13): 1882-6.
[http://dx.doi.org/10.1002/cbic.201402166] [PMID: 25044912]

[73] Sakai K, Hamada N, Watanabe Y. Degradation mechanism of poly(vinyl alcohol) by successive reactions of secondary alcohol oxidase and β -Diketone Hydrolase from *Pseudomonas* sp. Agric Biol Chem 1986; 50(4): 989-96.
[http://dx.doi.org/10.1080/00021369.1986.10867494]

[74] Suzuki T. Oxidation of secondary alcohols by polyvinyl alcohol-degrading enzyme produced by pseudomonas O–3. Agric Biol Chem 1978; 42: 1187-94.

[75] Chen J, Wang J, Li Y, *et al.* Catalysis mechanism of oxidized polyvinyl alcohol by pseudomonas hydrolase: Insights from molecular dynamics and QM/MM analysis. Chem Phys Lett 2019; 721: 49-56.
[http://dx.doi.org/10.1016/j.cplett.2019.02.023]

[76] Jerves C, Neves RPP, Ramos MJ, da Silva S, Fernandes PA. Reaction mechanism of the PET degrading enzyme PETase studied with DFT/MM molecular dynamics simulations. ACS Catal 2021; 11(18): 11626-38.
[http://dx.doi.org/10.1021/acscatal.1c03700]

[77] Han X, Liu W, Huang JW, *et al.* Structural insight into catalytic mechanism of PET hydrolase. Nat Commun 2017; 8(1): 2106.
[http://dx.doi.org/10.1038/s41467-017-02255-z] [PMID: 29235460]

[78] Quesne MG, Ward RA, de Visser SP. Cysteine protease inhibition by nitrile-based inhibitors: A computational study. Front Chem 2013; 1: 39.
[http://dx.doi.org/10.3389/fchem.2013.00039] [PMID: 24790966]

[79] Turk V, Turk B, Turk D. New embo members' review: Lysosomal cysteine proteases: facts and opportunities. EMBO J 2001; 20(17): 4629-33.
[http://dx.doi.org/10.1093/emboj/20.17.4629] [PMID: 11532926]

[80] Olson OC, Joyce JA. Cysteine cathepsin proteases: Regulators of cancer progression and therapeutic response. Nat Rev Cancer 2015; 15(12): 712-29.
[http://dx.doi.org/10.1038/nrc4027] [PMID: 26597527]

[81] Baker PJ, Numata K. Chemoenzymatic synthesis of poly(L-alanine) in aqueous environment. Biomacromolecules 2012; 13(4): 947-51.
[http://dx.doi.org/10.1021/bm201862z] [PMID: 22380731]

[82] Yazawa K, Numata K. Recent advances in chemoenzymatic peptide syntheses. Molecules 2014; 19(9): 13755-74.
[http://dx.doi.org/10.3390/molecules190913755] [PMID: 25191871]

[83] Schwab LW, Kloosterman WMJ, Konieczny J, Loos K. Papain catalyzed (co)oligomerization of α-amino acids. Polymers 2012; 4(1): 710-40.
[http://dx.doi.org/10.3390/polym4010710]

[84]   Numata K. Poly(amino acid)s/polypeptides as potential functional and structural materials. Polym J 2015; 47(8): 537-45.
[http://dx.doi.org/10.1038/pj.2015.35]

[85]   Bordusa F. Proteases in organic synthesis. Chem Rev 2002; 102(12): 4817-68.
[http://dx.doi.org/10.1021/cr010164d] [PMID: 12475208]

[86]   Bergmann M, Fruton JS. Some synthetic and hydrolytic experiments with chymotrypsin. J Biol Chem 1938; 124(1): 321-9.
[http://dx.doi.org/10.1016/S0021-9258(18)74101-X]

[87]   Capellas M, Benaiges MD, Caminal G, Gonzalez G, Lopez-Santín J, Clapés P. Enzymatic synthesis of a CCK-8 tripeptide fragment in organic media. Biotechnol Bioeng 1996; 50(6): 700-8.
[http://dx.doi.org/10.1002/(SICI)1097-0290(19960620)50:6<700::AID-BIT11>3.0.CO;2-I]   [PMID: 18627079]

[88]   Koshland DE Jr. Kinetics of peptide bond formation. J Am Chem Soc 1951; 73(9): 4103-8.
[http://dx.doi.org/10.1021/ja01153a016]

[89]   Guzman F, Barberis S, Illanes A. Peptide synthesis: Chemical or enzymatic. Electron J Biotechnol 2007; 10-0–0.

[90]   Ageitos JM, Chuah JA, Numata K. Chemo-enzymatic synthesis of linear and branched cationic peptides: Evaluation as gene carriers. Macromol Biosci 2015; 15(7): 990-1003.
[http://dx.doi.org/10.1002/mabi.201400487] [PMID: 25828913]

[91]   Ageitos JM, Yazawa K, Tateishi A, Tsuchiya K, Numata K. The Benzyl ester group of amino acid monomers enhances substrate affinity and broadens the substrate specificity of the enzyme catalyst in chemoenzymatic copolymerization. Biomacromolecules 2016; 17(1): 314-23.
[http://dx.doi.org/10.1021/acs.biomac.5b01430] [PMID: 26620763]

[92]   Ettari R, Nizi E, Di Francesco ME, *et al.* Nonpeptidic vinyl and allyl phosphonates as falcipain-2 inhibitors. ChemMedChem 2008; 3(7): 1030-3.
[http://dx.doi.org/10.1002/cmdc.200800050] [PMID: 18428116]

[93]   de Beer RJAC, Zarzycka B, Amatdjais-Groenen HIV, *et al.* Papain-catalyzed peptide bond formation: Enzyme-specific activation with guanidinophenyl esters. ChemBioChem 2011; 12(14): 2201-7.
[http://dx.doi.org/10.1002/cbic.201100267] [PMID: 21826775]

[94]   de Beer RJAC, Zarzycka B, Mariman M, *et al.* Papain-specific activating esters in aqueous dipeptide synthesis. ChemBioChem 2012; 13(9): 1319-26.
[http://dx.doi.org/10.1002/cbic.201200017] [PMID: 22615272]

[95]   Tsuchiya K, Numata K. Chemoenzymatic synthesis of polypeptides containing the unnatural amino acid 2-aminoisobutyric acid. Chem Commun 2017; 53(53): 7318-21.
[http://dx.doi.org/10.1039/C7CC03095A] [PMID: 28485427]

[96]   Drenth J, Jansonius JN, Koekoek R, Swen HM, Wolthers BG. Structure of papain. Nature 1968; 218: 929-32.

[97]   Ménard R, Carrière J, Laflamme P, *et al.* Contribution of the glutamine 19 side chain to transition-state stabilization in the oxyanion hole of papain. Biochemistry 1991; 30(37): 8924-8.
[http://dx.doi.org/10.1021/bi00101a002] [PMID: 1892809]

[98]   Vernet T, Tessier DC, Chatellier J, *et al.* Structural and functional roles of asparagine 175 in the cysteine protease papain. J Biol Chem 1995; 270(28): 16645-52.
[http://dx.doi.org/10.1074/jbc.270.28.16645] [PMID: 7622473]

[99]   Tsuchiya K, Numata K. Papain-catalyzed chemoenzymatic synthesis of telechelic polypeptides using bis(leucine ethyl ester) initiator. Macromol Biosci 2016; 16(7): 1001-8.
[http://dx.doi.org/10.1002/mabi.201600005] [PMID: 26947148]

[100] Serveau C, Juliano L, Bernard P, Moreau T, Mayer R, Gauthier F. New substrates of papain, based on

the conserved sequence of natural inhibitors of the cystatin family. Biochimie 1994; 76(2): 153-8.
[http://dx.doi.org/10.1016/0300-9084(94)90007-8] [PMID: 8043651]

[101] Gimenez-Dejoz J, Tsuchiya K, Numata K. Insights into the stereospecificity in papain-mediated chemoenzymatic polymerization from quantum mechanics/molecular mechanics simulations. ACS Chem Biol 2019; 14(6): 1280-92.
[http://dx.doi.org/10.1021/acschembio.9b00259] [PMID: 31063345]

[102] WHO. Malaria. Available from: https://www.who.int/news-room/fact-sheets/detail/malaria (Accessed Dec 15, 2022).

[103] da Silva Neto GJ, Silva LR, de Omena RJM, *et al.* Dual quinoline-hybrid compounds with antimalarial activity against *Plasmodium falciparum* parasites. New J Chem 2022; 46(14): 6502-18.
[http://dx.doi.org/10.1039/D1NJ05598D]

[104] Miller LH, Baruch DI, Marsh K, Doumbo OK. The pathogenic basis of malaria. Nature 2002; 415(6872): 673-9.
[http://dx.doi.org/10.1038/415673a] [PMID: 11832955]

[105] Singh A, Rosenthal PJ. Selection of cysteine protease inhibitor-resistant malaria parasites is accompanied by amplification of falcipain genes and alteration in inhibitor transport. J Biol Chem 2004; 279(34): 35236-41.
[http://dx.doi.org/10.1074/jbc.M404235200] [PMID: 15192087]

[106] Shenai BR, Sijwali PS, Singh A, Rosenthal PJ. Characterization of native and recombinant falcipain-2, a principal trophozoite cysteine protease and essential hemoglobinase of Plasmodium falciparum. J Biol Chem 2000; 275(37): 29000-10.
[http://dx.doi.org/10.1074/jbc.M004459200] [PMID: 10887194]

[107] Arafet K, Ferrer S, Martí S, Moliner V. Quantum mechanics/molecular mechanics studies of the mechanism of falcipain-2 inhibition by the epoxysuccinate E64. Biochemistry 2014; 53(20): 3336-46.
[http://dx.doi.org/10.1021/bi500060h] [PMID: 24811524]

[108] Powers JC, Asgian JL, Ekici ÖD, James KE. Irreversible inhibitors of serine, cysteine, and threonine proteases. Chem Rev 2002; 102(12): 4639-750.
[http://dx.doi.org/10.1021/cr010182v] [PMID: 12475205]

[109] Clayton J. Chagas disease 101. Nature 2010; 465(S7301): S4-5.
[http://dx.doi.org/10.1038/nature09220] [PMID: 20571553]

[110] Moncayo SAC. Current Epidemiological Trends of Chagas Disease in Latin America and Futur Challenges: Epidemiology, Surveillance, and Health Policies. 2nd., Elsevier Inc. 2017.

[111] Castro JA, deMecca MM, Bartel LC. Toxic side effects of drugs used to treat Chagas' disease (American trypanosomiasis). Hum Exp Toxicol 2006; 25(8): 471-9.
[http://dx.doi.org/10.1191/0960327106het653oa] [PMID: 16937919]

[112] Filardi LS, Brener Z. Susceptibility and natural resistance of *Trypanosoma cruzi* strains to drugs used clinically in Chagas disease. Trans R Soc Trop Med Hyg 1987; 81(5): 755-9.
[http://dx.doi.org/10.1016/0035-9203(87)90020-4] [PMID: 3130683]

[113] Campetella O, Henriksson J, Åslund U, Frasch ACC, Pettersson U, Cazzulo JJ. The major cysteine proteinase (cruzipain) from *Trypanosoma cruzi* is encoded by multiple polymorphic tandemly organized genes located on different chromosomes. Mol Biochem Parasitol 1992; 50(2): 225-34.
[http://dx.doi.org/10.1016/0166-6851(92)90219-A] [PMID: 1311053]

[114] Fujii N, Mallari JP, Hansell EJ, *et al.* Discovery of potent thiosemicarbazone inhibitors of rhodesain and cruzain. Bioorg Med Chem Lett 2005; 15(1): 121-3.
[http://dx.doi.org/10.1016/j.bmcl.2004.10.023] [PMID: 15582423]

[115] Schnapp AR, Eickhoff CS, Sizemore D, Curtiss R III, Hoft DF. Cruzipain induces both mucosal and systemic protection against *Trypanosoma cruzi* in mice. Infect Immun 2002; 70(9): 5065-74.
[http://dx.doi.org/10.1128/IAI.70.9.5065-5074.2002] [PMID: 12183554]

[116] Scott CJ, Taggart CC. Biologic protease inhibitors as novel therapeutic agents. Biochimie 2010; 92(11): 1681-8.
[http://dx.doi.org/10.1016/j.biochi.2010.03.010] [PMID: 20346385]

[117] Brocklehurst K. Specific covalent modification of thiols: Applications in the study of enzymes and other biomolecules. Int J Biochem 1979; 10(4): 259-74.
[http://dx.doi.org/10.1016/0020-711X(79)90088-0] [PMID: 456716]

[118] Polgár L. On the mode of activation of the catalytically essential sulfhydryl group of papain. Eur J Biochem 1973; 33(1): 104-9.
[http://dx.doi.org/10.1111/j.1432-1033.1973.tb02660.x] [PMID: 4691346]

[119] Polgár L. Mercaptide-imidazolium ion-pair: The reactive nucleophile in papain catalysis. FEBS Lett 1974; 47(1): 15-8.
[http://dx.doi.org/10.1016/0014-5793(74)80415-1] [PMID: 4426388]

[120] Lewis SD, Johnson FA, Shafer JA. Potentiometric determination of ionizations at the active site of papain. Biochemistry 1976; 15(23): 5009-17.
[http://dx.doi.org/10.1021/bi00668a010] [PMID: 10964]

[121] Lewis SD, Johnson FA, Shafer JA. Effect of cysteine-25 on the ionization of histidine-159 in papain as determined by proton nuclear magnetic resonance spectroscopy. Evidence for a histidine-159-cystei-e-25 ion pair and its possible role in catalysis. Biochemistry 1981; 20(1): 48-51.
[http://dx.doi.org/10.1021/bi00504a009] [PMID: 7470479]

[122] Mladenovic M, Junold K, Fink RF, Thiel W, Schirmeister T, Engels B. Atomistic insights into the inhibition of cysteine proteases: First QM/MM calculations clarifying the regiospecificity and the inhibition potency of epoxide- and aziridine-based inhibitors. J Phys Chem B 2008; 112(17): 5458-69.
[http://dx.doi.org/10.1021/jp711287c] [PMID: 18393547]

[123] Arafet K, Ferrer S, Moliner V. First quantum mechanics/molecular mechanics studies of the inhibition mechanism of cruzain by peptidyl halomethyl ketones. Biochemistry 2015; 54(21): 3381-91.
[http://dx.doi.org/10.1021/bi501551g] [PMID: 25965914]

[124] Arafet K, Świderek K, Moliner V. Computational study of the michaelis complex formation and the effect on the reaction mechanism of cruzain cysteine protease. ACS Omega 2018; 3(12): 18613-22.
[http://dx.doi.org/10.1021/acsomega.8b03010]

[125] McKerrow J, Engel JC, Caffrey CR. Cysteine protease inhibitors as chemotherapy for parasitic infections. Bioorg Med Chem 1999; 7(4): 639-44.
[http://dx.doi.org/10.1016/S0968-0896(99)00008-5] [PMID: 10353643]

[126] Oanca G, Asadi M, Saha A, Ramachandran B, Warshel A. Exploring the catalytic reaction of cysteine proteases. J Phys Chem B 2020; 124(50): 11349-56.
[http://dx.doi.org/10.1021/acs.jpcb.0c08192] [PMID: 33264018]

[127] Silva LR, Guimarães AS, do Nascimento J, *et al.* Computer-aided design of 1,4-naphthoquinone-based inhibitors targeting cruzain and rhodesain cysteine proteases. Bioorg Med Chem 2021; 41: 116213.
[http://dx.doi.org/10.1016/j.bmc.2021.116213] [PMID: 33992862]

[128] Nascimento IJ dos S, Mendonça de Aquino T. Molecular modeling applied to design of cysteine protease inhibitors : A powerful tool for the identification of hit compounds against neglected tropical diseases. Front Comput Chem 2020; 5: 63-110.

[129] Santos Nascimento IJ, Silva-Júnior EF, Aquino TM. Repurposing FDA-approved drugs targeting SARS-CoV2 3CL [pro] : A study by applying virtual screening, molecular dynamics, MM-PBSA calculations and covalent docking. Lett Drug Des Discov 2022; 19(7): 637-53.
[http://dx.doi.org/10.2174/1570180819666220106110133]

[130] dos Santos Nascimento IJ, de Aquino TM, da Silva-Júnior EF. Drug Repurposing: A strategy for discovering inhibitors against emerging viral infections. Curr Med Chem 2021; 28(15): 2887-942.
[http://dx.doi.org/10.2174/1875533XMTA5rMDYp5] [PMID: 32787752]

[131] Santos-Júnior PF da S, Nascimento IJ dos S, Aquino TM, Araújo-júnior JX, da Silva-Júnior EF. Drug discovery strategies against emerging coronaviruses : A global threat. Front Anti-Inf Drug Disco 2020; 56: 35-90.

[132] Zhang L, Lin D, Sun X, *et al.* Crystal structure of SARS-CoV-2 main protease provides a basis for design of improved a-ketoamide inhibitors. Science 2020; 368: 409-12.

[133] Jin Z, Du X, Xu Y, *et al.* Structure of M^{pro} from SARS-CoV-2 and discovery of its inhibitors. Nature 2020; 582(7811): 289-93.
[http://dx.doi.org/10.1038/s41586-020-2223-y] [PMID: 32272481]

[134] Jin Z, Zhao Y, Sun Y, *et al.* Structural basis for the inhibition of SARS-CoV-2 main protease by antineoplastic drug carmofur. Nat Struct Mol Biol 2020; 27(6): 529-32.
[http://dx.doi.org/10.1038/s41594-020-0440-6] [PMID: 32382072]

[135] Dai W, Zhang B, Jiang XM, *et al.* Structure-based design of antiviral drug candidates targeting the SARS-CoV-2 main protease. Science 368: 1331-5.

[136] Huang C, Wei P, Fan K, Liu Y, Lai L. 3C-like proteinase from SARS coronavirus catalyzes substrate hydrolysis by a general base mechanism. Biochemistry 2004; 43(15): 4568-74.
[http://dx.doi.org/10.1021/bi036022q] [PMID: 15078103]

[137] Connolly KM, Smith BT, Pilpa R, Ilangovan U, Jung ME, Clubb RT. Sortase from *Staphylococcus aureus* does not contain a thiolate-imidazolium ion pair in its active site. J Biol Chem 2003; 278(36): 34061-5.
[http://dx.doi.org/10.1074/jbc.M305245200] [PMID: 12824164]

[138] Świderek K, Moliner V. Revealing the molecular mechanisms of proteolysis of SARS-CoV-2 M^{pro} by QM/MM computational methods. Chem Sci 2020; 11(39): 10626-30.
[http://dx.doi.org/10.1039/D0SC02823A] [PMID: 34094317]

[139] Mondal D, Warshel A. Exploring the mechanism of covalent inhibition: Stimulating the binding free energy of α-ketoamide inhibitors of the main protease of SARS-CoV-2. Biochemistry 2020; 59(48): 4601-8.
[http://dx.doi.org/10.1021/acs.biochem.0c00782] [PMID: 33205654]

[140] Eno EA, Louis H, Unimuke TO, *et al.* Modeling of Re(I) tricarbonyl complexes against SARS-CoV-2 receptor *via* DFT, *in-silico* molecular docking, and QSAR. Chemical Physics Impact 2022; 5: 100105.
[http://dx.doi.org/10.1016/j.chphi.2022.100105]

[141] Karges J, Kalaj M, Gembicky M, Cohen SM, Re I. Re ^{I} tricarbonyl complexes as coordinate covalent inhibitors for the SARS-CoV-2 main cysteine protease. Angew Chem Int Ed 2021; 60(19): 10716-23.
[http://dx.doi.org/10.1002/anie.202016768] [PMID: 33606889]

[142] Lintuluoto M, Lintuluoto JM. DFT study on enzyme turnover including proton and electron transfers of copper-containing nitrite reductase. Biochemistry 2016; 55(33): 4697-707.
[http://dx.doi.org/10.1021/acs.biochem.6b00423] [PMID: 27455866]

[143] Pourmand F, Zahedi M, Safari N. Theoretical (DFT) study on the hydroxylation mechanism of Sn(IV)porphyrin: How does Sn(IV)porphyrin inhibit heme oxygenase catalysis function. J Mol Struct 2022; 1253: 132097.
[http://dx.doi.org/10.1016/j.molstruc.2021.132097]

[144] Trewick SC, Henshaw TF, Hausinger RP, Lindahl T, Sedgwick B. Oxidative demethylation by *Escherichia coli* AlkB directly reverts DNA base damage. Nature 2002; 419(6903): 174-8.
[http://dx.doi.org/10.1038/nature00908] [PMID: 12226667]

[145] Liu H, Llano J, Gauld JW. A DFT study of nucleobase dealkylation by the DNA repair enzyme AlkB. J Phys Chem B 2009; 113(14): 4887-98.
[http://dx.doi.org/10.1021/jp810715t] [PMID: 19338370]

[146] Ruiz-Pernía JJ, Alves CN, Moliner V, Silla E, Tuñón I. A QM/MM study of the reaction mechanism

for the 3'-processing step catalyzed by HIV-1 integrase. J Mol Struct THEOCHEM 2009; 898(1-3): 115-20.
[http://dx.doi.org/10.1016/j.theochem.2008.08.005]

[147] Rungrotmongkol T, Mulholland AJ, Hannongbua S. Active site dynamics and combined quantum mechanics/molecular mechanics (QM/MM) modelling of a HIV-1 reverse transcriptase/DNA/dTTP complex. J Mol Graph Model 2007; 26(1): 1-13.
[http://dx.doi.org/10.1016/j.jmgm.2006.09.004] [PMID: 17046299]

[148] Maekawa S, Matsui T, Hirao K, Shigeta Y. Theoretical study on reaction mechanisms of nitrite reduction by copper nitrite complexes: Toward understanding and controlling possible mechanisms of copper nitrite reductase. J Phys Chem B 2015; 119(17): 5392-403.
[http://dx.doi.org/10.1021/acs.jpcb.5b01356] [PMID: 25845517]

[149] Dokmanovic M, Clarke C, Marks PA. Histone deacetylase inhibitors: Overview and perspectives. Mol Cancer Res 2007; 5(10): 981-9.
[http://dx.doi.org/10.1158/1541-7786.MCR-07-0324] [PMID: 17951399]

[150] Hsu KC, Liu CY, Lin TE, *et al.* Novel class IIa-selective histone deacetylase inhibitors discovered using an *in silico* virtual screening approach. Sci Rep 2017; 7(1): 3228.
[http://dx.doi.org/10.1038/s41598-017-03417-1] [PMID: 28607401]

[151] Kollar J, Frecer V. Diarylcyclopropane hydroxamic acid inhibitors of histone deacetylase 4 designed by combinatorial approach and QM/MM calculations. J Mol Graph Model 2018; 85: 97-110.
[http://dx.doi.org/10.1016/j.jmgm.2018.08.008] [PMID: 30145395]

[152] Zhu L, Zhou J, Zhang R, *et al.* Degradation mechanism of biphenyl and 4-4'-dichlorobiphenyl cis-dihydroxylation by non-heme 2,3 dioxygenases BphA: A QM/MM approach. Chemosphere 2020; 247: 125844.
[http://dx.doi.org/10.1016/j.chemosphere.2020.125844] [PMID: 32069708]

[153] Bassan A, Blomberg MRA, Siegbahn PEM. A theoretical study of the cis-dihydroxylation mechanism in naphthalene 1,2-dioxygenase. J Biol Inorg Chem 2004; 9(4): 439-52.
[http://dx.doi.org/10.1007/s00775-004-0537-0] [PMID: 15042436]

[154] Polatoğlu İ, Karataş D. Modeling of molecular interaction between catechol and tyrosinase by DFT. J Mol Struct 2020; 1202: 127192.
[http://dx.doi.org/10.1016/j.molstruc.2019.127192]

[155] Stein M, Lubitz W. Relativistic DFT calculation of the reaction cycle intermediates of [NiFe] hydrogenase: A contribution to understanding the enzymatic mechanism. J Inorg Biochem 2004; 98(5): 862-77.
[http://dx.doi.org/10.1016/j.jinorgbio.2004.03.002] [PMID: 15134933]

[156] Fan HJ, Hall MB. A capable bridging ligand for Fe-only hydrogenase: Density functional calculations of a low-energy route for heterolytic cleavage and formation of dihydrogen. J Am Chem Soc 2001; 123(16): 3828-9.
[http://dx.doi.org/10.1021/ja004120i] [PMID: 11457119]

[157] Montet Y, Amara P, Volbeda A, *et al.* Gas access to the active site of Ni-Fe hydrogenases probed by X-ray crystallography and molecular dynamics. Nat Struct Biol 1997; 4(7): 523-6.
[http://dx.doi.org/10.1038/nsb0797-523] [PMID: 9228943]

CHAPTER 9

Free Energy Estimation for Drug Discovery: Background and Perspectives

Fernando D. Prieto-Martínez[1,*] and **Yelzyn Galván-Ciprés**[2]

[1] *Instituto de Química, Universidad Nacional Autónoma de México, Ciudad de México, México*

[2] *Escuela Nacional de Ciencias Biológicas, Instituto Politécnico Nacional, Ciudad de México, México*

Abstract: Drug development is a remarkably complex subject, with potency and specificity being the desired traits in the early stages of research. Yet, these need careful thought and rational design, which has led to the inclusion of multidisciplinary efforts and non-chemistry methods in the ever-changing landscape of medicinal chemistry. Computational approximation of protein-ligand interactions is the main goal of the so-called structure-based methods. Over the years, there has been a notable improvement in the predictive power of approaches like molecular force fields. Mainstream applications of these include molecular docking, a well-known method for high-throughput virtual screening. Still, even with notable success cases, the search for accurate and efficient methods for free energy estimation remains a major goal in the field. Recently, with the advent of technology, more exhaustive simulations are possible in a reasonable time. Herein, we discuss free energy predictions and applications of perturbation theory, with emphasis on their role in molecular design and drug discovery. Our aim is to provide a concise but comprehensive view of current trends, best practices, and overall perspectives in this maturing field of computational chemistry.

Keywords: Alchemistry, Computer-aided Drug Design, Free Energy Methods and Simulation.

INTRODUCTION

As of today, drug development is a multidisciplinary field where the areas of competence go beyond pharmacology or organic synthesis. Within such a context, the first question we must answer is; what exactly do we mean by the drug? The International Union for Pure and Applied Chemistry (IUPAC) has the following definition:

---

* **Corresponding author Fernando D. Prieto-Martínez:** Instituto de Química, Universidad Nacional Autónoma de México, Ciudad de México, México; E-mail: ferdpm4@hotmail.com

Igor José dos Santos Nascimento (Ed.)

"Any substance which, when absorbed into a living organism, may modify one or more of its functions" [1].

Starting from it, we may add that for our intent and purposes, a drug shall be any "small" molecule (*i.e.*, with a molecular weight < 650 Da) with an intended target, rationally designed or optimized and with the clear goal of having a therapeutic use.

Since the dawn of civilization, along with agriculture, mankind has indirectly learned of the medicinal potential of some plant species. In a historical context, it has been a long and slow transition from mystical to therapeutic. A paragon example is perhaps salicylic acid; a natural product which led to the development of one of the most well-known drugs: aspirin. It is very remarkable that this rather "simple" structure is, in fact, a prodrug from a metabolite present in the willow tree bark, which came to be associated with a Nobel Prize in physiology while also being one of the most commercially successful drugs of all time [2].

At first, drug design relied on mimicking endogenous ligands of known targets. This may seem rather straightforward, but quite the opposite is true. Returning to the aspirin example, its active ingredient, salicylic acid, had its mode of action identified and described until the latter half of the XX century. Thus, since the development and maturing of pharmacology, approaches towards drug design have become more systematic and less form of art and chance. A prime example to consider is the development of angiotensin converting enzyme (ACE) inhibitors. It was between the decades of 1960 and 1970 that John Vane's group actively researched the cause of hypertension. Studying the effect of the venom from a Brazilian viper (*Bothrops jararaca*) *in vitro*, Vane recognized the importance of ACE as a major regulator of blood pressure [3]. This led to the involvement of Squibb, specifically David Cushman and Miguel Ondetti, who characterized several peptides as antihypertensive agents and hypothesized that ACE was a zinc metallopeptidase [4]. From here, the main challenge to overcome was oral bioavailability, so the group turned to recently described carboxypeptidase A inhibitors, hypothesizing structural similarities towards ACE (Fig. **1**). This rationale led to the eventual synthesis of captopril [5]. Nonetheless, such a decision proved to be fortuitous, as the crystal structures of ACE (published almost 30 years later) showed that its catalytic domain is actually unrelated to that of carboxypeptidase A [6].

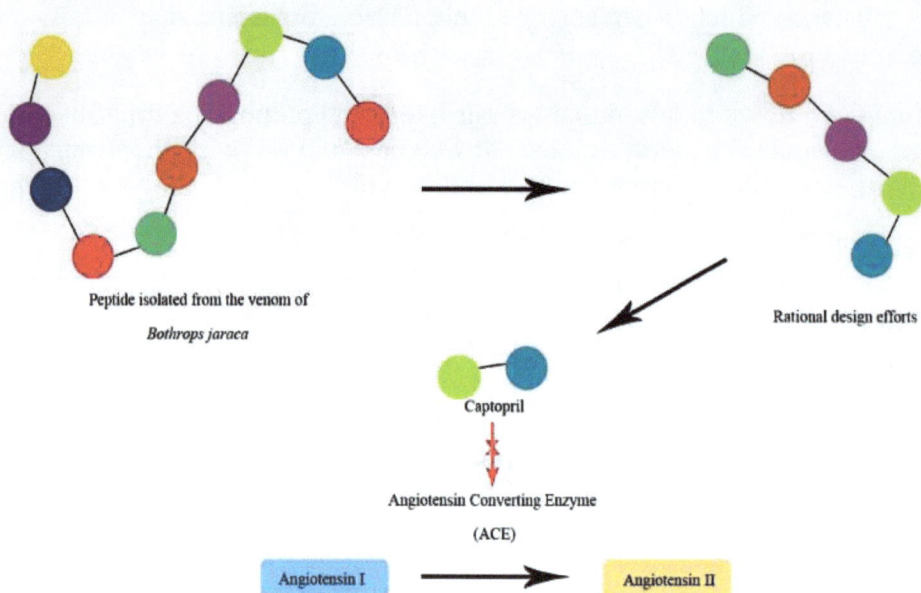

Fig. (1). Schematic view of the optimization process leading to the development of captopril.

From here, it becomes clear that drug development endeavours involve a great deal of complexity. For instance, during the 1980s, combinatorial chemistry was seen as a promising venue to tackle molecular diversity. The proposal of synthesizing hundreds of compounds in a systematic and rather efficient way was very appealing, leading to the development of high-throughput screening methods. A framework where thousands of compounds could be quickly evaluated using a combination of biochemical assays and robotic machinery [7]. While HTS has had notable success cases, the overall rate of developed drugs from it is rather discrete. Additionally, HTS campaigns gave rise to a phenomenon known as frequent hitters or pan-assay interference compounds (PAINS). Said designation is given to compounds showing "promiscuity" to a broad range of proteins or more generally to false positives due to interference with assay elements [8]. Indeed, this situation revealed that drug development cannot be solved by brute force, as it demands both critical and creative thinking [9].

Thus, the industry turned to state-of-the-art methodologies. Parallel to the development of HTS, there was a significant shift towards other computational modelling techniques. Early examples of this include the pharmacophore elucidation studies during the late 1960s and 1970s [10]. Nonetheless, pharmacophore models proved insufficient tools, as no information on the target is obtained. The pressing need for methods capable of predicting the binding affinity ushered in the development of molecular docking, leading the way to the establishment of structure-based drug design (SBDD). With the passage of time, SBDD techniques quickly settled in. There is no denying that with them, drug discovery and design have seen significant progress. However, critical points such as robustness, "realism" and overall construction of said tools need to be objectively assessed to determine the best way forward [11].

In the following sections, we will discuss the context of free energy estimation by computational means. Beginning with an overview of the theoretical aspects, with a bird's eye view on perturbation theory. Followed by current developments in the field and general guidelines for best practices. Next, we present an outlook on future directions and perspectives. We also present a brief case study on free energy methods and their role in drug design. Finally, we comment on some recommendations for yielding newcomers to the field.

THEORETICAL BACKGROUND

As previously discussed in Chapter 8, molecular recognition arises from pairwise interactions. This process can be quite complex, however, by making some assumptions, it is possible to make a theoretical estimate of the associated free energy. A major concession that must be made is that equilibrium conditions have been reached. Therefore, it follows that:

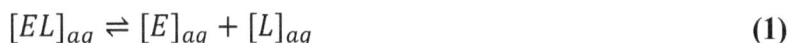

$$[EL]_{aq} \rightleftharpoons [E]_{aq} + [L]_{aq} \tag{1}$$

In this framework, affinity can be expressed as a rate constant, thus:

$$K_A = \frac{[E + L]_{aq}}{[E]_{aq}[L]_{aq}} \tag{2}$$

Which in turn can be effectively related to free energy:

$$\Delta G = -RT \ln K_A \tag{3}$$

Hence, free energy can be computed from physical parameters such as electrostatics, weak interactions, etc. For a protein-ligand complex, recognition has been described by two main approaches: the lock and key (rigid) or induced fit models. If we take the former, the free energy of binding may be simplified to:

$$\Delta G = E_{inter} + E_{intra} + \Delta S_{conf} \tag{4}$$

Similar reasoning is the basis of molecular docking, which in early implementations used force field-based estimation of free energy. For example:

$$U = W_{vdw} \sum_{i,j} \left(\frac{A_{ij}}{r_{ij}^{12}} - \frac{B_{ij}}{r_{ij}^6}\right) + W_{Hbond} \sum_{i,j} E(t) \left(\frac{C_{ij}}{r_{ij}^{12}} - \frac{D_{ij}}{r_{ij}^{10}}\right) + W_{elec} \sum_{i,j} \frac{q_i q_j}{e(r_{ij})r_{ij}}$$
$$+ W_{sol} \sum_{i,j} (S_i V_j + S_j V_i)e^{\left(-r_{ij}^2/2\sigma^2\right)} \tag{5}$$

The previous equation is implemented in the well-known program Autodock 4 [12]. It can be seen that the main parameters for affinity estimation include pairwise evaluation of van der Waals forces, hydrogen bonding energy, long-range interaction and a desolvation based on volume terms.

These calculations offer a good starting point, as the computational time is reasonable even on consumer-grade hardware. Nevertheless, there are some significant problems: there is no description of the entropic component, the interactions and potentials are limited by atom typing, and any modification to the underlying scheme is not trivial. With the development of protein crystallization and repositories such as the Protein Data Bank, another possibility emerged. Regression models trained from crystallographic data are known as empirical scoring functions [13]. Just like molecular mechanics estimates, there is a crucial assumption: binding affinity is an additive phenomenon. Thus, the general equation for such methods comes from the following:

$$\Delta G = \sum w_i \Delta G_i \tag{6}$$

Empirical approaches had more success when compared to force field scoring [14 - 17]. With the added benefit of faster computing time. Still, the disadvantages are also evident; due to the additive nature, empirical functions have a strong bias for bigger and polyfunctionalized molecules. Such is the case of some natural products, which tend to be overestimated as strong binders. Moreover, the overall robustness of these functions depends on the available data and the training set [18, 19]. Hence, it follows that empirical scoring functions are hardly universal.

As time went by, it became clear that molecular docking is a good tool for virtual screening; nevertheless, its affinity estimates are quite poor indeed, as shown in several studies [20 - 22]. It was during this time, two disciplines of computational chemistry gained prominence: artificial intelligence and molecular dynamics simulations. This led to a divergence in the pursuit of more accurate methods for free energy estimation.

Some research groups saw the renaissance of machine learning methods as an opportunity, given the notable success rate it achieved in fields like QSAR modelling [23]. Thus came the development of descriptor-based scoring functions, with notable examples being: RF-score [24], NNScore 2.0 [25] and GNINA [26]. This novel generation of affinity estimators held strong promise, with significant improvements over physics-based or regression-based models [27]. Yet, the black-box nature of some methods in artificial intelligence makes it difficult to truly assess the performance of these functions. Recently, the knowledge gained with these models has been questioned, showing that in the majority of cases, machine-learning scoring functions are indeed more accurate, but this may be due to unexpected behavior [28 - 30].

In parallel, other groups turned to the enhanced capabilities offered by molecular dynamics simulations. Using the sampled conformations, affinity could be computed iteratively. Such techniques are known as end-point methods, which include: Linear Response Approximation (LRA), Linear Interaction Energy (LIE) and MM-PBSA/GBSA.

End-point Methods

LIE methods were originally proposed by Åqvist, Medina and Samuelsson [31], similar to free energy perturbations (FEP); these are based on a thermodynamic cycle (Fig. **2**). However, in contrast to perturbative methods, binding energetics are obtained from "single points" pertaining to relevant states; *i.e.*, bound and unbound. This is possible if the potential energy is transformed into free energy, to do this a linear response approximation can be used to evaluate the polar contribution [32]:

$$\Delta G_{solv}^{elec} = \frac{1}{2}\{\langle U_{l-s}^{elec}\rangle_{on} + \langle U_{l-s}^{elec}\rangle_{off}\} \tag{7}$$

A similar reasoning can be applied to handle non-polar contributions, with the added need for a constant representing "cavity creation". Therefore, to compute ΔG_{bind} values, ensemble averages of van der Waals and electrostatic interactions are scaled and summed, yielding the LIE equation:

$$\Delta G_{bind} = \alpha\Delta\langle U_{l-s}^{vdW}\rangle + \beta\Delta\langle U_{l-s}^{elec}\rangle + \gamma \tag{8}$$

Where α, β are empirical coefficients and γ is the aforementioned constant, which was originally assigned to zero. It follows that the scaling values have a high influence on the estimated values of ΔG_{bind}. Despite this, the initial parametrization of this equation returned reasonably robust results in different systems [33]. In particular, β has been studied by several groups to automate the method for virtual screening campaigns based on the nature of ligands and their functional groups [34]. For these reasons, LIE offers notable advantages over FEP methods, the main one being its computational cost. Moreover, the method is better suited for structurally diverse ligands, such is not the case for FEP (*vide infra*). Due to this, LIE is particularly suited for lead optimization stages.

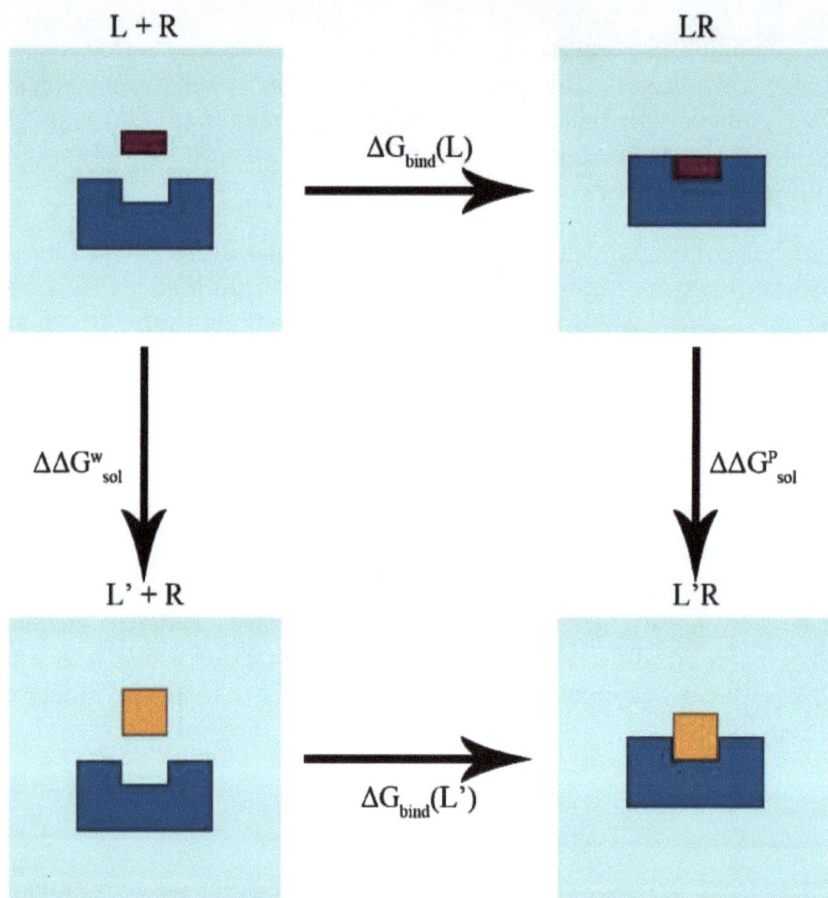

Fig. (2). Thermodynamic cycle is delimited by free energy calculation in the context of drug binding.

Since its publication, the LIE method has been tested in several systems and different force fields [35]. Initial observations from these suggested the need for additional constants and significant dependence on force field parameters. Brandsdal *et al.* [35] suggested that additional parametrization is needed in cases with strong hydrophobic contribution; while also dismissing any force field dependence, still they remarked that LIE methods are sensitive to the overall protocol of the analyzed simulations.

Of course, further improvements to the method have been proposed too. A good portion of these enhancements are shifted towards the parametrization of the LIE equation. For instance, Linder, Ranganathan and Brinck [36] developed an adapted version of LIE. In their approach, the empirical coefficients for the polar and non-polar terms are scaled with the aid of empirical descriptors of several protein-ligand complexes. Analogous ideas were pursued before to enhance the overall description of specific cases, such as solvation energies [37]. Extended descriptions have proven useful, yet they seem to inherit the problem of empirical scoring functions as extensive fitting of quality data is required [38]. Others have searched for improvements in the thermodynamics described by the method. Either by including data from multiple simulations [39], using filtering schemes to identify configurational transitions [40] or by substituting the non-polar term of the equation with values obtained directly from FEP methods of ligand solvation [41].

Perhaps one of the most interesting combinations to improve LIE was the one proposed by Zhou *et al.* [42]. Using a continuum solvent method in order to reduce computing overhead, improvements of almost one order of magnitude were observed. Let this mention serve as an introduction to the other half of end-point methods: MM/PB(GB)SA.

On the other hand, MM-implicit solvent techniques are the quintessential implementation of end-point calculations. This is due to the use of a more robust description of the solvation term, either from the Poisson-Boltzmann or Generalized-Born approaches for the surface solvation area (*vide infra*). The former was originally proposed by Srinivasan *et al.* [43], but perhaps its most notable success came in 2001. Then, a combination of molecular docking and MM/PBSA as a post-docking refinement protocol was used to accurately predict the binding mode of efavirenz to HIV-1 reverse transcriptase [44]. A noteworthy milestone indeed, considering that it was made prior to co-crystallization of the efavirenz-RT complex and the overall deviation of the predicted arrangement was 1.1 A [45].

Briefly, the MM-PBSA approach computes affinity using some of the previously discussed assumptions, with the energy terms coming from molecular mechanics parameters to approximate the enthalpic contributions of binding. Hence, affinity comes from the following equations:

$$\Delta G_{bind} \approx \Delta E_{MM} + \Delta G_{bind,solv} - T\Delta S \tag{9}$$

$$\Delta G_{bind,solv} = \Delta G_{polar} + \Delta G_{non-polar} \tag{10}$$

As mentioned, these calculations come from molecular dynamics simulations, usually from the last portion of a given trajectory, *i.e.*, 5-10 ns. From a practical point of view, these approaches may be seen as an enhancement of force field-based scoring. Therefore, it also inherits its pitfalls and limitations, with the main difficulty being the computation of entropy contributions. Generally speaking, the entropic term is configurational in nature. Due to this, it can be approached using several methods, with initial proposals including normal mode analysis. This is possible because low frequency vibrations dominate motion along the protein backbone. Using the multidimensional treatment of multiple harmonic oscillators, the vibrational motion can be computed. Thus, using molecular mechanics, frequencies for normal modes may be obtained using the Hessian matrix of energetic terms for molecular interaction [46]. Similar reasoning gave rise to "essential" dynamics analysis, identifying anharmonic motions of proteins which can be associated with folding or allosteric phenomena [47]. A notable disadvantage of such computations is their complexity and resources needed, given the number of atoms and the resulting matrix. Moreover, it has been shown that normal mode analysis does fail to estimate the absolute vibrational entropy; nonetheless it provides a reasonable estimate for free energy calculations [48].

Additionally, the entropic contribution is system dependent, making it harder to assess its overall impact for every case [49]. As an example, consider the epigenetic reader BRD4. Bromodomains are small "finger" proteins present in several eukaryotic species. Recently, these have been validated as druggable targets whose inhibition offers a wide array of therapeutic phenotypes [50 - 54]. Due to their conserved structure and overall features of its binding site, bromodomains have been slowly adopted as test systems for molecular modelling. In 2017, Aldeghi *et al.* [55] presented a comprehensive analysis of the statistical performance of several free energy estimations in these proteins. In this case, the addition of entropy contributions had a low impact, with the authors observing that the loss of the water network may be the reason for this. Said hypothesis can be supported by the fact that a number of conserved water molecules condition an enthalpy-driven binding; a trend present in most BRD4-ligands [56].

In recent years, novel methods for entropy computation have been suggested. The first approach is known as interaction entropy, proposed by Duan *et al.* [57]. This magnitude is defined as:

$$-T\Delta S = k_B T ln \langle e^{\beta \Delta E_{pl}^{int}} \rangle \tag{11}$$

Where ΔE_{pl}^{int} comes from the electrostatic and van der Waals interactions, this ensemble average may be obtained from molecular dynamics snapshots, suppressing the need for normal node calculations. This approach has performed significantly better than normal MM/PBSA in proteins with a high polarization effect [58].

Contemporary to this, Ben-Shalom *et al.* [59] proposed binding entropy estimation of rotation and translation (BEERT). With it, the configurational entropy is approximated with the difference of rotational and translational volumes for the ligand in bounded and unbounded states.

Now, we will discuss the development in the differences between Poisson-Boltzmann and Generalized Born approaches. Implicit solvation makes use of continuum electrostatics, meaning that the solvent is represented as a dielectric medium. For a macroscopic media, the Poisson equation for electrostatics takes the following form:

$$\nabla[\epsilon(r)\nabla\phi(r)] = -4\pi\rho(r) \tag{12}$$

Where r represents the positional vector, ϕ is the electrostatic potential, p is the fixed charged density (of the solute) and ϵ is the dielectric constant. A numerical solution to this equation is available if the equation is linearized, *i.e.*, assuming that both the electric field and ionic strength are weak, yielding:

$$\nabla[\epsilon(r)\nabla\phi(r)] = -4\pi\rho(r) + \kappa^2\epsilon(r)\phi(r); \kappa^2 = \frac{8\pi e^2 I}{\epsilon k_B T} \tag{13}$$

Such delimiting is needed to ensure free energy ΔG_{elec} is independent of the path in which the system is formed. This is a well-known issue, yet its impact on practical implementations has not been fully explored [60]. Another relevant aspect of the numerical solution is the geometric model used for interphase description, as it defines the discontinuity in the system. In practice, several solvers have been proposed, such as van der Waals or Connolly surfaces or solvent accessible area with the latter offering a good compromise between accuracy and computational cost. Nevertheless, it has been shown that it remains insufficient for a robust description of hydrophobic phenomena [61].

On the other hand, the linearity in continuum electrostatics offers an advantage for further simplification. Using analytical functions, the free energy in continuum media can be generalized to:

$$\Delta G_{elec} = \frac{1}{2}\sum_{i,j} q_i q_j F(x_i, x_j) \tag{14}$$

Where $F(x_i, x_j)$ is a geometry dependent function, a candidate for such function is the Born equation:

$$\frac{1}{2}\left(\frac{1}{\epsilon}-1\right)\frac{q^2}{a} \tag{15}$$

For an N-atom system, equation **15** takes the following form:

$$\Delta G_{elec} = \frac{1}{2}\left(\frac{1}{\epsilon}-1\right)\sum_{i}\frac{q_i^2}{a_i} + \frac{1}{2}\left(\frac{1}{\epsilon}-1\right)\sum_{i}^{N}\sum_{i\neq j}^{N}\frac{q_iq_j}{r_{ij}} \tag{16}$$

Where r_{ij} comes from the distance between atom centers. To offer a unified model, GB approaches compute the following:

$$\Delta G_{elec} = \frac{1}{2}\left(\frac{1}{\epsilon}-1\right)\sum_{ij}\frac{q_iq_j}{f_{GB}} \tag{17}$$

Where f_{GB} is a geometric factor and should be "effective" Born radius. The intrinsic assumption is a sphere to which f_{GB} comes from [62]:

$$f_{GB} = \sqrt{r_{ij}^2 + R_iR_je^{-\left(\frac{r_{ij}^2}{4}R_iR_j\right)}} \tag{18}$$

Overall, the Generalized Born approach is very affordable when compared to PBSA. Such is the case, that it has been implemented for molecular dynamics simulation and not just endpoint calculations [63]. With the added benefit of similar performance on different systems. However, it should be mentioned that GBSA methods come in "families" which have different performances and applicability. This goes to show that while simpler, the use of semi-analytical approaches is not universal nor straightforward [64].

Compared to LIE, MM/implicit solvent methods are more accessible, and easier to conduct and analyze. Currently, their main weakness remains the calculation of entropy contribution and its overall effect on correlation with experimental results. Moreover, it has been argued that the choice of the dielectric parameter is crucial to the results but also system dependent; thus, these approaches may not be robust enough and lean towards educated guesses [65].

Even with the aforementioned drawbacks and limitations, implicit solvation methods coupled with molecular mechanics have been extensively reported in the context of drug design, often with successful results [66 - 68]. For this, they have become a staple in virtual screening as a means of molecular docking and scoring refinement [66]. Also, considering their performance and lesser computational cost, MM/GBSA has been proposed as the superior choice for batch use [69, 70]. In this sense, the versatility of the analytical functions in GBSA has been exploited to further refine the model. For example, last year, Wang *et al.* presented a variable atomic dielectric GBSA model [71]. This approach made use of molecular descriptors to obtain an optimal dielectric factor, using machine learning algorithms for optimization. The authors state that the obtained model is significantly superior to conventional GBSA. In an akin study, Dong *et al.* [72] reported GXLE, a hybrid strategy also based on machine learning models to improve MM/GBSA accuracy; the method showed notable performance on the CASF-2016 set [72].

Just as LIE, these methods have room for improvement; this, combined with the strong pace at which FEP continues to rise, may lead to the conclusion that MM/PB(GB)SA should be abandoned. Despite this, the development of novel protocols and models shows that implicit solvent models and their application in affinity prediction stay (and will continue to be) strong for years on end [73].

With this, we have presented a background on the main methods for affinity prediction in SBDD. Next, we offer an overview of perturbation methods. Recently, these have attracted significant attention due to positive results and high correlation with experimental results. In reality, the theoretical foundation for FEP has been around since the latter half of the last century.

FREE ENERGY PERTURBATION THEORY

Generally speaking, the starting point is a transformation between two thermodynamic states. For our particular case, these can refer to unbound protein and drug A and the protein-drug complex B. Considering Helmholtz free energy, the difference between A and B comes from:

$$\Delta A_{AB} = A_B - A_A = -kT ln(\frac{Q_B}{Q_A}) \tag{19}$$

With this equation, it is possible to obtain the Zwanzig formula for free energy perturbation. The said equation states that a set of configurations of A may be used to sample the canonical distribution of B:

$$\Delta A_{AB} = -kT ln \langle e^{-\beta(U_B - U_A)} \rangle_A \tag{20}$$

This method is also known as exponential averaging and serves as a simple and straightforward introduction to free energy perturbations. Nevertheless, this approach has notable weaknesses. For example, consider what happens if A and B are very different (*i.e.,* their phase spaces do not overlap). To mitigate this problem, we could introduce a small perturbation along the path from A to B. Usually, this coordinate is known as λ and can be used to construct a "gradient" of sorts to improve the overlap between states. With it, the Zwanzig equation may be expressed as:

$$e^{-\beta\Delta G} = \prod_{i=0}^{n-1} \langle e^{-\beta(U(\lambda_{i+1})-U(\lambda_i))} \rangle_{\lambda_i} \tag{21}$$

This reasoning can be used to introduce thermodynamic integration (TI). Returning to the drug binding problem, the phenomenon can be described by a cycle like the one discussed above. Based on this, the overall transition can be approximated by adding the remaining legs in the cycle. Hence, the general equation for this method states:

$$\Delta G = \int_{\lambda=0}^{1} \langle \partial H / \partial \lambda \rangle_{\lambda} d\lambda \tag{22}$$

Thermodynamic integration is another elegant scheme to estimate free energy. However, it also comes with certain drawbacks. A major one is that, just like in exponential averaging, numerous simulations are needed to perform the numerical approximation [74] with the added disadvantage that the intermediate stages have unphysical potentials. Moreover, either for FEP or TI, the selection of λ values may have undesired effects, mostly due to the unknown dependence of λ_i and ΔG. While the stratification approach may seem straightforward, its implementation is far from trivial due to uncertainties along the alchemical path [75]. The worst case scenario is an endpoint catastrophe that occurs due to strong perturbations. At the very least, abrupt changes in λ lead to errors in the estimation of free energy. This phenomenon has been observed in calculations of solvation energies, when highly polar solutes are used [76].

In summary, the overarching difficulty of FEP methods is to determine if the simulation time for each stratum is reasonable enough to obtain a significant sample. All of this combined with a robust estimate on the number of windows needed for this. To overcome the limitations of the stratification scheme, non-equilibrium work (NEW) can be used. This is possible due to the Jarzynski's equality and the Crooks theorem, which states that the free energy ΔF can be generalized to:

$$e^{-\beta\Delta F} = \frac{\langle f(W) \rangle_F}{\langle f(-W)e^{-\beta W} \rangle_R} \tag{23}$$

Therefore, equilibrium free energy differences may be computed from many independent simulations where the initial state is rapidly converted to the end state [77]. A proposal that is appealing for parallel schemes, yielding faster computer times. In practice, all that is needed for these protocols is an enhanced sampling method and fast adiabatic switching. Still, the main obstacle to the widespread implementation of NEW schemes is the lack of proper software for accessible setup and analysis [78].

This leaves us with yet another challenge, what would serve as a proper estimator of the rate between forward and backward work values? The first proposal came from Charles Bennett, a notable alumnus of Berni Alder; currently known as Bennett's acceptance ratio (BAR) [79]. Initially, the estimator is two-sided whose performance depends on the following ratio:

$$r = \frac{n_R}{n_F} \qquad (24)$$

Where n_R and n_F are the numbers of reverse and forward work values, respectively. Since its publication, the ratio has been extensively studied and validated in the context of free energy calculations [80, 81]. For instance, it has been shown that BAR is a robust estimator due to its mean-squared error convexity [82]. Plus, the method can be further extended to weight data from multiple thermodynamic states yielding a multi-state generalization (MBAR) [83]. This is a very desirable trait, considering the underlying problem in computational simulations: sampling.

As stated in Chapter 8, one of the persistent problems with simulations is the sampling of rare events. Enhanced sampling methods involve the production of multistate data, which aims to identify and focus on relevant degrees of freedom [84]. While helpful, enhanced sampling adds another layer of complexity, due to the need of additional treatment prior to any analysis. Umbrella sampling is an often cited example, in such instances weighted histogram analysis method (WHAM) may be used to obtain the potential of mean force in a robust manner [85]. MBAR, on the other hand, persists as a statistically optimal alternative as it is a bindless method, and biases are avoided with the added benefit of lower computational bottlenecks [86].

In addition to this, MBAR equations can be generalized to fit concrete cases such as perturbation, scaling, accumulation, and full potential energy [87]. Wu *et al.* [88] explored an alternative generalization of the method, arriving at transition-based reweighting analysis method (TRAM) which can be used to construct Markov state models from a combination of biased and non-biased simulations, which proves useful to obtain either thermodynamic or kinetic information.

Lastly, it must be stated that the previous discussion implies the so-called absolute binding. Another variant of FEP is alchemy, where the binding is relative and the perturbation involves a mutation or change in the ligand or ligand. For this case, two alternatives are possible: a single or dual topology for the transformation. The main difference is how the transformation evolves and of course the steps needed for setup. To date, there is no consensus on which alternative is better. On the one hand, single topology methods converge quickly but introduce strong perturbations, whereas dual topology schemes may experience strong clashes as the scaled moieties vanish and grow [89]. Such a situation could arise in alchemical transformations involving the introduction of bulky moieties [90]. Commonly, these approaches often follow a protocol known as FEP/REST with the latter referring to replica exchange with solute tempering. This method delimits the system into two main regions: a cold one (being the solvent) and a hot one (the complex). This is quite effective as it reduces the number of needed replicas with the added benefit of increased exploration of the conformational space [89]. Additionally, relative methods have seen recent success in the study of scaffold hopping [91]. It has been shown that relative binding methods generally present lower errors when compared to their absolute counterparts [92]. Hence, these have become a regular choice for lead optimization, nevertheless, just like the absolute variants, notable difficulties persist [93].

METADYNAMICS

A family of enhanced sampling methods proposed by the Parrinello research group [94 - 96].

Metadynamics simulations are based on the construction of significant reaction coordinates, also known as collective variables (CVs). Practically speaking, this concept is comparable to umbrella sampling. Nonetheless, metadynamics protocols do not use attractive potentials, but rather a history-dependent bias to avoid low energy regions. This bias comprises a collection of d-dimensional Gaussians, whose height and width determine the accuracy and efficiency of the free energy reconstruction [97]. The main assumption behind metadynamics is that given enough time, these Gaussian potentials will guide the system through collective-variable space, effectively filling the free energy surface. This assumption was not formally derived but rather heuristically proposed from the observed behavior of functionals with known form [94]. This strong point of metadynamics is also its weakness, as it can lead to very high energy states that are mostly irrelevant. Moreover, it often leads to an oscillation in free energy values impeding convergence [95]. To overcome said issue, one can use well-tempered metadynamics which includes a scaling factor ΔT which prevents the systems from spending time on high energy regions.

It can be seen that the performance and results obtained with metadynamics are strongly dependent on the choice of CVs. This is not trivial as metastable basins and saddle points must be considered [98]. Also, to improve the description of the surface metadynamics algorithms can be parallelized or include weighting schemes. For instance, on the fly probability enhanced sampling (OPES) has been implemented to optimize the chosen biases toward efficient results [99].

In our given context, metadynamics algorithms may be used to enhance sampling prior to FEP methods. Such an approach has been successfully used by Tanida and Matsuura [100], in this study the theophylline-RNA aptamer was used as a model. The protocol consisted of metadynamics simulations to assess binding poses, followed by reweighting and clustering of results. In it, six metastable binding modes were chosen as the starting points of absolute free energy calculation. Using simulations of 20 ns, their best estimation was on par with experimental measurements.

Software and Workflows

There are many obstacles one can encounter when choosing the appropriate software to perform free-energy calculations on, and we can note that among these, the main culprits are availability of computer resources, time of calculation, software compatibility, researcher's level of expertise, step automation, etcetera. In lieu of this, Table **1** summarizes some of the most relevant attributes of available tools or workflows focusing on FEP calculations.

Table 1. Common workflows for FEP calculations.

Workflow	Type of FEP Calculation	Post-analysis	Force-field	Software	GPU	Refs.
BAC	Alchemical and endpoint	Yes	AMBER, CHARMM	GROMACS, NAMD, OpenMM	No	[101]
BLaDE	Alchemical	No	CHARMM	CHARMM	Yes	[102]
FEP+	Alchemical and endpoint	Yes	OPLS	Desmond	Yes	[103]
FESetup	Alchemical	No	AMBER	AMBER, CHARMM, DL_POLY, GROMACS, NAMD	No	[104]
FEW	Alchemical and endpoint	Yes	AMBER	AMBER	No	[105]
Flare	Alchemical	Yes	AMBER	OpenMM	No	[106]
STaGE	Alchemical	No	AMBER, CHARMM, OPLS	GROMACS	No	[107]

(Table 1) cont.....

Workflow	Type of FEP Calculation	Post-analysis	Force-field	Software	GPU	Refs.
pmx	Alchemical	Yes	AMBER, CHARMM, OPLS	GROMACS	No	[108]
PyAutoFEP	Alchemical and endpoint	Yes	AMBER, CHARMM, OPLS	GROMACS	No	[109]
YANK	Alchemical	Yes	AMBER, CHARMM	OpenMM	No	[110]

Current Developments

While protein force fields have reached notable descriptive power and overall robustness [111, 112]; the same cannot be said for small molecules. Parametrization of organic and more specifically; drug-like molecules has been attempted since the 1990s. A staple of these is the MMFF94 force field, developed by scientists at Merck [113, 114].

Since its publication, the AMBER and CHARMM force fields have been generalized in an attempt to describe organic molecules [115, 116]. Similarly, for the OPLS family of force fields, the Jorgensen group developed LigparGen server [117] to provide enhanced parameters for organic ligands [118]. On the commercially-licensed side, this family of force fields has been revisited to improve the parameter coverage of the chemical space [91, 119 - 120]. The most recent iteration: OPLS4 has been optimized in an attempt to reduce the root-mean-squared error on FEP estimations [121]. This is a persistent issue, suggesting the need for improvement and reparametrization, as noted by previous studies [122]. Other problems involving molecular mechanics descriptions include conformer generation. As it has been shown that competent, effective prediction of biologically relevant arrangements is far from being a solved problem [123].

Yet, there has been a notable improvement in the descriptive power of these tools for scientific applications [124]. Recently, endeavors such as the Open Force Field Consortium are on track to provide better tools for the development and validation of novel force fields [125 - 128]. The OpenFF family of force fields derives from the parm@Frosst initiative, which comprises parameters derived by Merck scientists to use with AMBER force fields within drug designing problems (http://www.ccl.net/cca/data/parm_at_Frosst/). Using this data and the concept of chemical perception [127], the precursor to OpenFF: SMINORFF99Frosst has comparable performance to that of GAFF for binding thermodynamics [129]. As of today, OpenFF is in its second major release "Sage" which has been benchmarked and compared to other major options for MM minimization and optimization of molecules [130]. Summing up, the initiative has shown high promise, with important milestones being expected for its next major releases "Rosemary" and "Thyme". However, for this reason this is yet another case where "mainstream" usage stays on the horizon, at least for the time being.

In this regard, other developments towards this end have been found in QM and machine-learning potentials, with the latter being trained on QM but with the added advantage of shorter computing times. A notable example of machine learning potential is ANI, a potential trained with a neural network which aims to provide DFT accuracy with MM computing cost [131]. In its most recent version, ANI has been successfully applied in molecular dynamics simulations [132]. Recently, Wieder, Fass and Chodera reported the use of ANI to predict tautomer ratios in solution [133]. While preliminary, this study shows how machine learning potentials can be expanded upon and fitted towards practical applications with discrete but promising results. The rise of machine learning potential has led to the development of dedicated libraries that allow their implementation and testing on molecular dynamics engines. A notable example of this is TorchMD, a framework for machine learning training using PyTorch for molecular dynamics potentials [134].

The role and effect of water molecules is also an unsolved issue in FEP. The main reason is that the transitions from bulk to the protein pocket are slower or even worse, as water may be trapped in the pocket when it should reach the bulk liquid [135]. This is especially critical when waters are part of a hydrogen bond network, but these are not properly resolved. In general, the recommendation is to conserve the water as it may add stabilization to the complete structure, but to resolve kinetically trapped water is no easy task. Many methods such as 3D-RISM, Aquaalta, Consolv, Waterdock, WaterMap and Water-score [136] have been proposed to predict water positions, but this issue continues to be under development. A putative solution has been found using the grand canonical ensemble; either in Monte Carlo simulations or in replica exchange protocols where the chemical potential of water is considered as an additional factor of the alchemical path [137].

Best Practices

A few details need to be taken into consideration before performing FEP calculations for these to be as accurate as possible and of high relevance in order to aid in accelerating the drug discovery process [138]. In this regard, any prior knowledge on the complex is very valuable. With such foresight, inherent difficulties can be tackled before the FEP protocol leading to accurate results [139]. Another important aspect to bear in mind is the FEP protocol to be used, either absolute or relative variants. As each has significant critical aspects that affect performance and overall issues. For example, for absolute binding estimation, simulations should be bidirectional whenever possible [140]. Such data can be easily analyzed with the BAR estimator. As stated, the use of the ratio is always superior to the unidirectional approach. Relative calculations, on the other hand, need careful construction of the transformation cycle. Commonly, this involves the use of similarity measures as weighting schemes to propose pairs for FEP. It has been shown that such methods are reasonable for low error estimates, nonetheless, additional steps, such as the mathematical construction of perturbation maps, also improve accuracy [141].

There are some general and practical recommendations regarding the initial structure from which the calculation will be performed, such as having the selected crystal falling within a resolution range of 2.5 - 3.0 Å, which have yielded decent results. Despite this, users must be aware that $\Delta\Delta G$ values may significantly change between alternative structures [142]. All the same, when a crystal structure is not available and homology models need to be used, although there has been some controversy on the topic when these models are accurately predicted, the results one can obtain from this are acceptable [143].

With the initial structure selected, we can now take a closer look at the preparation of the system. Many small details, when overlooked, can become a big enough issue which could lead to an abrupt ending of the simulation as FEP calculations are more sensible to protein preparation than others [144]. Some of these details include tautomer states, ionization states, modeled loops, unresolved or kinetically trapped waters and inaccurate binding poses, all of which contribute to reaching the global minima. Particularly discussing hydrogen atoms, these also are required to be placed as accurately as possible and to model hydrogen bonds in a correct manner, as these can heavily contribute to pivotal interactions with the ligand of interest. For charged species, it has been suggested that pKa and tautomer distribution can be included as part of the FEP formalism, significantly improving accuracy [145, 146].

For pose placement, it is important to recognize that a proper starting position is the key to obtaining robust predictions. When no crystal experimental structure is available, the binding mode will be modeled by docking. For relative binding energies, stable binding modes are required, the main drawback is that, in most cases, the dominant binding mode is unknown [147]. In this sense, docking poses often are non-ideal, and a preliminary round of equilibration with conventional molecular dynamics may help to discard unstable binding modes [148]. Next is the magnitude of the perturbation, which needs not to be above 10 heavy atoms so as to not compromise accuracy [149]. However, when this is of interest is a good practice to keep a close eye on the cycle values. Contrasting methodologies should be used whenever possible. In this way, information on the sensitivity and convergence of estimates is attained [150].

Restraints in itself present another issue that has been extensively covered by different research groups [151, 152]. In practice, one can choose one of the many types of restraints available that range from harmonic distance to flat bottom. The difference between the two resides in that the first allows for the ligand to explore multiple binding modes in a binding site, whereas the second allows the ligand to explore multiple binding sites. Regarding the protein, adding harmonic restraints to the protein backbone can help take the global minimum of both the homology models as well as structures obtained from crystals with more consistency. Furthermore, results that include a proper set of restraints fall more in accordance with what is seen experimentally [147]. Another type of restraint was proposed by Boresch in which all six rigid body degrees of freedom are restrained [153].

Other general recommendations for FEP methods include the following [154]:

• Pathways should maximize the similarity between states whenever possible.

• Transformations involving rings should avoid opening or closing perturbations. In such cases, the addition or removal of the whole ring works best.

• Uncertainty between neighboring states should be equal, this way the overall variance can be kept low.

• State prototype should be encouraged, even in short simulations, an estimate of variance can be assessed.

• Simulation must be at equilibrium to avoid any outsized contributions to free energy. Moreover, equilibrium must be reached at each stratum (λ).

• Beware of simulation parameters, do not assume that conventional choices for molecular dynamics are applicable to FEP. In most cases, such settings may affect the potential energy surface,

• Samples must be uncorrelated in time, as most analysis methods assume this property.

Finally, for the analysis of the results, there needs to be an appropriate evaluation of the convergence of the system and the final error obtained as to judge whether or not these predictions hold enough relevance for them to be used in the decision-making stage of the drug discovery process. This generally involves the identification of non-equilibrated regions, some guidelines for this have been described previously [155] and are currently implemented in the pyMBAR library [156].

BOCEPREVIR AS A CASE STUDY: A RELEVANT ANTIVIRAL FOR THE TREATMENT OF HEP C

Hepatitis C (Hep C) has become a prominent infectious disease worldwide. Furthermore, Hep C virus is the leading cause of the development of cirrhosis and hepatocellular carcinoma. It was during the 90's that more novel therapy research approaches started to emerge as a response to this health emergency, which included interferon-based strategies. However, drug discovery was not left behind and in combination with the rising presence of biomolecular assays, the discovery of novel compounds was accelerated.

Hep C virus is a RNA-based virus that belongs to the Flaviviridae family. It is now known that the NS3 protease plays a major role in the virus replication process, which positions it as a major target for direct antiviral mechanisms. In short, to date there are a few possible mechanisms that have been proposed by which the NS3 protease may regulate replication; by assisting polyprotein processing, aiding RNA replication and packaging RNA into virions [157]. NS3 forms a complex with its coenzyme 4A and a catalytic amino acid triad has already been described at the active site of this enzyme. With this information taken together, the search for a small inhibitor for NS3-4A complex began with high-throughput screening, but without much success. It wasn't until structure-based design efforts were made based on the enzyme's substrate that promising compounds started to emerge. This led to the discovery of boceprevir, the first ever direct-acting antiviral drug to be approved. Initially, it was approved in 2012 but withdrawn in 2015 after being linked with hepatic injury when taken in combination with other antiviral agents. Although it is known that boceprevir is a CYP4A4 isoenzyme inhibitor and a pgp substrate, the exact mechanism by which it could lead to hepatic injury is not known.

As it was mentioned previously, the search for small compounds started by HTS, however, without any leads from this attempt, the first efforts to rationally design a bioactive compound began by functionalizing an undecapeptide based on the protease substrate with a ketoamide that could covalently bond to Ser139 which was recognized as a pivotal residue. As potent as it was (k_i = 1.9 nmol), this effort did not meet any required pharmacokinetic properties, and so the search continued focusing on the reduction of the molecular weight. This led to the discovery of a pentapeptide that retained activity within the nanomolar range. Further optimization continued, as it is shown in Fig. (**3**), where a pharmacophore became apparent.

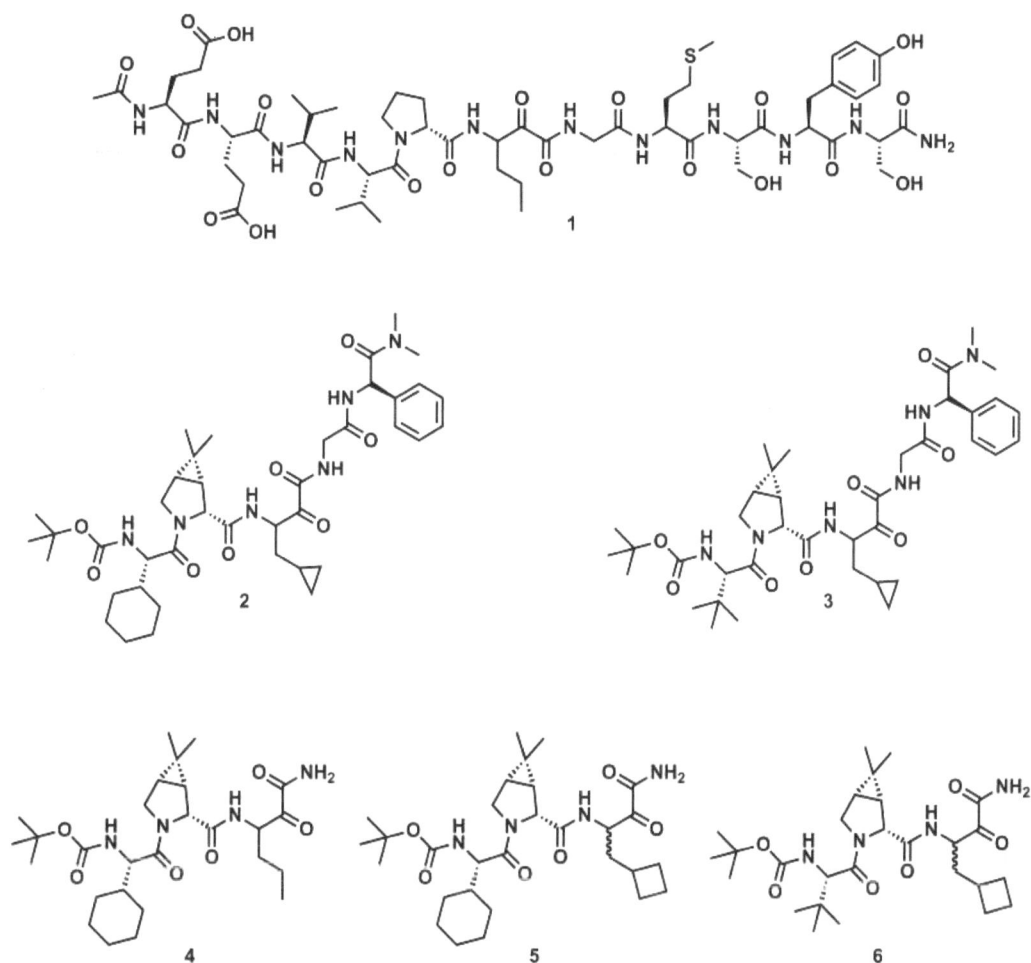

Fig. (3). Discovery of boceprevir.

The main caveat for drug design targeting NS3 is its shallow and solvent-exposed recognition site that is mainly regulated by weak lipophilic and electrostatic interactions. This has two main complications for rational design: molecular size and the inclusion of liable functional groups. Thus, combinatorial efforts, including docking, molecular dynamics, and automated in vitro assays provide a more appropriate strategy for drug discovery [158].

Efforts to understand exactly how this small compound can be used as an example for further discovery are currently underway. Boceprevir (Fig. **4**) has been recently taken into consideration against SARS-CoV-2 and once again, it is through in silico strategies that this potential repurposing is possible. Traditional de novo drug design has a major disadvantage as it is very time-consuming, and during the COVID-19 pandemic, time was of the essence. In this sense, drug repurposing or drug re-profiling offers a time-saving approach as the toxicity and pharmacokinetic profiles are already described.

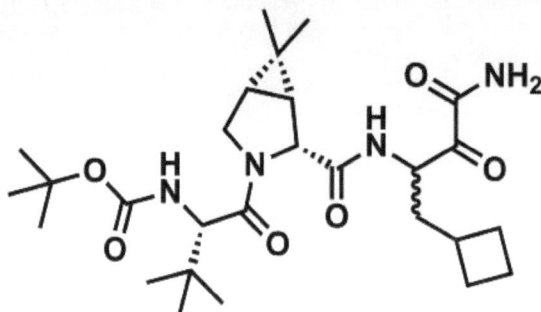

Fig. (4). Chemical structure of boceprevir.

In 2020, different research groups simultaneously proposed boceprevir repurposing as a potential ligand for two proteins of special interest for SARS-CoV-2 targeting; the spike protein and Mpro. However, it wasn't until 2021 that more in-depth studies were performed. Arooj and collaborators [159] described the free energy landscapes of both boceprevir and dexamethasone bound to the spike protein as well as Mpro through umbrella sampling and steered molecular dynamics. After running each drug-protein complex through MDS of one microsecond each, the representative complex of the top cluster was selected to perform steered molecular dynamics using the center of mass pulling. Every complex was placed in an elongated water box along the x-axis and the corresponding drug was then forced to leave the binding site with a pulling force of 100 kJ/mol.nm^2 and this was carried out for 500 ps thus obtaining 500 conformations each sampling window of varying drug-protein distances. Of these, only 90 sampling windows were selected to perform 10 ns simulations. Although both drugs showed interesting results, boceprevir had a slightly better binding mode with the selected proteins.

CONCLUSION

There is no denying that free energy prediction is the most outstanding challenge for structure-based design. The diversity of methods to tackle it is a silent testament to the inherent complexity and overall elusiveness. In this regard, free energy perturbations have acquired a leading role in the last decade. However, the current state of maturity of these methods is a subject of debate. On one side, there are confident and overall optimistic views on recent developments [160, 161]. On the other hand, there have been some conservative and critical voices warning of the pitfalls still unsolved [162].

Objectively, the estimation of affinity values based on molecular simulation has come a long way. There are numerous case studies and success cases that showcase its potential and merits. Of course, it can be argued that such results arise from a proper combination of user proficiency and test systems. In this sense, another subject that emerges is force field accuracy. For example, in a recent paper, Mondal, Florian, and Warshel benchmarked the relative binding of FEP to a series of thrombin inhibitors [163]. Their results showed that the overestimation of affinity is systematically persistent even with rigorous enhanced sampling protocols. This suggests that the convergence issues could be due to force field accuracy and not the sampling protocol.

Moreover, artificial intelligence has also been slowly rising in the field. At first, this was limited to scoring function development. Giving a steady rise to such studies where accuracy seemed to be notably improved [164], this has led to the adoption of such methodologies for structure-based design either by training the identification of collective variables [165, 166], or to improve free energy estimation [29]. Nevertheless, critical assessment of the apparent success rates of artificial intelligence quickly showed that the results are indeed more accurate, but this may be due to wrong reasons or even hidden biases [30]. Plus, this phenomenon is not exclusive to the chemical information field but a general trend. Such a scenario presses the need for robust tools and the establishment of proper benchmarks [167].

So, back to the matter, are free energy perturbations ready for widespread use? To really answer this question, we must draw a line. If it involves active usage in drug discovery and rational design, then sure. By all means, the methods and tools are a vantage point to help with decision making. Yet, if the widespread use involves non-experts, the answer is not straightforward. Even with the promising examples in the literature and the overall improvement in hardware capabilities; the truth of the matter is that FEP methods require notable effort, as previously suggested [149]. Furthermore, there are still methodological gaps and tough cases that often require a savvy user on the controls. For example, FEP affinity can be robust for diverse and complex ligands, even in transformations involving several heteroatoms. Still, values involving protonated ligands, metal interactions, or water bridges often show high variance [90]; all of which are common-place in drug designing campaigns. Moreover, current software implementations are limited to a handful of options, which could increase the learning curve significantly.

For the time being, newcomers to the field should be patient and get a grasp of the theory and its implications. Our opinion is that every medicinal chemist should know and understand the fundamentals of free energy calculations. To avoid trivialization, as it has been the case with molecular docking and machine learning to some extent. Free energy estimation will continue to evolve and gain refinement, perhaps in the following decade, a turning point will be reached. We also hold an optimistic perspective on the new grounds that lie before us; but for the time being we are not quite there yet.

To close this chapter, we would like to paraphrase Shirts and Mobley. Just as mentioned in the abstract, we attempted to provide a concise overview on free energy methods. However, we are aware that such a task escapes the scope of a single review or chapter. Therefore, we would like to point sharp and curious readers to the following literature [106, 135, 150, 161, 168 - 175].

ACKNOWLEDGEMENTS

Fernando D. Prieto-Martínez is thankful to CONACyT for a postdoctoral fellowship No. 31146; granted to the project FORDECYT-PRONACES, No. 1561802.

REFERENCES

[1]   IUPAC. The IUPAC Compendium of Chemical Terminology. 2nd., Research Triangle Park, NC: International Union of Pure and Applied Chemistry (IUPAC), 2019.
[http://dx.doi.org/10.1351/goldbook]

[2]   Beck H, Härter M, Haß B, Schmeck C, Baerfacker L. Small molecules and their impact in drug discovery: A perspective on the occasion of the 125th anniversary of the bayer chemical research laboratory. Drug Discov Today 2022; 27(6): 1560-74.
[http://dx.doi.org/10.1016/j.drudis.2022.02.015] [PMID: 35202802]

[3]   Smith CG, Vane JR. The discovery of captopril. FASEB J 2003; 17(8): 788-9.
[http://dx.doi.org/10.1096/fj.03-0093life] [PMID: 12724335]

[4]   Cushman DW, Ondetti MA. History of the design of captopril and related inhibitors of angiotensin converting enzyme. Hypertension 1991; 17(4): 589-92.
[http://dx.doi.org/10.1161/01.HYP.17.4.589] [PMID: 2013486]

[5]   Erdös EG. The ACE and I: How ACE inhibitors came to be. FASEB J 2006; 20(8): 1034-8.
[http://dx.doi.org/10.1096/fj.06-0602ufm] [PMID: 16770001]

[6]   Hooper NM, Turner AJ. An ACE structure. Nat Struct Mol Biol 2003; 10(3): 155-7.
[http://dx.doi.org/10.1038/nsb0303-155] [PMID: 12605218]

[7]   Harding D, Banks M, Fogarty S, Binnie A. Development of an automated high-throughput screening system: A case history. Drug Discov Today 1997; 2(9): 385-90.
[http://dx.doi.org/10.1016/S1359-6446(97)01082-9]

[8]   Yang ZY, He JH, Lu AP, Hou TJ, Cao DS. Frequent hitters: Nuisance artifacts in high-throughput screening. Drug Discov Today 2020; 25(4): 657-67.
[http://dx.doi.org/10.1016/j.drudis.2020.01.014] [PMID: 31987936]

[9]   Murcko MA. What makes a great medicinal chemist? a personal perspective. J Med Chem 2018; 61(17): 7419-24.
[http://dx.doi.org/10.1021/acs.jmedchem.7b01445] [PMID: 29745657]

[10]   Güner OF, Bowen JP. Setting the record straight: The origin of the pharmacophore concept. J Chem

Inf Model 2014; 54(5): 1269-83.
[http://dx.doi.org/10.1021/ci5000533] [PMID: 24745881]

[11] Michael E, Simonson T. How much can physics do for protein design? Curr Opin Struct Biol 2022; 72: 46-54.
[http://dx.doi.org/10.1016/j.sbi.2021.07.011] [PMID: 34461593]

[12] Morris GM, Huey R, Lindstrom W, *et al.* Autodock4 and autodocktools4: Automated docking with selective receptor flexibility. J Comput Chem 2009; 30(16): 2785-91.
[http://dx.doi.org/10.1002/jcc.21256] [PMID: 19399780]

[13] Charifson PS, Corkery JJ, Murcko MA, Walters WP. Consensus scoring: A method for obtaining improved hit rates from docking databases of three-dimensional structures into proteins. J Med Chem 1999; 42(25): 5100-9.
[http://dx.doi.org/10.1021/jm990352k] [PMID: 10602695]

[14] McInnes C. Virtual screening strategies in drug discovery. Curr Opin Chem Biol 2007; 11(5): 494-502.
[http://dx.doi.org/10.1016/j.cbpa.2007.08.033] [PMID: 17936059]

[15] Wang R, Lai L, Wang S. Further development and validation of empirical scoring functions for structure-based binding affinity prediction. J Comput Aided Mol Des 2002; 16(1): 11-26.
[http://dx.doi.org/10.1023/A:1016357811882] [PMID: 12197663]

[16] Guedes IA, Pereira FSS, Dardenne LE. Empirical scoring functions for structure-based virtual screening: Applications, critical aspects, and challenges. Front Pharmacol 2018; 9: 1089.
[http://dx.doi.org/10.3389/fphar.2018.01089] [PMID: 30319422]

[17] Ren X, Shi YS, Zhang Y, *et al.* Novel consensus docking strategy to improve ligand pose prediction. J Chem Inf Model 2018; 58(8): 1662-8.
[http://dx.doi.org/10.1021/acs.jcim.8b00329] [PMID: 30044626]

[18] Spyrakis F, Cozzini P, Eugene Kellogg G. Applying computational scoring functions to assess biomolecular interactions in food science: Applications to the estrogen receptors. Nucl Receptor Res 2016; 3
[http://dx.doi.org/10.11131/2016/101202]

[19] Huang SY, Grinter SZ, Zou X. Scoring functions and their evaluation methods for protein–ligand docking: Recent advances and future directions. Phys Chem Chem Phys 2010; 12(40): 12899-908.
[http://dx.doi.org/10.1039/c0cp00151a] [PMID: 20730182]

[20] Anighoro A, Bajorath J. Three-dimensional similarity in molecular docking: Prioritizing ligand poses on the basis of experimental binding modes. J Chem Inf Model 2016; 56(3): 580-7.
[http://dx.doi.org/10.1021/acs.jcim.5b00745] [PMID: 26918284]

[21] Kalinowsky L, Weber J, Balasupramaniam S, Baumann K, Proschak E. A diverse benchmark based on 3D matched molecular pairs for validating scoring functions. ACS Omega 2018; 3(5): 5704-14.
[http://dx.doi.org/10.1021/acsomega.7b01194] [PMID: 31458770]

[22] Pinzi L, Rastelli G. Molecular docking: Shifting paradigms in drug discovery. Int J Mol Sci 2019; 20(18): 4331.
[http://dx.doi.org/10.3390/ijms20184331] [PMID: 31487867]

[23] Cherkasov A, Muratov EN, Fourches D, *et al.* QSAR modeling: Where have you been? Where are you going to? J Med Chem 2014; 57(12): 4977-5010.
[http://dx.doi.org/10.1021/jm4004285] [PMID: 24351051]

[24] Ballester PJ, Mitchell JBO. A machine learning approach to predicting protein–ligand binding affinity with applications to molecular docking. Bioinformatics 2010; 26(9): 1169-75.
[http://dx.doi.org/10.1093/bioinformatics/btq112] [PMID: 20236947]

[25] Durrant JD, McCammon JA. NNScore 2.0: A neural-network receptor-ligand scoring function. J Chem Inf Model 2011; 51(11): 2897-903.

[http://dx.doi.org/10.1021/ci2003889] [PMID: 22017367]

[26] McNutt AT, Francoeur P, Aggarwal R, *et al.* GNINA 1.0: Molecular docking with deep learning. J Cheminform 2021; 13(1): 43.
[http://dx.doi.org/10.1186/s13321-021-00522-2] [PMID: 34108002]

[27] Wójcikowski M, Ballester PJ, Siedlecki P. Performance of machine-learning scoring functions in structure-based virtual screening. Sci Rep 2017; 7(1): 46710.
[http://dx.doi.org/10.1038/srep46710] [PMID: 28440302]

[28] Liu J, Wang R. Classification of current scoring functions. J Chem Inf Model 2015; 55(3): 475-82.
[http://dx.doi.org/10.1021/ci500731a] [PMID: 25647463]

[29] Chen L, Cruz A, Ramsey S, *et al.* Hidden bias in the DUD-E dataset leads to misleading performance of deep learning in structure-based virtual screening. PLoS One 2019; 14(8): e0220113.
[http://dx.doi.org/10.1371/journal.pone.0220113] [PMID: 31430292]

[30] Volkov M, Turk JA, Drizard N, *et al.* On the frustration to predict binding affinities from protein–ligand structures with deep neural networks. J Med Chem 2022; 65(11): 7946-58.
[http://dx.doi.org/10.1021/acs.jmedchem.2c00487] [PMID: 35608179]

[31] Åqvist J, Medina C, Samuelsson JE. A new method for predicting binding affinity in computer-aided drug design. Protein Eng Des Sel 1994; 7(3): 385-91.
[http://dx.doi.org/10.1093/protein/7.3.385] [PMID: 8177887]

[32] Gutiérrez-de-Terán H, Åqvist J. Linear interaction energy: Method and applications in drug design. Methods Mol Biol 2012; 819: 305-823.
[http://dx.doi.org/10.1007/978-1-61779-465-0_20]

[33] Åqvist J, Luzhkov VB, Brandsdal BO. Ligand binding affinities from MD simulations. Acc Chem Res 2002; 35(6): 358-65.
[http://dx.doi.org/10.1021/ar010014p] [PMID: 12069620]

[34] Rifai EA, van Dijk M, Geerke DP. Recent developments in linear interaction energy based binding free energy calculations. Front Mol Biosci 2020; 7: 114.
[http://dx.doi.org/10.3389/fmolb.2020.00114] [PMID: 32626725]

[35] Brandsdal BO, Österberg F, Almlöf M, Feierberg I, Luzhkov VB, Åqvist J. Free energy calculations and ligand binding. Adv Protein Chem 2003; 123-58.
[http://dx.doi.org/10.1016/S0065-3233(03)66004-3]

[36] Linder M, Ranganathan A, Brinck T. "Adapted linear interaction energy": A structure-based lie parametrization for fast prediction of protein–ligand affinities. J Chem Theory Comput 2013; 9(2): 1230-9.
[http://dx.doi.org/10.1021/ct300783e] [PMID: 26588766]

[37] Åqvist J, Hansson T. On the validity of electrostatic linear response in polar solvents. J Phys Chem 1996; 100(22): 9512-21.
[http://dx.doi.org/10.1021/jp953640a]

[38] van Dijk M, ter Laak AM, Wichard JD, Capoferri L, Vermeulen NPE, Geerke DP. Comprehensive and automated linear interaction energy based binding-affinity prediction for multifarious cytochrome P450 aromatase inhibitors. J Chem Inf Model 2017; 57(9): 2294-308.
[http://dx.doi.org/10.1021/acs.jcim.7b00222] [PMID: 28776988]

[39] Stjernschantz E, Oostenbrink C. Improved ligand-protein binding affinity predictions using multiple binding modes. Biophys J 2010; 98(11): 2682-91.
[http://dx.doi.org/10.1016/j.bpj.2010.02.034] [PMID: 20513413]

[40] Vosmeer CR, Kooi DP, Capoferri L, Terpstra MM, Vermeulen NPE, Geerke DP. Improving the iterative linear interaction energy approach using automated recognition of configurational transitions. J Mol Model 2016; 22(1): 31.
[http://dx.doi.org/10.1007/s00894-015-2883-y] [PMID: 26757914]

[41]   Rifai EA, Ferrario V, Pleiss J, Geerke DP. Combined linear interaction energy and alchemical solvation free-energy approach for protein-binding affinity computation. J Chem Theory Comput 2020; 16(2): 1300-10.
[http://dx.doi.org/10.1021/acs.jctc.9b00890] [PMID: 31894691]

[42]   Zhou R, Friesner RA, Ghosh A, Rizzo RC, Jorgensen WL, Levy RM. New linear interaction method for binding affinity calculations using a continuum solvent model. J Phys Chem B 2001; 105(42): 10388-97.
[http://dx.doi.org/10.1021/jp011480z]

[43]   Srinivasan J, Miller J, Kollman PA, Case DA. Continuum solvent studies of the stability of RNA hairpin loops and helices. J Biomol Struct Dyn 1998; 16(3): 671-82.
[http://dx.doi.org/10.1080/07391102.1998.10508279] [PMID: 10052623]

[44]   Wang J, Morin P, Wang W, Kollman PA. Use of MM-PBSA in reproducing the binding free energies to HIV-1 RT of TIBO derivatives and predicting the binding mode to HIV-1 RT of efavirenz by docking and MM-PBSA. J Am Chem Soc 2001; 123(22): 5221-30.
[http://dx.doi.org/10.1021/ja003834q] [PMID: 11457384]

[45]   Halperin I, Ma B, Wolfson H, Nussinov R. Principles of docking: An overview of search algorithms and a guide to scoring functions. Proteins 2002; 47(4): 409-43.
[http://dx.doi.org/10.1002/prot.10115] [PMID: 12001221]

[46]   Xu B, Shen H, Zhu X, Li G. Fast and accurate computation schemes for evaluating vibrational entropy of proteins. J Comput Chem 2011; 32(15): 3188-93.
[http://dx.doi.org/10.1002/jcc.21900] [PMID: 21953554]

[47]   Amadei A, Linssen ABM, Berendsen HJC. Essential dynamics of proteins. Proteins 1993; 17(4): 412-25.
[http://dx.doi.org/10.1002/prot.340170408] [PMID: 8108382]

[48]   Carrington BJ, Mancera RL. Comparative estimation of vibrational entropy changes in proteins through normal modes analysis. J Mol Graph Model 2004; 23(2): 167-74.
[http://dx.doi.org/10.1016/j.jmgm.2004.05.003] [PMID: 15363458]

[49]   Sun H, Duan L, Chen F, *et al.* Assessing the performance of MM/PBSA and MM/GBSA methods. 7. Entropy effects on the performance of end-point binding free energy calculation approaches. Phys Chem Chem Phys 2018; 20(21): 14450-60.
[http://dx.doi.org/10.1039/C7CP07623A] [PMID: 29785435]

[50]   Smith SG, Zhou MM. The bromodomain: A new target in emerging epigenetic medicine. ACS Chem Biol 2016; 11(3): 598-608.
[http://dx.doi.org/10.1021/acschembio.5b00831] [PMID: 26596782]

[51]   Kougnassoukou Tchara PE, Filippakopoulos P, Lambert JP. Emerging tools to investigate bromodomain functions. Methods 2020; 184: 40-52.
[http://dx.doi.org/10.1016/j.ymeth.2019.11.003] [PMID: 31726225]

[52]   Bechter O, Schöffski P. Make your best BET: The emerging role of BET inhibitor treatment in malignant tumors. Pharmacol Ther 2020; 208: 107479.
[http://dx.doi.org/10.1016/j.pharmthera.2020.107479] [PMID: 31931101]

[53]   Fioravanti R, Mautone N, Rovere A, Rotili D, Mai A. Targeting histone acetylation/deacetylation in parasites: An update (2017–2020). Curr Opin Chem Biol 2020; 57: 65-74.
[http://dx.doi.org/10.1016/j.cbpa.2020.05.008] [PMID: 32615359]

[54]   Acharya A, Kutateladze TG, Byrareddy SN. Combining antiviral drugs with BET inhibitors is beneficial in combatting SARS☐CoV☐2 infection. Clin Transl Discov 2022; 2(2): e66.
[http://dx.doi.org/10.1002/ctd2.66] [PMID: 35633739]

[55]   Aldeghi M, Bodkin MJ, Knapp S, Biggin PC. Statistical analysis on the performance of molecular mechanics poisson–boltzmann surface area versus absolute binding free energy calculations:

bromodomains as a case study. J Chem Inf Model 2017; 57(9): 2203-21.
[http://dx.doi.org/10.1021/acs.jcim.7b00347] [PMID: 28786670]

[56]   Shadrick WR, Slavish PJ, Chai SC, *et al.* Exploiting a water network to achieve enthalpy-driven, bromodomain-selective BET inhibitors. Bioorg Med Chem 2018; 26(1): 25-36.
[http://dx.doi.org/10.1016/j.bmc.2017.10.042] [PMID: 29170024]

[57]   Duan L, Liu X, Zhang JZH. Interaction entropy: A new paradigm for highly efficient and reliable computation of protein–ligand binding free energy. J Am Chem Soc 2016; 138(17): 5722-8.
[http://dx.doi.org/10.1021/jacs.6b02682] [PMID: 27058988]

[58]   Duan L, Feng G, Wang X, Wang L, Zhang Q. Effect of electrostatic polarization and bridging water on CDK2–ligand binding affinities calculated using a highly efficient interaction entropy method. Phys Chem Chem Phys 2017; 19(15): 10140-52.
[http://dx.doi.org/10.1039/C7CP00841D] [PMID: 28368432]

[59]   Ben-Shalom IY, Pfeiffer-Marek S, Baringhaus KH, Gohlke H. Efficient approximation of ligand rotational and translational entropy changes upon binding for use in MM-PBSA calculations. J Chem Inf Model 2017; 57(2): 170-89.
[http://dx.doi.org/10.1021/acs.jcim.6b00373] [PMID: 27996253]

[60]   Roux B, Simonson T. Implicit solvent models. Biophys Chem 1999; 78(1-2): 1-20.
[http://dx.doi.org/10.1016/S0301-4622(98)00226-9] [PMID: 17030302]

[61]   Decherchi S, Masetti M, Vyalov I, Rocchia W. Implicit solvent methods for free energy estimation. Eur J Med Chem 2015; 91: 27-42.
[http://dx.doi.org/10.1016/j.ejmech.2014.08.064] [PMID: 25193298]

[62]   Bashford D, Case DA. Generalized born models of macromolecular solvation effects. Annu Rev Phys Chem 2000; 51(1): 129-52.
[http://dx.doi.org/10.1146/annurev.physchem.51.1.129] [PMID: 11031278]

[63]   Onufriev A. Implicit solvent models in molecular dynamics simulations: A brief overview.Annual Reports in Computational Chemistry. Elsevier 2008; pp. 125-37.
[http://dx.doi.org/10.1016/S1574-1400(08)00007-8]

[64]   Onufriev AV, Case DA. Generalized born implicit solvent models for biomolecules. Annu Rev Biophys 2019; 48(1): 275-96.
[http://dx.doi.org/10.1146/annurev-biophys-052118-115325] [PMID: 30857399]

[65]   de Ruiter A, Oostenbrink C. Free energy calculations of protein–ligand interactions. Curr Opin Chem Biol 2011; 15(4): 547-52.
[http://dx.doi.org/10.1016/j.cbpa.2011.05.021] [PMID: 21684797]

[66]   Kuhn B, Gerber P, Schulz-Gasch T, Stahl M. Validation and use of the MM-PBSA approach for drug discovery. J Med Chem 2005; 48(12): 4040-8.
[http://dx.doi.org/10.1021/jm049081q] [PMID: 15943477]

[67]   Genheden S, Ryde U. The MM/PBSA and MM/GBSA methods to estimate ligand-binding affinities. Expert Opin Drug Discov 2015; 10(5): 449-61.
[http://dx.doi.org/10.1517/17460441.2015.1032936] [PMID: 25835573]

[68]   Wang C, Greene DA, Xiao L, Qi R, Luo R. Recent developments and applications of the MMPBSA method. Front Mol Biosci 2018; 4: 87.
[http://dx.doi.org/10.3389/fmolb.2017.00087] [PMID: 29367919]

[69]   Xu L, Sun H, Li Y, Wang J, Hou T. Assessing the performance of MM/PBSA and MM/GBSA methods. 3. The impact of force fields and ligand charge models. J Phys Chem B 2013; 117(28): 8408-21.
[http://dx.doi.org/10.1021/jp404160y] [PMID: 23789789]

[70]   Sun H, Li Y, Shen M, *et al.* Assessing the performance of MM/PBSA and MM/GBSA methods. 5. Improved docking performance using high solute dielectric constant MM/GBSA and MM/PBSA

rescoring. Phys Chem Chem Phys 2014; 16(40): 22035-45.
[http://dx.doi.org/10.1039/C4CP03179B] [PMID: 25205360]

[71] Wang E, Fu W, Jiang D, *et al.* VAD-MM/GBSA: A variable atomic dielectric MM/GBSA model for improved accuracy in protein–ligand binding free energy calculations. J Chem Inf Model 2021; 61(6): 2844-56.
[http://dx.doi.org/10.1021/acs.jcim.1c00091] [PMID: 34014672]

[72] Dong L, Qu X, Zhao Y, Wang B. Prediction of binding free energy of protein–ligand complexes with a hybrid molecular mechanics/generalized born surface area and machine learning method. ACS Omega 2021; 6(48): 32938-47.
[http://dx.doi.org/10.1021/acsomega.1c04996] [PMID: 34901645]

[73] Tuccinardi T. What is the current value of MM/PBSA and MM/GBSA methods in drug discovery? Expert Opin Drug Discov 2021; 16(11): 1233-7.
[http://dx.doi.org/10.1080/17460441.2021.1942836] [PMID: 34165011]

[74] Ryde U. How many conformations need to be sampled to obtain converged QM/MM energies? the curse of exponential averaging. J Chem Theory Comput 2017; 13(11): 5745-52.
[http://dx.doi.org/10.1021/acs.jctc.7b00826] [PMID: 29024586]

[75] Procacci P. Solvation free energies *via* alchemical simulations: let's get honest about sampling, once more. Phys Chem Chem Phys 2019; 21(25): 13826-34.
[http://dx.doi.org/10.1039/C9CP02808K] [PMID: 31211310]

[76] Jorgensen WL, Thomas LL. Perspective on free-energy perturbation calculations for chemical equilibria. J Chem Theory Comput 2008; 4(6): 869-76.
[http://dx.doi.org/10.1021/ct800011m] [PMID: 19936324]

[77] Michel J, Essex JW. Prediction of protein–ligand binding affinity by free energy simulations: Assumptions, pitfalls and expectations. J Comput Aided Mol Des 2010; 24(8): 639-58.
[http://dx.doi.org/10.1007/s10822-010-9363-3] [PMID: 20509041]

[78] Procacci P. Methodological uncertainties in drug-receptor binding free energy predictions based on classical molecular dynamics. Curr Opin Struct Biol 2021; 67: 127-34.
[http://dx.doi.org/10.1016/j.sbi.2020.08.001] [PMID: 33220532]

[79] Bennett CH. Efficient estimation of free energy differences from Monte Carlo data. J Comput Phys 1976; 22(2): 245-68.
[http://dx.doi.org/10.1016/0021-9991(76)90078-4]

[80] Shirts MR, Bair E, Hooker G, Pande VS. Equilibrium free energies from nonequilibrium measurements using maximum-likelihood methods. Phys Rev Lett 2003; 91(14): 140601.
[http://dx.doi.org/10.1103/PhysRevLett.91.140601] [PMID: 14611511]

[81] Gutiérrez M, Vallejos GA, Cortés MP, Bustos C. Bennett acceptance ratio method to calculate the binding free energy of BACE1 inhibitors: Theoretical model and design of new ligands of the enzyme. Chem Biol Drug Des 2019; 93(6): 1117-28.
[http://dx.doi.org/10.1111/cbdd.13456] [PMID: 30693676]

[82] Hahn AM, Then H. Characteristic of bennett's acceptance ratio method. Phys Rev E Stat Nonlin Soft Matter Phys 2009; 80(3): 031111.
[http://dx.doi.org/10.1103/PhysRevE.80.031111] [PMID: 19905066]

[83] Procacci P. Multiple Bennett acceptance ratio made easy for replica exchange simulations. J Chem Phys 2013; 139(12): 124105.
[http://dx.doi.org/10.1063/1.4821814] [PMID: 24089748]

[84] Chen H, Chipot C. Enhancing sampling with free-energy calculations. Curr Opin Struct Biol 2022; 77: 102497.
[http://dx.doi.org/10.1016/j.sbi.2022.102497] [PMID: 36410221]

[85] Kumar S, Rosenberg JM, Bouzida D, Swendsen RH, Kollman PA. THE weighted histogram analysis

method for free-energy calculations on biomolecules. I. The method. J Comput Chem 1992; 13(8): 1011-21.
[http://dx.doi.org/10.1002/jcc.540130812]

[86]   Shirts MR, Chodera JD. Statistically optimal analysis of samples from multiple equilibrium states. J Chem Phys 2008; 129(12): 124105.
[http://dx.doi.org/10.1063/1.2978177] [PMID: 19045004]

[87]   Matsunaga Y, Kamiya M, Oshima H, Jung J, Ito S, Sugita Y. Use of multistate Bennett acceptance ratio method for free-energy calculations from enhanced sampling and free-energy perturbation. Biophys Rev 2022; 14(6): 1503-12.
[http://dx.doi.org/10.1007/s12551-022-01030-9] [PMID: 36659993]

[88]   Wu H, Paul F, Wehmeyer C, Noé F. Multiensemble Markov models of molecular thermodynamics and kinetics. Proc Natl Acad Sci 2016; 113(23): E3221-30.
[http://dx.doi.org/10.1073/pnas.1525092113] [PMID: 27226302]

[89]   Cournia Z, Allen BK, Beuming T, Pearlman DA, Radak BK, Sherman W. Rigorous free energy simulations in virtual screening. J Chem Inf Model 2020; 60(9): 4153-69.
[http://dx.doi.org/10.1021/acs.jcim.0c00116] [PMID: 32539386]

[90]   Decherchi S, Cavalli A. Thermodynamics and kinetics of drug-target binding by molecular simulation. Chem Rev 2020; 120(23): 12788-833.
[http://dx.doi.org/10.1021/acs.chemrev.0c00534] [PMID: 33006893]

[91]   Shivakumar D, Harder E, Damm W, Friesner RA, Sherman W. Improving the prediction of absolute solvation free energies using the next generation opls force field. J Chem Theory Comput 2012; 8(8): 2553-8.
[http://dx.doi.org/10.1021/ct300203w] [PMID: 26592101]

[92]   Azimi S, Khuttan S, Wu JZ, Pal RK, Gallicchio E. Relative binding free energy calculations for ligands with diverse scaffolds with the alchemical transfer method. J Chem Inf Model 2022; 62(2): 309-23.
[http://dx.doi.org/10.1021/acs.jcim.1c01129] [PMID: 34990555]

[93]   Bhati AP, Wan S, Hu Y, Sherborne B, Coveney PV. Uncertainty quantification in alchemical free energy methods. J Chem Theory Comput 2018; 14(6): 2867-80.
[http://dx.doi.org/10.1021/acs.jctc.7b01143] [PMID: 29678106]

[94]   Laio A, Parrinello M. Escaping free-energy minima. Proc Natl Acad Sci 2002; 99(20): 12562-6.
[http://dx.doi.org/10.1073/pnas.202427399] [PMID: 12271136]

[95]   Barducci A, Bussi G, Parrinello M. Well-tempered metadynamics: A smoothly converging and tunable free-energy method. Phys Rev Lett 2008; 100(2): 020603.
[http://dx.doi.org/10.1103/PhysRevLett.100.020603] [PMID: 18232845]

[96]   Bonomi M, Barducci A, Parrinello M. Reconstructing the equilibrium Boltzmann distribution from well-tempered metadynamics. J Comput Chem 2009; 30(11): 1615-21.
[http://dx.doi.org/10.1002/jcc.21305] [PMID: 19421997]

[97]   Laio A, Parrinello M. Computing Free Energies and Accelerating Rare Events with Metadynamics.Computer Simulations in Condensed Matter Systems: From Materials to Chemical Biology. Berlin, Heidelberg: Springer Berlin Heidelberg 2006; 1: pp. 315-47.
[http://dx.doi.org/10.1007/3-540-35273-2_9]

[98]   Bussi G, Branduardi D. Free-energy calculations with metadynamics: Theory and practice. Rev Comput Chem 2015; 28: 1-49.
[http://dx.doi.org/10.1002/9781118889886.ch1]

[99]   Invernizzi M, Parrinello M. Rethinking metadynamics: From bias potentials to probability distributions. J Phys Chem Lett 2020; 11(7): 2731-6.
[http://dx.doi.org/10.1021/acs.jpclett.0c00497] [PMID: 32191470]

[100] Tanida Y, Matsuura A. Alchemical free energy calculations *via* metadynamics: Application to the THEOPHYLLINE☐RNA aptamer complex. J Comput Chem 2020; 41(20): 1804-19.
[http://dx.doi.org/10.1002/jcc.26221] [PMID: 32449538]

[101] Sadiq SK, Wright D, Watson SJ, Zasada SJ, Stoica I, Coveney PV. Automated molecular simulation based binding affinity calculator for ligand-bound HIV-1 proteases. J Chem Inf Model 2008; 48(9): 1909-19.
[http://dx.doi.org/10.1021/ci8000937] [PMID: 18710212]

[102] Hayes RL, Buckner J, Brooks CL III. BLaDE: A basic lambda dynamics engine for gpu-accelerated molecular dynamics free energy calculations. J Chem Theory Comput 2021; 17(11): 6799-807.
[http://dx.doi.org/10.1021/acs.jctc.1c00833] [PMID: 34709046]

[103] Fratev F, Sirimulla S. An improved free energy perturbation fep+ sampling protocol for flexible ligand-binding domains. Sci Rep 2019; 9(1): 16829.
[http://dx.doi.org/10.1038/s41598-019-53133-1] [PMID: 31728038]

[104] Loeffler HH, Michel J, Woods C. FESetup: Automating setup for alchemical free energy simulations. J Chem Inf Model 2015; 55(12): 2485-90.
[http://dx.doi.org/10.1021/acs.jcim.5b00368] [PMID: 26544598]

[105] Homeyer N, Gohlke H. FEW: A workflow tool for free energy calculations of ligand binding. J Comput Chem 2013; 34(11): 965-73.
[http://dx.doi.org/10.1002/jcc.23218] [PMID: 23288722]

[106] Kuhn M, Firth-Clark S, Tosco P, Mey ASJS, Mackey M, Michel J. Assessment of binding affinity *via* alchemical free-energy calculations. J Chem Inf Model 2020; 60(6): 3120-30.
[http://dx.doi.org/10.1021/acs.jcim.0c00165] [PMID: 32437145]

[107] Lundborg M, Lindahl E. Automatic gromacs topology generation and comparisons of force fields for solvation free energy calculations. J Phys Chem B 2015; 119(3): 810-23.
[http://dx.doi.org/10.1021/jp505332p] [PMID: 25343332]

[108] Gapsys V, Michielssens S, Seeliger D, de Groot BL. pmx: Automated protein structure and topology generation for alchemical perturbations. J Comput Chem 2015; 36(5): 348-54.
[http://dx.doi.org/10.1002/jcc.23804] [PMID: 25487359]

[109] Carvalho Martins L, Cino EA, Ferreira RS. PyAutoFEP: An automated free energy perturbation workflow for gromacs integrating enhanced sampling methods. J Chem Theory Comput 2021; 17(7): 4262-73.
[http://dx.doi.org/10.1021/acs.jctc.1c00194] [PMID: 34142828]

[110] Wang K, Chodera JD, Yang Y, Shirts MR. Identifying ligand binding sites and poses using GPU-accelerated Hamiltonian replica exchange molecular dynamics. J Comput Aided Mol Des 2013; 27(12): 989-1007.
[http://dx.doi.org/10.1007/s10822-013-9689-8] [PMID: 24297454]

[111] Vanommeslaeghe K, MacKerell AD Jr. CHARMM additive and polarizable force fields for biophysics and computer-aided drug design. Biochim Biophys Acta, Gen Subj 2015; 1850(5): 861-71.
[http://dx.doi.org/10.1016/j.bbagen.2014.08.004] [PMID: 25149274]

[112] Tian C, Kasavajhala K, Belfon KAA, *et al.* ff19SB: Amino-acid-specific protein backbone parameters trained against quantum mechanics energy surfaces in solution. J Chem Theory Comput 2020; 16(1): 528-52.
[http://dx.doi.org/10.1021/acs.jctc.9b00591] [PMID: 31714766]

[113] Halgren TA. Merck molecular force field. II. MMFF94 van der Waals and electrostatic parameters for intermolecular interactions. J Comput Chem 1996; 17(5-6): 520-52.
[http://dx.doi.org/10.1002/(SICI)1096-987X(199604)17:5/6<520::AID-JCC2>3.0.CO;2-W]

[114] Halgren TA. Merck molecular force field. I. Basis, form, scope, parameterization, and performance of MMFF94. J Comput Chem 1996; 17(5-6): 490-519.

[http://dx.doi.org/10.1002/(SICI)1096-987X(199604)17:5/6<490::AID-JCC1>3.0.CO;2-P]

[115]   Wang J, Wolf RM, Caldwell JW, Kollman PA, Case DA. Development and testing of a general amber force field. J Comput Chem 2004; 25(9): 1157-74.
[http://dx.doi.org/10.1002/jcc.20035] [PMID: 15116359]

[116]   Vanommeslaeghe K, MacKerell AD Jr. Automation of the charmm general force field (CGenFF) I: Bond perception and atom typing. J Chem Inf Model 2012; 52(12): 3144-54.
[http://dx.doi.org/10.1021/ci300363c] [PMID: 23146088]

[117]   Dodda LS, Cabeza de Vaca I, Tirado-Rives J, Jorgensen WL. LigParGen web server: An automatic OPLS-AA parameter generator for organic ligands. Nucleic Acids Res 2017; 45(W1): W331-6.
[http://dx.doi.org/10.1093/nar/gkx312] [PMID: 28444340]

[118]   Dodda LS, Vilseck JZ, Tirado-Rives J, Jorgensen WL. 1.14*CM1A-LBCC: Localized bond-charge corrected cm1a charges for condensed-phase simulations. J Phys Chem B 2017; 121(15): 3864-70.
[http://dx.doi.org/10.1021/acs.jpcb.7b00272] [PMID: 28224794]

[119]   Harder E, Damm W, Maple J, *et al.* OPLS3: A force field providing broad coverage of drug-like small molecules and proteins. J Chem Theory Comput 2016; 12(1): 281-96.
[http://dx.doi.org/10.1021/acs.jctc.5b00864] [PMID: 26584231]

[120]   Roos K, Wu C, Damm W, *et al.* OPLS3e: Extending force field coverage for drug-like small molecules. J Chem Theory Comput 2019; 15(3): 1863-74.
[http://dx.doi.org/10.1021/acs.jctc.8b01026] [PMID: 30768902]

[121]   Lu C, Wu C, Ghoreishi D, *et al.* OPLS4: Improving force field accuracy on challenging regimes of chemical space. J Chem Theory Comput 2021; 17(7): 4291-300.
[http://dx.doi.org/10.1021/acs.jctc.1c00302] [PMID: 34096718]

[122]   Zhu S. Validation of the generalized force fields GAFF, CGenFF, OPLS-AA, and PRODRGFF by testing against experimental osmotic coefficient data for small drug-like molecules. J Chem Inf Model 2019; 59(10): 4239-47.
[http://dx.doi.org/10.1021/acs.jcim.9b00552] [PMID: 31557024]

[123]   Friedrich NO, de Bruyn Kops C, Flachsenberg F, Sommer K, Rarey M, Kirchmair J. Benchmarking commercial conformer ensemble generators. J Chem Inf Model 2017; 57(11): 2719-28.
[http://dx.doi.org/10.1021/acs.jcim.7b00505] [PMID: 28967749]

[124]   Vassetti D, Pagliai M, Procacci P. Assessment of GAFF2 and OPLS-AA General Force Fields in Combination with the Water Models TIP3P, SPCE, and OPC3 for the Solvation Free Energy of Druglike Organic Molecules. J Chem Theory Comput 2019; 15(3): 1983-95.
[http://dx.doi.org/10.1021/acs.jctc.8b01039] [PMID: 30694667]

[125]   Boothroyd S, Wang LP, Mobley DL, Chodera JD, Shirts MR. Open force field evaluator: An automated, efficient, and scalable framework for the estimation of physical properties from molecular simulation. J Chem Theory Comput 2022; 18(6): 3566-76.
[http://dx.doi.org/10.1021/acs.jctc.1c01111] [PMID: 35507313]

[126]   Ehrman JN, Lim VT, Bannan CC, Thi N, Kyu DY, Mobley DL. Improving small molecule force fields by identifying and characterizing small molecules with inconsistent parameters. J Comput Aided Mol Des 2021; 35(3): 271-84.
[http://dx.doi.org/10.1007/s10822-020-00367-1] [PMID: 33506360]

[127]   Zanette C, Bannan CC, Bayly CI, *et al.* Toward learned chemical perception of force field typing rules. J Chem Theory Comput 2019; 15(1): 402-23.
[http://dx.doi.org/10.1021/acs.jctc.8b00821] [PMID: 30512951]

[128]   Wang LP, Martinez TJ, Pande VS. Building force fields: An automatic, systematic, and reproducible approach. J Phys Chem Lett 2014; 5(11): 1885-91.
[http://dx.doi.org/10.1021/jz500737m] [PMID: 26273869]

[129]   Slochower DR, Henriksen NM, Wang LP, Chodera JD, Mobley DL, Gilson MK. Binding

thermodynamics of host–guest systems with SMIRNOFF99FROSST 1.0.5 from the open force field initiative. J Chem Theory Comput 2019; 15(11): 6225-42.
[http://dx.doi.org/10.1021/acs.jctc.9b00748] [PMID: 31603667]

[130] D'Amore L, Hahn DF, Dotson DL, *et al.* Collaborative assessment of molecular geometries and energies from the open force field. J Chem Inf Model. 2022; 62(23): 6094-104. Epub 2022 Nov 26.
[http://dx.doi.org/10.1021/acs.jcim.2c01185] [PMID: 36433835] [PMCID: PMC9873353]

[131] Smith JS, Isayev O, Roitberg AE. ANI-1: An extensible neural network potential with DFT accuracy at force field computational cost. Chem Sci 2017; 8(4): 3192-203.
[http://dx.doi.org/10.1039/C6SC05720A] [PMID: 28507695]

[132] Devereux C, Smith JS, Huddleston KK, *et al.* Extending the applicability of the ANI deep learning molecular potential to sulfur and halogens. J Chem Theory Comput 2020; 16(7): 4192-202.
[http://dx.doi.org/10.1021/acs.jctc.0c00121] [PMID: 32543858]

[133] Wieder M, Fass J, Chodera JD. Fitting quantum machine learning potentials to experimental free energy data: predicting tautomer ratios in solution. Chem Sci 2021; 12(34): 11364-81.
[http://dx.doi.org/10.1039/D1SC01185E] [PMID: 34567495]

[134] Doerr S, Majewski M, Pérez A, *et al.* TorchMD: A deep learning framework for molecular simulations. J Chem Theory Comput 2021; 17(4): 2355-63.
[http://dx.doi.org/10.1021/acs.jctc.0c01343] [PMID: 33729795]

[135] Limongelli V. Ligand binding free energy and kinetics calculation in 2020. Wiley Interdiscip Rev Comput Mol Sci 2020; 10(4)
[http://dx.doi.org/10.1002/wcms.1455]

[136] Cournia Z, Allen B, Sherman W. Relative binding free energy calculations in drug discovery: Recent advances and practical considerations. J Chem Inf Model 2017; 57(12): 2911-37.
[http://dx.doi.org/10.1021/acs.jcim.7b00564] [PMID: 29243483]

[137] de Ruiter A, Oostenbrink C. Advances in the calculation of binding free energies. Curr Opin Struct Biol 2020; 61: 207-12.
[http://dx.doi.org/10.1016/j.sbi.2020.01.016] [PMID: 32088376]

[138] Mobley DL, Klimovich PV. Perspective: Alchemical free energy calculations for drug discovery. J Chem Phys 2012; 137(23): 230901.
[http://dx.doi.org/10.1063/1.4769292] [PMID: 23267463]

[139] Zara L, Efrém NL, van Muijlwijk-Koezen JE, de Esch IJP, Zarzycka B. Progress in free energy perturbation: Options for evolving fragments. Drug Discov Today Technol 2021; 40: 36-42.
[http://dx.doi.org/10.1016/j.ddtec.2021.10.001] [PMID: 34916020]

[140] Pohorille A, Jarzynski C, Chipot C. Good practices in free-energy calculations. J Phys Chem B 2010; 114(32): 10235-53.
[http://dx.doi.org/10.1021/jp102971x] [PMID: 20701361]

[141] Yang Q, Burchett W, Steeno GS, *et al.* Optimal designs for pairwise calculation: An application to free energy perturbation in minimizing prediction variability. J Comput Chem 2020; 41(3): 247-57.
[http://dx.doi.org/10.1002/jcc.26095] [PMID: 31721260]

[142] Pérez-Benito L, Casajuana-Martin N, Jiménez-Rosés M, van Vlijmen H, Tresadern G. Predicting activity cliffs with free-energy perturbation. J Chem Theory Comput 2019; 15(3): 1884-95.
[http://dx.doi.org/10.1021/acs.jctc.8b01290] [PMID: 30776226]

[143] Cappel D, Hall ML, Lenselink EB, *et al.* Relative binding free energy calculations applied to protein homology models. J Chem Inf Model 2016; 56(12): 2388-400.
[http://dx.doi.org/10.1021/acs.jcim.6b00362] [PMID: 28024402]

[144] Lee TS, Allen BK, Giese TJ, *et al.* Alchemical binding free energy calculations in AMBER20: Advances and best practices for drug discovery. J Chem Inf Model 2020; 60(11): 5595-623.
[http://dx.doi.org/10.1021/acs.jcim.0c00613] [PMID: 32936637]

[145] de Oliveira C, Yu HS, Chen W, Abel R, Wang L. Rigorous free energy perturbation approach to estimating relative binding affinities between ligands with multiple protonation and tautomeric states. J Chem Theory Comput 2019; 15(1): 424-35.
[http://dx.doi.org/10.1021/acs.jctc.8b00826] [PMID: 30537823]

[146] Chen W, Deng Y, Russell E, Wu Y, Abel R, Wang L. Accurate calculation of relative binding free energies between ligands with different net charges. J Chem Theory Comput 2018; 14(12): 6346-58.
[http://dx.doi.org/10.1021/acs.jctc.8b00825] [PMID: 30375870]

[147] Mobley DL, Chodera JD, Dill KA. On the use of orientational restraints and symmetry corrections in alchemical free energy calculations. J Chem Phys 2006; 125(8): 084902.
[http://dx.doi.org/10.1063/1.2221683] [PMID: 16965052]

[148] Heinzelmann G, Gilson MK. Automation of absolute protein-ligand binding free energy calculations for docking refinement and compound evaluation. Sci Rep 2021; 11(1): 1116.
[http://dx.doi.org/10.1038/s41598-020-80769-1] [PMID: 33441879]

[149] Abel R, Wang L, Harder ED, Berne BJ, Friesner RA. Advancing drug discovery through enhanced free energy calculations. Acc Chem Res 2017; 50(7): 1625-32.
[http://dx.doi.org/10.1021/acs.accounts.7b00083] [PMID: 28677954]

[150] Hansen N, van Gunsteren WF. Practical aspects of free-energy calculations: A Review. J Chem Theory Comput 2014; 10(7): 2632-47.
[http://dx.doi.org/10.1021/ct500161f] [PMID: 26586503]

[151] Ebrahimi M, Hénin J. Symmetry-adapted restraints for binding free energy calculations. J Chem Theory Comput 2022; 18(4): 2494-502.
[http://dx.doi.org/10.1021/acs.jctc.1c01235] [PMID: 35230113]

[152] Menzer WM, Xie B, Minh DDL. On restraints in end□point protein–ligand binding free energy calculations. J Comput Chem 2020; 41(6): 573-86.
[http://dx.doi.org/10.1002/jcc.26119] [PMID: 31821590]

[153] Leitgeb M, Schröder C, Boresch S. Alchemical free energy calculations and multiple conformational substates. J Chem Phys 2005; 122(8): 084109.
[http://dx.doi.org/10.1063/1.1850900] [PMID: 15836022]

[154] Shirts MR, Mobley DL. An Introduction to best practices in free energy calculations. Methods Mol Biol 2013; 924: 271-311.
[http://dx.doi.org/10.1007/978-1-62703-017-5_11]

[155] Klimovich PV, Shirts MR, Mobley DL. Guidelines for the analysis of free energy calculations. J Comput Aided Mol Des 2015; 29(5): 397-411.
[http://dx.doi.org/10.1007/s10822-015-9840-9] [PMID: 25808134]

[156] Chodera JD. A simple method for automated equilibration detection in molecular simulations. J Chem Theory Comput 2016; 12(4): 1799-805.
[http://dx.doi.org/10.1021/acs.jctc.5b00784] [PMID: 26771390]

[157] Belon CA, Frick DN. Helicase inhibitors as specifically targeted antiviral therapy for hepatitis C. Future Virol 2009; 4(3): 277-93.
[http://dx.doi.org/10.2217/fvl.09.7] [PMID: 20161209]

[158] Arooj M, Shehadi I, Nassab CN, Mohamed AA. Computational insights into binding mechanism of drugs as potential inhibitors against SARS-CoV-2 targets. Chem Zvesti 2022; 76(1): 111-21.
[http://dx.doi.org/10.1007/s11696-021-01843-0] [PMID: 34483461]

[159] Fu H, Zhou Y, Jing X, Shao X, Cai W. Meta-analysis reveals that absolute binding free-energy calculations approach chemical accuracy. J Med Chem 2022; 65(19): 12970-8.
[http://dx.doi.org/10.1021/acs.jmedchem.2c00796] [PMID: 36179112]

[160] Feng M, Heinzelmann G, Gilson MK. Absolute binding free energy calculations improve enrichment

of actives in virtual compound screening. Sci Rep 2022; 12(1): 13640.
[http://dx.doi.org/10.1038/s41598-022-17480-w] [PMID: 35948614]

[161] Wan S, Bhati AP, Zasada SJ, Coveney PV. Rapid, accurate, precise and reproducible ligand–protein binding free energy prediction. Interface Focus 2020; 10(6): 20200007.
[http://dx.doi.org/10.1098/rsfs.2020.0007] [PMID: 33178418]

[162] Mondal D, Florian J, Warshel A. Exploring the Effectiveness of Binding Free Energy Calculations. J Phys Chem B 2019; 123(42): 8910-5.
[http://dx.doi.org/10.1021/acs.jpcb.9b07593] [PMID: 31560539]

[163] Cheng T, Li X, Li Y, Liu Z, Wang R. Comparative assessment of scoring functions on a diverse test set. J Chem Inf Model 2009; 49(4): 1079-93.
[http://dx.doi.org/10.1021/ci9000053] [PMID: 19358517]

[164] Bonati L, Rizzi V, Parrinello M. Data-driven collective variables for enhanced sampling. J Phys Chem Lett 2020; 11(8): 2998-3004.
[http://dx.doi.org/10.1021/acs.jpclett.0c00535] [PMID: 32239945]

[165] Noé F, Tkatchenko A, Müller KR, Clementi C. Machine learning for molecular simulation. Annu Rev Phys Chem 2020; 71(1): 361-90.
[http://dx.doi.org/10.1146/annurev-physchem-042018-052331] [PMID: 32092281]

[166] Wider M, Fass J, Chodera JD. Teaching free energy calculations to learn from exprimental data. bioRxiv 2021.
[http://dx.doi.org/10.1101/2021.08.24.457513]

[167] Chipot C, Pearlman DA. Free energy calculations. the long and winding gilded road. Mol Simul 2002; 28(1-2): 1-12.
[http://dx.doi.org/10.1080/08927020211974]

[168] Barbu A, Zhu S-C. Monte Carlo Methods. Singapore: Springer Singapore 2020.
[http://dx.doi.org/10.1007/978-981-13-2971-5]

[169] Khalak Y, Tresadern G, Aldeghi M, *et al.* Alchemical absolute protein–ligand binding free energies for drug design. Chem Sci 2021; 12(41): 13958-71.
[http://dx.doi.org/10.1039/D1SC03472C] [PMID: 34760182]

[170] Schindler CEM, Baumann H, Blum A, *et al.* Large-scale assessment of binding free energy calculations in active drug discovery projects. J Chem Inf Model 2020; 60(11): 5457-74.
[http://dx.doi.org/10.1021/acs.jcim.0c00900] [PMID: 32813975]

[171] Loeffler HH, Bosisio S, Duarte Ramos Matos G, *et al.* Reproducibility of free energy calculations across different molecular simulation software packages. J Chem Theory Comput 2018; 14(11): 5567-82.
[http://dx.doi.org/10.1021/acs.jctc.8b00544] [PMID: 30289712]

[172] Song LF, Merz KM Jr. Evolution of alchemical free energy methods in drug discovery. J Chem Inf Model 2020; 60(11): 5308-18.
[http://dx.doi.org/10.1021/acs.jcim.0c00547] [PMID: 32818371]

[173] Tobias DJ, Brooks CL III. Calculation of free energy surfaces using the methods of thermodynamic perturbation theory. Chem Phys Lett 1987; 142(6): 472-6.
[http://dx.doi.org/10.1016/0009-2614(87)80646-2]

[174] Armacost KA, Riniker S, Cournia Z. Novel directions in free energy methods and applications. J Chem Inf Model 2020; 60(1): 1-5.
[http://dx.doi.org/10.1021/acs.jcim.9b01174] [PMID: 31983210]

[175] Rizzi A, Jensen T, Slochower DR, *et al.* The SAMPL6 sampling challenge: Assessing the reliability and efficiency of binding free energy calculations. J Comput Aided Mol Des 2020; 34(5): 601-33.
[http://dx.doi.org/10.1007/s10822-020-00290-5] [PMID: 31984465]

SUBJECT INDEX